Problems during the Design and Testing of Instrument, Special and Power Transformers

Problems during the Design and Testing of Instrument, Special and Power Transformers

Editors

Elzbieta Lesniewska
Xose Lopez-Fernandez
Pawel Witczak

MDPI

Basel • Beijing • Wuhan • Barcelona • Belgrade • Novi Sad • Cluj • Manchester

Editors
Elzbieta Lesniewska
Institute of Mechatronics
and Information Systems
Lodz University
of Technology
Lodz
Poland

Xose Lopez-Fernandez
Electrical Engineering
Universidad de Vigo
Vigo
Spain

Pawel Witczak
Institute of Mechatronics
and Information Systems
Lodz University
of Technology
Lodz
Poland

Editorial Office
MDPI
St. Alban-Anlage 66
4052 Basel, Switzerland

This is a reprint of articles from the Special Issue published online in the open access journal *Energies* (ISSN 1996-1073) (available at: www.mdpi.com/journal/energies/special_issues/43U7U74X2M).

For citation purposes, cite each article independently as indicated on the article page online and as indicated below:

Lastname, A.A.; Lastname, B.B. Article Title. *Journal Name* **Year**, *Volume Number*, Page Range.

ISBN 978-3-7258-0252-4 (Hbk)
ISBN 978-3-7258-0251-7 (PDF)
doi.org/10.3390/books978-3-7258-0251-7

Contents

About the Editors

Elzbieta Lesniewska

Elzbieta Lesniewska was born in Torun, Poland. She obtained her M.Sc. in Electrical Engineering at the Technical University of Lodz in 1974 and her Ph.D. in 1983. The title D.Sc. was received in 1997. Since 1999, she has worked as a professor at the Technical University of Lodz. She obtained the Full Professor title from the Polish President in 2012.

In 1974, she began working in the Institute of Electrical Engineering and, from 1991, in the Department of Applied Electrical Engineering and Instrument Transformers. She specializes in electrical engineering, especially in the theory of electromagnetic fields and instrument transformers. For many years, she cooperated with constructors of instrument transformers and worked in industry.

She is the author or joint author of over 100 scientific articles, 3 patents, and 2 printed course lectures.

She is a member of the International Compumag Society and the Polish Society of Applied Electromagnetism.

Xose Lopez-Fernandez

Prof. Dr. Xose M. Lopez-Fernandez received his M.Sc. degree in Electrical Engineering and obtained his Ph.D. degree (with first class honors) from Vigo University, Vigo, Spain, in 1992 and 1997, respectively. He is currently Professor in the Department of Electrical Engineering, University of Vigo, and was Visiting Professor at the University of Artois, Bethune, France. He was Director of the R&D Department for EFACEC Transformers (Porto-Portugal and Savannah-USA) from 2012 to 2014. He is the main editor and co-author of the monograph "Transformers: Analysis, Design, and Measurement".

Prof. Dr. Lopez-Fernandez is the Founder and General Chairman of the International Advanced Research Workshop on Transformers. He is leading research projects and consulting for utilities and power transformer manufacturers. His research interests are in the design aspects of electrical machines. Prof. Dr. Lopez-Fernandez received the Alfons Hoffmann's Medal from the Polish Power Engineering Society in 2004 for his contribution to promoting knowledge on Power Transformer. Xose is the recipient of the 2023 IEEE PES Outstanding Volunteer Award for inspiring and outstanding leadership as President of the PES Spain Chapter.

Prof. Dr. Xose is a senior member of IEEE, a member of the IEEE-PES Transformers Committee, and an active member of CIGRE Working Groups.

Prof. Dr. Xose is the Founder Member and President of the Spanish Association for Transformers Innovation (RedInTransf), Editorial Board Member of *IET Science, Measurement and Technology* (*IET SMT Journal*), Editor of the IEEE Transactions on Power Delivery (TPWRD), IEEE PES Transformers Committee's TPWRD Editor Liaison, Vice-Chair IEEE-PES Spain Chapter (2017–2019), Chair IEEE-PES Spain Chapter (2019–2021/2022–2023), and Past-Chair IEEE-PES Spain Chapter (2024–2025).

Pawel Witczak

Pawel Witczak received his Ph.D. and D.Sc. in Electrical Engineering from Lodz University of Technology, Poland, in 1981 and 1996, respectively. He is a Professor at the Institute of Mechatronics and Information Systems, Lodz University of Technology. He was a Visiting Professor at the University of Artois, Bethune, France, from 1996 to 2006. Professor Witczak was the leader of many projects awarded by the industry, including Crompton Greaves, Mumbai, India, 1998–2000; OTIS

United Technologies, USA, 2001–2012; and Hitachi Energy (formerly ABB), Poland and Germany, from 2017. His main research interests are designing and measuring electrical machines and power transformers, particularly emphasizing the numerical modeling of magnetic vibrations and acoustic emissions.

Preface

This Special Issue aims to present and disseminate the latest developments in the theory, design, modeling, and laboratory testing of all types of transformers. The scope includes design calculations and diagnostics of various types and structures of 1- and 3-phase power transformers, special transformers, and measurement transformers with various applications. Design requirements concern not only the exploitation but also the optimal design methodologies, advanced device modeling methods, and technical research methods. This article is addressed primarily to producers and users of power, special, and measurement transformers and scientists interested in transformer-related topics. The authors of the articles included in the Special Issue are employees of technological universities or research institutes with professorial and doctoral degrees. In their work, they deal with specific technical problems, most often cooperating with industry. During their everyday research, they usually work alongside technicians, to whom they are grateful for their help.

Elzbieta Lesniewska, Xose Lopez-Fernandez, and Pawel Witczak
Editors

energies

Editorial

Problems during the Design and Testing of Instrument, Special and Power Transformers: The Outlook

Elzbieta Lesniewska * **and Pawel Witczak**

Institute of Mechatronics and Information Systems, Lodz University of Technology, 90-924 Lodz, Poland; pawel.witczak@p.lodz.pl
* Correspondence: elzbieta.lesniewska-komeza@p.lodz.pl; Tel.: +48-426312694

In 1831, Michael Faraday discovered the phenomenon of electromagnetic induction, which allowed for the construction of previously unknown electrical devices and significantly impacted economic and social development. Transformers work on the principle of Faraday's law and are used to transform electrical energy (changing voltage and current values). Ottó Bláthy, Miksa Déri, and Károly Zipernowsky from Austria-Hungary initiated the design and use of transformers for both experimental and commercial purposes and patented the single-phase transformer in 1885.

The pioneer of alternating current technology was Tesla, but it was Michał Doliwo-Dobrowolski who invented the 1891 squirrel cage motor and three-phase power transformer and energy transmission system. In Europe, the first experimental transmission of energy via a three-phase 15 kV high-voltage line took place in 1891 (from a hydroelectric power plant to Frankfurt am Main via a 178 km long line) under the leadership of Doliwo-Dobrowolski.

As the transmitted power and voltages of transmission lines increased, instrument transformers became an essential element of the measurement system, and the requirements for their accuracy also began to rise.

In 1895, transformers began to be used to measure voltage, and in 1899, Siemens patented the current transformer. Increasing powers and voltages increased the role of instrument transformers as measuring and protection devices, resulting in increased requirements for the accuracy of instrument transformers, especially when measuring the amount of energy produced. Instrument transformers are part of power systems [1]. They aim to transmit operating current and voltage values for measurement or protection applications. So, the design problems of instrument transformers differ from those of power transformers problems [2]. Power transformer standards IEC/EN 60076, IEC/EN 60726, IEC/EN 61558, and IEC/EN 60551 apply to both 1- and 3-phase transformers, voltage values, losses in transformers, and noise levels. However, although the principle of operation of instrument transformers is the same as that of a power transformer, the requirements placed on them concern the conversion of currents and voltages with high accuracy [3,4]. The standards vary depending on the type of instrument transformer; current, voltage, combined, measurement, insurance, for steady, and transient state but are also very restrictive ICE/EN 61896-1 and etc. For example, classes of 0.1, 0.2, and 0.5 are required for measuring current transformers, i.e., the amplitude error in percent and phase shift errors in minutes.

Currently, designs other than inductive instrument transformers have appeared, such as electronic and optoelectronic instrument transformers based on the Faraday magneto-optical phenomenon [5] and the Pockels effect using optical fibers [6], as well as transformers using microwaves and the so-called Bluetooth technology. However, induction instrument transformers are still most often used due to their reliability, and problems related to the design of more and more accurate structures still exist.

Citation: Lesniewska, E.; Witczak, P. Problems during the Design and Testing of Instrument, Special and Power Transformers: The Outlook. *Energies* **2024**, *17*, 363. https://doi.org/10.3390/en17020363

Received: 14 November 2023
Accepted: 10 December 2023
Published: 11 January 2024

Special transformers are often single phase and built for various particular purposes, e.g., short circuits, welding, testing, semiconductor, and safety transformers [7,8].

In this Special Issue of *Energies*, in addition to issues related to three-phase power transformers (contribution 1–4), problems of measurement transformers are addressed in articles (contribution 5,6) and a review article (contribution 7). However, the authors of (contribution 5,8–10) analyzed special transformers. In addition, there is also a review article (contribution 11), which presents exciting transformer models implemented using the SPICE program, considering the nonlinearity of magnetization characteristics and magnetic couplings, thermal compressions, and temperature differences [9].

The authors of (contribution 1) studied the physical phenomena occurring in a three-phase power transformer during the first period of its power supply. The problem of limiting inrush current surges in 3-phase power transformers is a current one and has also been the subject of other articles, e.g., [10]. The research used a nonlinear model based on classical modeling, which does not consider hysteresis. The model is described using a system of stiff, nonlinear ordinary differential equations. Computational methods that considered the influence of residual magnetism in various columns of the transformer core and the influence of the time moment determined in the voltage waveform at which the indicated voltage is applied to a given transformer winding were examined.

The authors of (contribution 2) compared 3D and 2D finite element models of a power transformer designed for reactive power compensation stations. The results of numerical 2D and 3D calculations of no-load current and losses in the transformer core were obtained and compared with data measured at the test station. Correctly estimating power losses is essential for the correct operating conditions of the transformer. Even moderate excesses of rated oil temperatures may release excessive amounts of dissolved gases [11]. The impact of considering the hysteresis loop phenomenon on the calculation of core losses was investigated using the Jiles–Atherton model [12]. The equivalence of two- and three-dimensional numerical models has been described in a previous publication [13]. The results presented in this paper also show that the numerical model should consider the area of core overlapping.

In turn, the authors of (contribution 3) considered the forces that can appear in conductors connecting the transformer windings and on-load tap changers during short-circuit conditions, which may lead to extremely dangerous electromechanical effects [14].

A mechanical transient simulation was analyzed in [15] to determine the deformation and mechanical stresses in the busbar system, which has a simple geometry and linear material properties.

Through repeated 3D numerical experiments, standard analytical formulas were extended to consider the presence of a ferromagnetic plate and proximity effects introduced by eddy currents inside the conductors. It makes it possible to overcome the limitations of the so-called multi-physics approach [16] and calculate the realistic structures of transformer leads.

The authors of (contribution 4) presented a step-by-step methodology for calculating the vibrations of the transformer tank caused by electromagnetic forces acting in the phase windings. In their approach, they use three-dimensional finite element models to calculate the distribution of forces and the acoustic pressure field in an oil-filled tank. A key factor in this analysis is the cross-domain data connection algorithm. The calculation results are verified by measuring the vibrations of the tank walls. The detailed model of the transformer, including the winding structure reduced to its orthotropic equivalent, was analyzed in [17] using the coupled magneto-mechanical approach. A similar treatment but extended to the real coil distribution in space may be found in [18]. In both of these works, a two-stage 2D-3D transformer model was used with considering construction details.

Instrument transformers are an element of the power measurement system; these devices process signals but add their transformation errors to the measured values. International Electro-technical Commission (IEC) standards require designers to identify and limit these errors. Moreover, by coupling the primary and secondary circuits, instrument transformers enable the propagation of electromagnetic disturbances generated in their primary circuit [19]. Problems with measurement accuracy are becoming more important due to the use of non-linear receivers and distributed generation units with uneven electricity production depending on external factors that cause deformation of the voltage shape in the power system. Also, due to their magnetic characteristic, instrument transformers operating in an overcurrent state may be an element introducing higher harmonics into measurement systems.

The authors of (contribution 6,7) studied the influence of higher harmonics of supply (up to the 100th harmonic) on the accuracy of measuring inductive instrument transformers. The authors of (contribution 7) present a review article that discusses advanced work devoted to characterizing the accuracy of the transformation of currents and voltages distorted by inductive voltage and current transformers.

Special current transformers deserve special attention.

Electronic instrument transformers using low signals from a Rogowski coil are known [20]. However, they have not found widespread use. Nevertheless, the authors of (contribution 10) developed an electronic DC transformer, where it is impossible to apply Faraday's law to change the voltage and current values. The work is innovative and interesting, mainly because electricity is sometimes transmitted via direct current lines. HVDC direct current transmission technology has been developed and improved since the first half of the 20th century, and now—especially in the face of energy transformation—is gaining importance [21]. A limitation of HVDC is also the complexity of the electrical energy conversion process and operation of HVDC stations—(this also applies to electronic DC current transformers). Lem's DC current transformers are known, but in (contribution 10), a different method of measuring current was proposed; this method involves using Hall sensors without iron cores, significantly reducing environmental interference during measurement. The authors focused on the accuracy of this method, i.e., the precise measurement and calibration of the delay time of an electronic DC current transformer under operating conditions.

In turn, the authors of (contribution 5) studied the known through-line primary conductor current transformer with a unique application and a particular design. Due to its application, it can be said to belong to both measuring current transformers and special transformers. Compared to classic through-line current transformers, the difference is that instead of the primary wire with a current of approximately several to several dozen kiloamps, a beam of particles passes through the center of the core.

This creates other measurement conditions, and it turns out that the non-uniform magnetic permeability of the ferromagnetic sheets around the core circumference influences the measurement accuracy. Therefore, what is new here is the design used by the authors to compensate for inaccuracies in the distribution of the core material. In this way, they obtained an amplitude error of 0.1%, i.e., a current transformer with very high accuracy. It enables precise measurements of the number of particles contained in the beams of the Large Hadron Collider at CERN [22]. The technology described and tested in the laboratory constitutes a significant advance in the state of knowledge in the instrumentation of super-accurate current transformers.

The authors of (contribution 8,9) special single-phase transformers, which have opposing design assumptions due to their operation. These are special high-power transformers that operate practically in the state of short circuits. The no-load condition for these types of transformers is an emergency, not an operating condition. Therefore, they are tested in the short-circuit condition, and the parameters of the transverse branch of the transformer equivalent circuit are irrelevant. The leakage inductance of these transformers is a critical design parameter.

The authors of (contribution 9) studied a semiconductor transformer operating in a system with electronic elements. For the proper operation of the entire system, obtaining the lowest possible transformer leakage inductance is necessary because the power that the transformer can transfer depends on it. The dissipation flow is unfavorable because it partially passes through the core, causes additional heat release, and thus limits the power transferred to the secondary circuit. This article considers the problem of selecting the type of transformer windings that will provide the lowest leakage inductance. Interleaved windings of the Litz type [23] and windings made of thin aluminum foil [24,25] were considered. The developed 3D model of a transformer with a GOES steel core considers the phenomena of current displacement and proximity in the windings. The results were compared with measurements made on the prototype.

In turn, the authors of (contribution 8) described the problems of designing a dissipation transformer with increased leakage reactance, powering systems operating in a near-short-circuit state, where there is a need to limit the short-circuit current. Such a device must ensure the continuity of the electric arc and prevent the flow of high current in the event of a short-circuit of the electrodes. These include car starting transformers, welding machines, soldering irons, and welding transformers [7]. The authors analyzed the magnetic field dissipation in a transformer with beveled edges and a magnetic shunt. The reactance of the fault transformer was calculated as a parameter for the selected position of the magnetic shunt. Point measurements of magnetic flux density vectors and winding reactance were carried out to verify the calculation results.

Author Contributions: Conceptualization: E.L.; methodology: E.L. and P.W.; validation: E.L. and P.W.; formal analysis: E.L. and P.W.; investigation: E.L.; resources: E.L. and P.W.; data curation: E.L. and P.W.; writing—original draft preparation: E.L. and P.W.; writing—review and editing: E.L.; supervision: E.L. All authors have read and agreed to the published version of the manuscript.

Conflicts of Interest: The authors declare no conflict of interest.

List of Contributions

1. Łukaniszyn, M.; Baron, B.; Kolańska-Płuska, J.; Majka, Ł. Inrush Current Reduction Strategy for a Three-Phase Dy Transformer Based on Pre-Magnetization of the Columns and Controlled Switching. *Energies* **2023**, *16*, 5238. https://doi.org/10.3390/en16135238.
2. Osinski, P.; Witczak, P. Analysis of Core Losses in Transformer Working at Static Var Compensator. *Energies* **2023**, *16*, 4584. https://doi.org/10.3390/en16124584.
3. Smoliński, M.; Witczak, P. Numerical and Analytical Determination of Steady-State Forces Acting on Cleats and Leads Conductor of the Power Transformer. *Energies* **2023**, *16*, 3600. https://doi.org/10.3390/en16083600.
4. Witczak, P.; Swiatkowski, M. Transmission of Vibrations from Windings to Tank in High Power Transformers. *Energies* **2023**, *16*, 2886. https://doi.org/10.3390/en16062886.
5. Krupa, M.; Gasior, M. A new wall current transformer for accurate beam intensity measurements in the Large Hadron Collider. *Energies* **2023**, *16*, 7442. https://doi.org/10.3390/en16217442.
6. Kaczmarek, M.; Stano, E. New Approach to Evaluate the Transformation Accuracy of Inductive CTs for Distorted Current. *Energies* **2023**, *16*, 3026. https://doi.org/10.3390/en16073026.
7. Kaczmarek, M.; Stano, E. Challenges of Accurate Measurement of Distorted Current and Voltage in the Power Grid by Conventional Instrument Transformers. *Energies* **2023**, *16*, 2648. https://doi.org/10.3390/en16062648.
8. Tomczuk, B.; Weber, D. Effect of Magnetic Shunts on Shell-Type Transformers Characteristics. *Energies* **2023**, *16*, 6814. https://doi.org/10.3390/en16196814.
9. Lesniewska, E.; Daniel Roger, D. Selection of the Winding Type of Solid-State Transformers in Terms of Transmitting the Greatest Possible Power in the Frequency Range from 500 Hz to 6000 Hz. *Energies* **2023**, *16*, 6528. https://doi.org/10.3390/en16186528.
10. Zhu, M.; Tang, H.; He, Z.; Liao, Y.; Tang, B.; Zhang, Q.; Shu, H.; Deng, Y.; Zeng, F.; Cao, P. Field Test Method and Equivalence Analysis of Delay Characteristics of DC Electronic Current Transformer. *Energies* **2023**, *16*, 5727. https://doi.org/10.3390/en16155727.
11. Górecki, K.; Detka, K. SPICE-Aided Models of Magnetic Elements—A Critical Review. *Energies* **2023**, *16*, 6568. https://doi.org/10.3390/en16186568.

References

1. Wiszniewski, A. *Instrument Transformers in the Power Industry*; WNT: Warsow, Poland, 1983; pp. 1–302. ISBN 8320403707. (In Polish)
2. Jezierski, E. *Power Transformer*; WNT: Warsow, Poland, 1983; pp. 1–650. ISBN 5905279053229. (In Polish)

3. Koszmider, A.; Olak, J.; Piotrowski, Z. *Current Transformers*; WNT: Warsow, Poland, 1985; pp. 1–283. ISBN 8320407109. (In Polish)
4. Lesniewska, E. Modern Methods of Construction Problem Solving in Designing Various Types of Instrument Transformers. *Energies* **2022**, *15*, 8199. [CrossRef]
5. Mihailovic, P.; Petricevic, S. Fiber Optic Sensors Based on the Faraday Effect. *Sensors* **2021**, *21*, 6564. [CrossRef] [PubMed]
6. Kurosawa, K.; Yoshida, S.; Mori, E.; Takahashi, G.; Saito, S. Development of an optical instrument transformer for DC voltage measurement. *IEEE Trans. Power Deliv.* **1993**, *8*, 1721–1726. [CrossRef]
7. Sakhno, L.; Sakhno, O.; Dubitsky, S. Field-circuit modelling of an advanced welding transformer with two parallel rectifiers. *Arch. Electr. Eng.* **2015**, *64*, 249–257. [CrossRef]
8. Lesniewska, E.; Olak, J. Analysis of the Operation of Cascade Current Transformers for Measurements of Short-Circuit Currents with a Non-Periodic Component with a Large Time Constant of Its Decay. *Energies* **2022**, *15*, 2925. [CrossRef]
9. Yi, R.; Chen, S.; Zhang, J.; Zhao, H.; Yu, Y.; Zhu, H. Research on Reverse optimized Control of Dual Active Bridge Based on PSpice-MATLAB/Simulink. In Proceedings of the 2022 IEEE 5th International Conference on Electronics Technology (ICET), Chengdu, China, 13–16 May 2022; pp. 243–248, ISBN 978-1-6654-8509-8.
10. Hamilton, R. Analysis of transformer inrush current and comparison of harmonic restraint methods in transformer protection. *IEEE Trans. Ind. Appl.* **2013**, *49*, 1890–1899. [CrossRef]
11. Girgis, R.; Nyenhuis, G.E. Hydrogen Gas Generation Due to Moderately Overheated Transformer Cores. In Proceedings of the IEEE Power & Energy Society General Meeting, Calgary, AB, Canada, 26–30 July 2009. [CrossRef]
12. Jiles, D.; Atherton, D. Theory of ferromagnetic hysteresis. *J. Magn. Magn. Mater.* **1986**, *61*, 48–60. [CrossRef]
13. Ciesielski, M.; Witczak, P. The use of finite elements modeling to analyze phase shifting transformer in steady-state service conditions. *COMPEL Int. J. Comput. Math. Electr. Electron. Eng.* **2022**, *41*, 1214–1222. [CrossRef]
14. Faiz, J.; Ebrahimi, B.M.; Noori, T. Three- and Two-Dimensional Finite-Element Computation of Inrush Current and Short-Circuit Electromagnetic Forces on Windings of a Three-Phase Core-Type Power Transformer. *IEEE Trans. Magn.* **2008**, *44*, 590–597. [CrossRef]
15. Szulborski, M.; Łapczynski, S.; Kolimas, Ł.; Kozarek, Ł.; Rasolomampionona, D.D. Calculations of Electrodynamic Forces in Three-Phase Asymmetric Busbar System with the Use of FEM. *Energies* **2020**, *13*, 5477. [CrossRef]
16. Kadkhodaei, G.; Sheshyekani, K.; Hamzeh, M.; Dadjo Tavakoli, S. Multiphysics analysis of busbars with various arrangements under short-circuit condition. *IET Electr. Syst. Transp.* **2016**, *6*, 250–260. [CrossRef]
17. Ertl, M.; Landes, H. Investigation of load noise generation of large power transformer by means of coupled 3D FEM analysis. *COMPEL-Int. J. Comput. Math. Electr. Electron. Eng.* **2007**, *26*, 788–799. [CrossRef]
18. Rausch, M.; Kaltenbacher, M.; Landes, H.; Lerch, R.; Anger, J.; Gerth, J.; Boss, P. Combination of finite and boundary element methods in investigation and prediction of load-controlled noise of power transformers. *J. Sound Vib.* **2002**, *250*, 323–338. [CrossRef]
19. Cataliotti, A.; Cosentino, V.; Crotti, G.; Giordano, D.; Modarres, M.; Di Cara, D.; Tinè, G.; Gallo, D.; Landi, C.; Luiso, M. Metrological performances of voltage and current instrument transformers in harmonics measurements. In Proceedings of the I2MTC—2018 IEEE International Instrumentation and Measurement Technology Conference: Discovering New Horizons in Instrumentation and Measurement, Houston, TX, USA, 14–17 May 2018; pp. 1–6. [CrossRef]
20. Tianshu, B.; Fei, S.; Hao, L.; Kenneth, E.M.; JInsong, L. Rogowski Electronic Transformer Performance Evaluation and Its Impact on PMUs. *FAC-PapersOnLine* **2016**, *49*, 96–102.
21. Bahmani, M.A.; Thiringer, T.; Kharezy, M. Design Methodology and Optimization of a Medium-Frequency Transformer for High-Power DC–DC Applications. *IEEE Trans. Ind. Applicat.* **2016**, *52*, 4225–4233. [CrossRef]
22. Castelnovo, P.; Florio, M.; Forte, S.; Rossi, L.; Sirtori, E. The economic impact of technological procurement for large-scale research infrastructures: Evidence from the Large Hadron Collider at CERN. *Res. Policy* **2018**, *47*, 1853–1867. [CrossRef]
23. Otomo, Y.; Igarashi, H.; Sano, H.; Yamada, T. Analysis of Litz Wire Losses Using Homogenization-Based FEM. *IEEE Trans. Magn.* **2021**, *57*, 7402409. [CrossRef]
24. Sullivan, C.R. Prospects for advances in power magnetics. In Proceedings of the CIPS 2016—9th International Conference on Integrated Power Electronics Systems, Nuremberg, Germany, 8–10 March 2016.
25. Gradinger, T.B.; Mogorovic, M. Foil-winding design for medium-frequency medium-voltage transformers. In Proceedings of the 2021 23rd European Conference on Power Electronics and Applications, Ghent, Belgium, 6–10 September 2021; pp. 1–10. [CrossRef]

energies

MDPI

Article

Inrush Current Reduction Strategy for a Three-Phase Dy Transformer Based on Pre-Magnetization of the Columns and Controlled Switching

Marian Łukaniszyn [1], Bernard Baron [1], Joanna Kolańska-Płuska [1] and Łukasz Majka [2,*]

[1] Department of Drive Automation and Robotics, Faculty of Electrical Engineering, Automatic Control and Informatics, Opole University of Technology, Prószkowska Street 76, 45-272 Opole, Poland; m.lukaniszyn@po.edu.pl (M.Ł.)

[2] Department of Electrical Engineering and Computer Science, Faculty of Electrical Engineering, Silesian University of Technology, Akademicka Street 10, 44-100 Gliwice, Poland

* Correspondence: lmajka@polsl.pl

Abstract: The methodology and test results of a three-phase three-column transformer with a Dy connection group are presented in this paper. This study covers the dynamics of events that took place in the first period of the transient state caused by the energizing of the transformer under no-load conditions. The origin of inrush currents was analyzed. The influence of factors accompanying the switch-on and the impact of the model parameters on the distribution and maximum values of these currents was studied. In particular, the computational methods of taking into account the influence of residual magnetism in different columns of the transformer core, as well as the impact of the time instant determined in the voltage waveform at which the indicated voltage is supplied to a given transformer winding, were examined. The study was carried out using a nonlinear model constructed on the basis of classical modeling, in which hysteresis is not taken into account. Such a formulated model requires simplification, which is discussed in this paper. The model is described using a system of stiff nonlinear ordinary differential equations. In order to solve the stiff differential state equations set for the transient states of a three-phase transformer in a no-load condition, a Runge–Kutta method, namely the Radau IIA method, with ninth-order quadrature formulas was applied. All calculations were carried out using the authors' own software, written in C#. A ready-made strategy for energizing a three-column three-phase transformer with a suitable pre-magnetization of its columns is given.

Keywords: stiff nonlinear ordinary differential equations; Runge–Kutta implicit methods; circuit model of a three-phase transformer; transient state; inrush currents of the unloaded transformer; the time instant specified in the waveform of the switched supply voltage; pre-magnetization

Citation: Łukaniszyn, M.; Baron, B.; Kolańska-Płuska, J.; Majka, Ł. Inrush Current Reduction Strategy for a Three-Phase Dy Transformer Based on Pre-Magnetization of the Columns and Controlled Switching. *Energies* **2023**, *16*, 5238. https://doi.org/10.3390/en16135238

Academic Editors: Elzbieta Lesniewska, Xose Lopez-Fernandez and Pawel Witczak

Received: 3 June 2023
Revised: 27 June 2023
Accepted: 29 June 2023
Published: 7 July 2023

1. Introduction

In transmission and distribution systems, transformer energization is a fairly routine operation. However, with the increase in distributed generation sources with inherent high intermittency resulting in more switching events, the transformers in service are becoming more and more vulnerable to electrical transients. A transient state during energizing can cause significant inrush currents [1].

Inrush currents generate forces comparable to those of short circuits. The problem is that they occur more frequently than short-circuiting and last longer. Thus, at every occurrence of an inrush, there is some degradation of the conductor and its insulation. After many such inrushes, local hotspots may emerge in the winding. The transformer inrush transient is not only dangerous because of its large current amplitude but also due to its rapid rise rate [2,3]. When a transformer is frequently exposed to transients, it will deteriorate due to severe mechanical and thermal stresses and may eventually fail [4–6]. The high inrush current may disturb or damage the operation of adjacent equipment in the circuit resulting in, e.g., the maloperations of power electronic converters [7,8] and protection relays [9]. Apart from affecting power quality in terms of temporary under-voltage (sagging) [10], the inrush currents contain many high frequency harmonics which can also lead to harmonic resonant over-voltage [1].

Inrush currents during transformer energization are widely studied and extensively described in the literature [1,7–21]. The survey of already documented research shows that the primary objective is to mitigate the impact of the transient phenomenon during transformer energization. The efforts of scientists and engineers are concentrated on creating an accurate model of the transformer to be used in the inrush transient studies [12,13,17] as well as investigating the nature (physics) of inrush current and the factors affecting this phenomenon [1,14,16,21].

One important issue is the correct selection of the mathematical description of the model (i.e., the choice of differential equations that define its dynamics) and method necessary to solve it [3,22–26]. In fact, this is a research matter, which leads to understanding the different phenomena involved and making suitable assumptions.

The research into transformer dynamics is a part of the much broader issue—the simulations of electromagnetic transients in power systems, which are essential for the adequate design of equipment and its protection [1,2]. Power transformers play important roles in power transmission and their performances have significant impact on power quality and the lifetime of power system apparatuses [9,11]. The major reasons for the failure of transformers are the thermal, electrical and mechanical stresses of transformer winding insulation. Therefore, the reliable models and methods associated with the representation of the transient states of transformers are required for investigation.

State of the Art

There are numerous mitigation techniques for transformer energization transients. These include introducing a pre-insertion resistor, controlling the switching time using the point-on-wave voltage at energization, varying the impedance of the power supply, and controlling the residual flux inside the transformer core during transformer energization [12–21].

A survey of scholarly knowledge on the topic also provides information about a number of controllable factors that are relevant to the theory of inrush currents.

The list of factors affecting inrush currents that are found in publications is given below:

- Starting/switching phase angle of voltage;
- Residual flux in core;
- Magnitude of voltage;
- Saturation flux;
- Core material;
- Supply/source impedance;
- Load and size of transformer.

This paper describes the simulation studies on the inrush current for the delta side energization of the transformer. This is a new research subject and seems essential in terms of practicality. It is also a continuation of research carried out by the authors [3,27–29].

2. Model of a Three-Phase Transformer with Connection Group Dy with a Non-Linear Magnetization Curve in the No-Load State

2.1. Model Concept

A research task of transient analysis was formulated for an unloaded three-phase transformer with primary side windings connected in delta. In order to carry it out, differential equations describing the adopted mathematical model were first formulated along with the assumptions.

The equivalent scheme of the transformer for the discussed model is shown in Figure 1, together with physical quantities that describe the model.

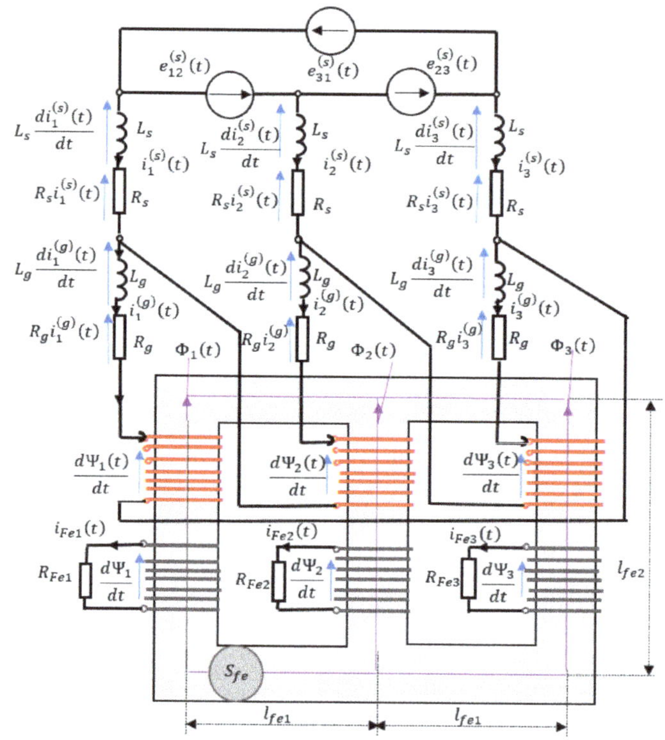

Figure 1. Equivalent scheme of the analyzed three-phase Dy transformer in no-load conditions.

A list of all the designations used in this work can be found in Nomenclature.

For the planned calculation process, it was assumed that the switching on of the transformer at time instant t_0 had the effect of the emergence of three voltages with instantaneous values in the following form:

$$\begin{aligned} e_{12}^{(s)}(t) &= E_{m12}\sin(\omega(t+t_0)), \\ e_{23}^{(s)}(t) &= E_{m23}\sin\left(\omega(t+t_0)+\tfrac{2}{3}\pi\right), \\ e_{31}^{(s)}(t) &= E_{m31}\sin\left(\omega(t+t_0)+\tfrac{4}{3}\pi\right). \end{aligned} \tag{1}$$

By making the parameter t_0 variable, it was possible to make the calculations dependent on the time instant specified in the waveform of the selected phase-to-phase voltage for the process of switching on the transformer, i.e., the start of the calculations.

In the investigation of the transient states, an invariant reference system was assumed. It was assumed that the reference point was the time instant at which the voltage $e_{12}^{(s)}(t)$ passes through zero (which occurs for $t = -t_0$).

We assume that the non-linear curve $H(B)$ of the core material is known in the case of discussed transformer. In our experiment, this task was implemented using a polynomial described by odd powers (2). The correctness of this method was verified experimentally [3].

$$H(B) = \sum_{k=1}^{11} a_k B^{2k-1} \tag{2}$$

Restricting the model to include only the nonlinear relationship $I_0 - \Psi$ was insufficient for model losses in the ferromagnetic material of the core. The total losses were taken into account according to the classical concept of transformer modeling. In each column of the three-phase transformer, an additional electromotive force was introduced in the form of a flux linkage derivative $\frac{d}{dt}\Psi_k(t)$ at the terminals of a coil with z_g turns. The coil with zero internal resistance was loaded with a resistance $R_{Fe,k}$, which modeled the total iron power losses in the $k-th$ column of the transformer (Figure 1).

Using the circuit description in modeling the magnetic circuit of the transformer, we can state that:

$$\frac{d}{dt}\Psi_k(t) = R_{Fe,k} i_{Fe,k}(t) \quad (3)$$

where $i_{Fe,k}(t)$ is a current of the coil under consideration.

It was assumed that the parameters $R_{Fe,k}$ could be determined by measuring the active power of a transformer operating at steady state under idling conditions.

The mathematical model of the transformer formulated in this way also offers the possibility of determining the steady state in the form of a limit cycle of the solution of the relevant differential equations. Details on the experimentally verified methodology of this subject may be found in [3,27].

2.2. Residual Magnetism

The formulated transformer model intentionally did not consider magnetic hysteresis. This was not necessary, since it was not intended to track the history of core magnetization, that is, the points at which reversals in the directions of changes in magnetic currents and fluxes occurred. Instead, the model offered the possibility of taking into account the magnetization of the transformer core at the particular time instant, when the transformer was switched on. The state of magnetization was taken into account in the model in the form of residual magnetism. If we know the amount of residual magnetism, then it is possible to set initial conditions for the state variables $\Psi_2(t_0), \Psi_3(t_0)$, which represent flux linkages.

As indicated by numerous literature sources, e.g., [11,14], the energizing of unloaded transformer is accompanied by the occurrence of current pulses. Their presence is limited to the first period determined by the supply voltage. The amplitudes of these currents, as a rule, exceed the rated values, and not infrequently reach values close to short-circuit values.

Basing on these findings, we argue that if the transient state analysis of the transformer is limited to a single period from the energizing instant, the proposed model will make it possible (if initial conditions are set) to investigate the impact of residual magnetism on the magnitudes of generated current pulses.

The analysis of the physics of the formation of current pulses has already established that they (the pulses) first appear as a result of locating the working point of the inductor ($I_0 - \Psi$ solution) far in the saturation region of transformer's core magnetization curve [3,12].

Therefore, the inclusion of the core's residual magnetism in the model must be supported by an accurate approximation of the core's nonlinear curve in the saturation region.

2.3. Dynamic Equations of the Unloaded Transformer Dy

In the developed mathematical model of the transformer, the magnetomotive forces (mmfs) generated by the coil currents $i_k^{(g)}(t)(k = 1, 2, 3)$ were supplemented by the magnetomotive forces generated by certain equivalent currents $i_{R_{Fe,k}}(t)$. These mmfs are expressed as $z_g i_{R_{Fe,k}}(t)$, and they corresponded to real power losses in the ferromagnetic core.

Such a concept for modeling a three-phase transformer with three columns and star-connected primary side windings has been comprehensively analyzed before, and the results of these studies were published in [27]. There, the flux linkages $\Psi_1(t), \Psi_2(t)$ were chosen as the state variables.

In the considered case, the windings of the primary side of the three-phase transformer were delta-connected. The flux linkages $\Psi_2(t), \Psi_3(t)$ were chosen as the state variables in the magnetic circuit equations.

Taking into account Kirchhoff's first and second laws for the magnetic circuit of a three-phase transformer with three columns (Figure 1) and Equation (3) modeling the total iron losses, the equations of state for the variables $\Psi_2(t), \Psi_3(t)$ were formulated:

$$\boldsymbol{G}\begin{bmatrix} \frac{\mathrm{d}}{\mathrm{d}t}\Psi_2(t) \\ \frac{\mathrm{d}}{\mathrm{d}t}\Psi_3(t) \end{bmatrix} = \begin{bmatrix} h_4(\boldsymbol{X}) \\ h_5(\boldsymbol{X}) \end{bmatrix} \tag{4}$$

where $\boldsymbol{G} = \begin{bmatrix} \frac{1}{R_{\mathrm{Fe1}}} + \frac{1}{R_{\mathrm{Fe2}}} & \frac{1}{R_{\mathrm{Fe1}}} \\ \frac{1}{R_{\mathrm{Fe2}}} & -\frac{1}{R_{\mathrm{Fe3}}} \end{bmatrix}$
and

$$\begin{aligned} h_4(\boldsymbol{X}) &= -i_1^{(\mathrm{g})}(t) + i_2^{(\mathrm{g})}(t) + \tfrac{1}{z_{\mathrm{g}}}U_{\mathrm{m},1}(-\Psi_2(t) - \Psi_3(t)) - \tfrac{1}{z_{\mathrm{g}}}U_{\mathrm{m},2}\Psi_2(t) \\ h_5(\boldsymbol{X}) &= -i_2^{(\mathrm{g})}(t) + i_3^{(\mathrm{g})}(t) - \tfrac{1}{z_{\mathrm{g}}}U_{\mathrm{m},2}\Psi_2(t) + \tfrac{1}{z_{\mathrm{g}}}U_{\mathrm{m},3}\Psi_3(t) \end{aligned}$$

When the system of Equation (4) is solved for the derivatives, the results are:

$$\begin{bmatrix} \frac{\mathrm{d}}{\mathrm{d}t}x_4(t) \\ \frac{\mathrm{d}}{\mathrm{d}t}x_5(t) \end{bmatrix} = \begin{bmatrix} \frac{\mathrm{d}}{\mathrm{d}t}\Psi_2(t) \\ \frac{\mathrm{d}}{\mathrm{d}t}\Psi_3(t) \end{bmatrix} = \boldsymbol{G}^{-1}\begin{bmatrix} h_4(\boldsymbol{X}) \\ h_5(\boldsymbol{X}) \end{bmatrix} = \begin{bmatrix} f_4(\boldsymbol{X}) \\ f_5(\boldsymbol{X}) \end{bmatrix} \tag{5}$$

where $\boldsymbol{C} = \boldsymbol{G}^{-1}\begin{bmatrix} c_{1,1} & c_{1,2} \\ c_{2,1} & c_{2,2} \end{bmatrix}$
and
$$c_{1,1} = \frac{1}{\frac{1}{R_{\mathrm{Fe1}}} + \frac{1}{R_{\mathrm{Fe2}}} + \frac{R_{\mathrm{Fe3}}}{R_{\mathrm{Fe1}}R_{\mathrm{Fe2}}}} \quad c_{1,2} = \frac{1}{\frac{1}{R_{\mathrm{Fe2}}} + \frac{1}{R_{\mathrm{Fe3}}} + \frac{R_{\mathrm{Fe1}}}{R_{\mathrm{Fe2}}R_{\mathrm{Fe3}}}}$$
$$c_{2,1} = \frac{1}{\frac{1}{R_{\mathrm{Fe3}}} + \frac{1}{R_{\mathrm{Fe1}}} + \frac{R_{\mathrm{Fe2}}}{R_{\mathrm{Fe3}}R_{\mathrm{Fe1}}}} \quad c_{2,2} = -\frac{1}{\frac{1}{R_{\mathrm{Fe3}}} + \frac{1}{R_{\mathrm{Fe1}} + R_{\mathrm{Fe2}}}}.$$

Equations for three independent circulations appeared as a result of applying Kirchhoff's second law to the modeled equivalent circuits of the primary side of the transformer (Figure 1):

$$\begin{aligned} &-L_{\mathrm{s}}\tfrac{\mathrm{d}}{\mathrm{d}t}i_1^{(\mathrm{s})}(t) - R_{\mathrm{s}}i_1^{(\mathrm{s})}(t) + L_{\mathrm{s}}\tfrac{\mathrm{d}}{\mathrm{d}t}i_2^{(\mathrm{s})}(t) + R_{\mathrm{s}}i_2^{(\mathrm{s})}(t) + L_{\mathrm{g}}\tfrac{\mathrm{d}}{\mathrm{d}t}i_2^{(\mathrm{g})}(t) + R_{\mathrm{g}}i_2^{(\mathrm{g})}(t) + \\ &\qquad + \tfrac{\mathrm{d}}{\mathrm{d}t}\Psi_2(t) = e_{12}^{(\mathrm{s})}(t) \\ &-L_{\mathrm{s}}\tfrac{\mathrm{d}}{\mathrm{d}t}i_2^{(\mathrm{s})}(t) - R_{\mathrm{s}}i_2^{(\mathrm{s})}(t) + L_{\mathrm{s}}\tfrac{\mathrm{d}}{\mathrm{d}t}i_3^{(\mathrm{s})}(t) + R_{\mathrm{s}}i_3^{(\mathrm{s})}(t) + L_{\mathrm{g}}\tfrac{\mathrm{d}}{\mathrm{d}t}i_3^{(\mathrm{g})}(t) + R_{\mathrm{g}}i_3^{(\mathrm{g})}(t) + \\ &\qquad + \tfrac{\mathrm{d}}{\mathrm{d}t}\Psi_3(t) = e_{23}^{(\mathrm{s})}(t) \\ &L_{\mathrm{g}}\tfrac{\mathrm{d}}{\mathrm{d}t}i_1^{(\mathrm{g})}(t) - R_{\mathrm{g}}i_1^{(\mathrm{g})}(t) + L_{\mathrm{g}}\tfrac{\mathrm{d}}{\mathrm{d}t}i_2^{(\mathrm{g})}(t) + R_{\mathrm{g}}i_2^{(\mathrm{g})}(t) + L_{\mathrm{g}}\tfrac{\mathrm{d}}{\mathrm{d}t}i_3^{(\mathrm{g})}(t) + R_{\mathrm{g}}i_3^{(\mathrm{g})}(t) + \\ &\qquad + \tfrac{\mathrm{d}}{\mathrm{d}t}\Psi_1(t) + \tfrac{\mathrm{d}}{\mathrm{d}t}\Psi_2(t) + \tfrac{\mathrm{d}}{\mathrm{d}t}\Psi_3(t) = e_{12}^{(\mathrm{s})}(t) \end{aligned} \tag{6}$$

where

$$\begin{aligned} i_1^{(\mathrm{s})} &= i_1^{(\mathrm{g})} - i_2^{(\mathrm{g})} \\ i_2^{(\mathrm{s})} &= i_2^{(\mathrm{g})} - i_3^{(\mathrm{g})} \\ i_3^{(\mathrm{s})} &= i_3^{(\mathrm{g})} - i_1^{(\mathrm{g})} \end{aligned}$$

The resulting system of equations was ordered in matrix form:

$$\boldsymbol{L}\begin{bmatrix} \frac{\mathrm{d}}{\mathrm{d}t}i_1^{(\mathrm{g})}(t) \\ \frac{\mathrm{d}}{\mathrm{d}t}i_2^{(\mathrm{g})}(t) \\ \frac{\mathrm{d}}{\mathrm{d}t}i_3^{(\mathrm{g})}(t) \end{bmatrix} = \boldsymbol{R}\begin{bmatrix} i_1^{(\mathrm{g})}(t) \\ i_2^{(\mathrm{g})}(t) \\ i_3^{(\mathrm{g})}(t) \end{bmatrix} - \begin{bmatrix} \frac{\mathrm{d}}{\mathrm{d}t}\Psi_2(t) \\ \frac{\mathrm{d}}{\mathrm{d}t}\Psi_3(t) \\ 0 \end{bmatrix} + \begin{bmatrix} e_{12}^{(\mathrm{s})}(t) \\ e_{12}^{(\mathrm{s})}(t) \\ 0 \end{bmatrix} \tag{7}$$

where

$$\boldsymbol{L} = \begin{bmatrix} -L_{\mathrm{s}} & 2L_{\mathrm{s}} + L_{\mathrm{g}} & -L_{\mathrm{s}} \\ -L_{\mathrm{s}} & -L_{\mathrm{s}} & 2L_{\mathrm{s}} + L_{\mathrm{g}} \\ L_{\mathrm{g}} & L_{\mathrm{g}} & L_{\mathrm{g}} \end{bmatrix}, \boldsymbol{R} = \begin{bmatrix} R_{\mathrm{s}} & -2R_{\mathrm{s}} - R_{\mathrm{g}} & -R_{\mathrm{s}} \\ R_{\mathrm{s}} & R_{\mathrm{s}} & -2R_{\mathrm{s}} - R_{\mathrm{g}} \\ -R_{\mathrm{g}} & -R_{\mathrm{g}} & -R_{\mathrm{g}} \end{bmatrix}$$

When the system of Equation (7) is solved for the derivatives, taking into account solution (5), the results are:

$$\begin{bmatrix} \frac{d}{dt}x_1(t) \\ \frac{d}{dt}x_2(t) \\ \frac{d}{dt}x_3(t) \end{bmatrix} = \begin{bmatrix} \frac{d}{dt}i_1^{(g)}(t) \\ \frac{d}{dt}i_2^{(g)}(t) \\ \frac{d}{dt}i_3^{(g)}(t) \end{bmatrix} = \begin{bmatrix} f_1(\boldsymbol{X}) \\ f_2(\boldsymbol{X}) \\ f_3(\boldsymbol{X}) \end{bmatrix} \tag{8}$$

where

$$\begin{bmatrix} f_1(\boldsymbol{X}) \\ f_2(\boldsymbol{X}) \\ f_3(\boldsymbol{X}) \end{bmatrix} = \boldsymbol{L}^{-1}\left(\boldsymbol{R}\begin{bmatrix} i_1^{(g)}(t) \\ i_2^{(g)}(t) \\ i_3^{(g)}(t) \end{bmatrix} + \begin{bmatrix} -f_4(\boldsymbol{X}) + e_{21}^{(s)}(t) \\ -f_5(\boldsymbol{X}) + e_{32}^{(s)}(t) \\ 0 \end{bmatrix} \right)$$

$$\begin{bmatrix} f_1(\boldsymbol{X}) \\ f_2(\boldsymbol{X}) \\ f_3(\boldsymbol{X}) \end{bmatrix} = \boldsymbol{L}^{-1}\boldsymbol{R}\begin{bmatrix} i_1^{(g)}(t) \\ i_2^{(g)}(t) \\ i_3^{(g)}(t) \end{bmatrix} + \boldsymbol{L}^{-1}\begin{bmatrix} -f_4(\boldsymbol{X}) + e_{21}^{(s)}(t) \\ -f_5(\boldsymbol{X}) + e_{32}^{(s)}(t) \\ 0 \end{bmatrix}$$

While the inverse matrix $\boldsymbol{L}^{-1}$ of matrix $\boldsymbol{L}$ assumes the following analytical form:

$$\boldsymbol{D} = \boldsymbol{L}^{-1} = \frac{1}{(3k+1)L_g}\begin{bmatrix} -1 & -1 & k+1 \\ 1 & 0 & k \\ 0 & 1 & k \end{bmatrix}$$

The relationships formulated as differential Equations (5) and (8) form a general system of equations of state for an unloaded transformer.

The model formulated in this way does not take into account magnetic hysteresis. However, its simplifications (Section 6) make it possible to use it in the analysis of the influence of residual magnetism (and the parameter $K_{Fe,k}$ of the model) on the maximum values of the inrush currents of the unloaded transformer appearing during the first period determined by the supply voltage.

3. Investigation

The transformer data adopted in the test calculations are included in Appendix A.

Powercore® H 105-30 electrical steel was used in the experiment, which is manufactured by the ThyssenKrupp company. Material characteristics: density 7.65 kg/dm^3, maximum specific loss at 1.7 T is 1.05 W/kg.

3.1. Definition of Supply Voltages Together with the Method That Takes into Account the Time Instant Determined (Set) in the Supply Voltage Waveform

To define the symmetrical supply voltages of the discussed transformer, it was assumed that they are sinusoidally alternating waveforms with pulsation $\omega = 2\pi f$ and arbitrarily set rms values of phase-to-phase voltages, E_{12}, E_{23}, E_{31}. It was also assumed that the initial phase angle for one (arbitrarily chosen) voltage was known. This was sufficient because the phase angles of other voltages were defined by the geometry of the triangle (Figure 2).

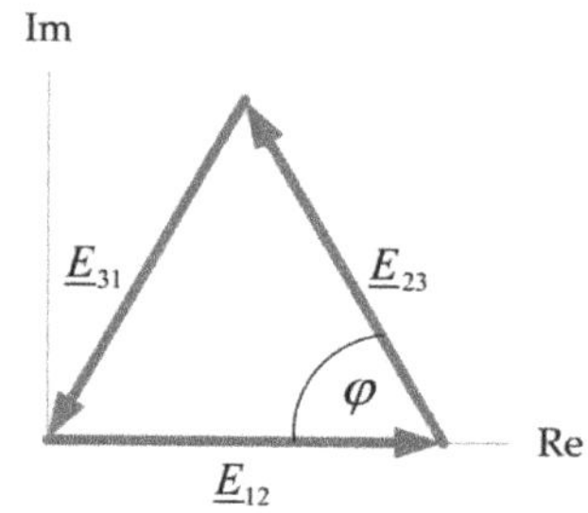

Figure 2. Phasor diagram of supply voltages.

The complex rms value of the reference voltage E_{12} assumed in the experiment was

$$\underline{E}_{12} = E_{12} \tag{9}$$

According to Figure 2, it was deduced that

$$\underline{E}_{23} = -a + \mathrm{j}k_{\mathrm{phsq}}E_{23}\sin\varphi$$

where

$$a = E_{23}\cos\varphi = \frac{E_{12}^2 + E_{23}^2 + E_{31}^2}{2E_{12}}, \ \varphi = \arccos\frac{a}{E_{23}}, \ k_{\mathrm{phsq}} = \begin{cases} -1_{\text{positive phase sequence}} \\ 1_{\text{negative phase sequence}} \end{cases} \tag{10}$$

The introduced parameter k_{phsq} made it possible to take into account the effect of the phase sequence on the calculated waveforms of transformer inrush currents.

The defined rms complex values of sinusoidally variable phase-to-phase voltages $\underline{E}_{12}, \underline{E}_{23}, \underline{E}_{31}$ constituted the basis for determining their instantaneous values in the form of the following functions:

$$\begin{aligned} e_{12}^{(\mathrm{s})}(t) &= \sqrt{2}|\underline{E}_{12}|\sin(\omega(t+t_0)+\arg(\underline{E}_{12})), \\ e_{23}^{(\mathrm{s})}(t) &= \sqrt{2}|\underline{E}_{23}|\sin(\omega(t+t_0)+\arg(\underline{E}_{23})), \\ e_{31}^{(\mathrm{s})}(t) &= \sqrt{2}|\underline{E}_{31}|\sin(\omega(t+t_0)+\arg(\underline{E}_{31})). \end{aligned} \tag{11}$$

Functions (11), representing input phase-to-phase voltages at the terminals of the primary winding of an unloaded transformer, met the requirements of the construction of Equations of states (5) and (8). The parameter t_0 present in the formula made it possible to define the time instant specified in the voltage waveform at which the indicated voltage appeared at the given transformer winding.

Since the argument of the reference voltage considered in the experiment $\underline{E}_{12}$ was assumed to be zero $\arg(\underline{E}_{12}) = 0$ (Figure 2), its function $e_{12}^{(\mathrm{s})}(t)$ took the form of

$$e_{12}^{(\mathrm{s})}(t) = \sqrt{2}|\underline{E}_{12}|\sin(\omega(t+t_0)) \tag{12}$$

The investigation of the transient state of the unloaded transformer was possible via the analysis of the solutions of the systems of differential Equations (5) and (8). In these equations, the initial point of integration was assumed to be zero ($t_0 = 0$). This meant that the starting point of the integral of the system of Equations (5) and (8) was t_0 away in time from the reference point determined by the passage of the voltage $e_{12}^{(\mathrm{s})}(t)$ through zero (Figure 3).

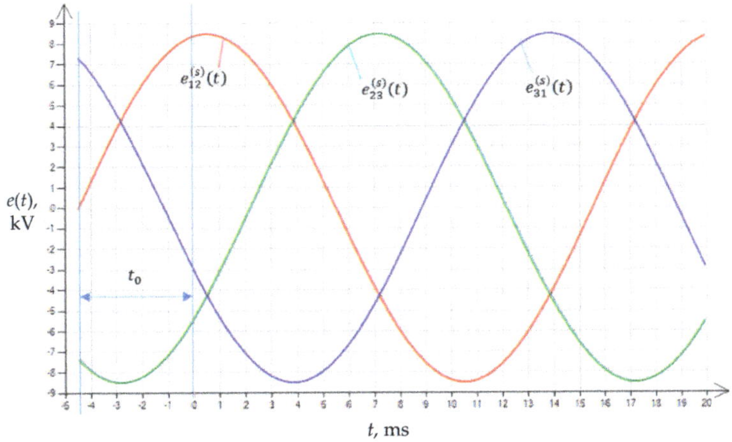

Figure 3. Visualization of how the t_0 parameter is set; the t_0 is the time shift from the reference point: the time instant when $e_{12}^{(s)}(t)$ voltage passes through zero.

The positive phase sequence $e_{12}^{(s)}(t), e_{23}^{(s)}(t), e_{31}^{(s)}(t)$ shown in Figure 3 was taken into account in further calculations as a coefficient with the value $k_{\text{phsq}} = -1$.

The presented method made it possible to set the parameter t_0 in the study of the effect of the time instant determined (defined) in the supply voltage waveform on the maximum values of the inrush current of the unloaded transformer in the first period determined by the supply voltage.

3.2. Study of the Influence of Residual Magnetism on the Maximum Values of Inrush Currents of an Unloaded Transformer

The study of the influence of residual magnetism on the maximum values of the current pulses accompanying the energizing of an unloaded transformer was carried out by setting initial conditions in the form of state variables

$$\begin{array}{l} x_4(0) = \Psi_2(0) \\ x_5(0) = \Psi_3(0) \end{array} \tag{13}$$

identical with flux linkages

$$\begin{array}{l} x_4(0) = \Psi_2(0) = B_2 z_g s_{\text{Fe}} \\ x_5(0) = \Psi_3(0) = B_3 z_g s_{\text{Fe}} \end{array} \tag{14}$$

where B_2, B_3 are the residual flux density of the transformer core columns, namely the center and outer columns (Figure 1).

For the purposes of the study, four scenarios (variants) were also established, with non-randomly selected sets of residual flux density values B_1, B_2, B_3 (Table 1) in successive columns of the transformer core. This procedure was carried out because of future applications, including the need to select the transformer core column whose winding will be supplied in the tests.

Table 1. Test sets of values (variants) of pre-magnetization in each column of the transformer core.

Variant	B_1,T	B_2,T	B_3,T
1	−0.6	1.2	−0.6
2	−1.2	0.6	0.6
3	−0.6	−0.6	1.2
4	0.6	−1.2	0.6

The remaining initial conditions of the system of Equations (5) and (8) were assumed to be zero in all calculation cycles $x_1(0) = i_1^{(g)}(0); x_2(0) = i_2^{(g)}(0); x_3(0) = i_3^{(g)}(0)$.

The influence of the time instant determined in the waveform of the supply voltage applied to the transformer was investigated by changing the parameter t_0 in functions (11) applied then to the equations of states (5) and (8).

3.3. Implementation of Calculations—Calculation Algorithm

Figure 4 shows the flowchart of the algorithm applied in both tests. The algorithm was input into the authors' own C# software [30]. The program allows the precise calculation and visualization of the distributions of the maximum values of the inrush currents of an unloaded transformer depending on the time instant t_0 characterizing the time instant in the waveform of the switched supply voltage and also the values and initial scenarios of the pre-magnetization of transformer's different columns. To integrate the stiff differential Equations (5) and (8) of the transformer model, the authors' own ninth-order Radau IIA method was used [31–33].

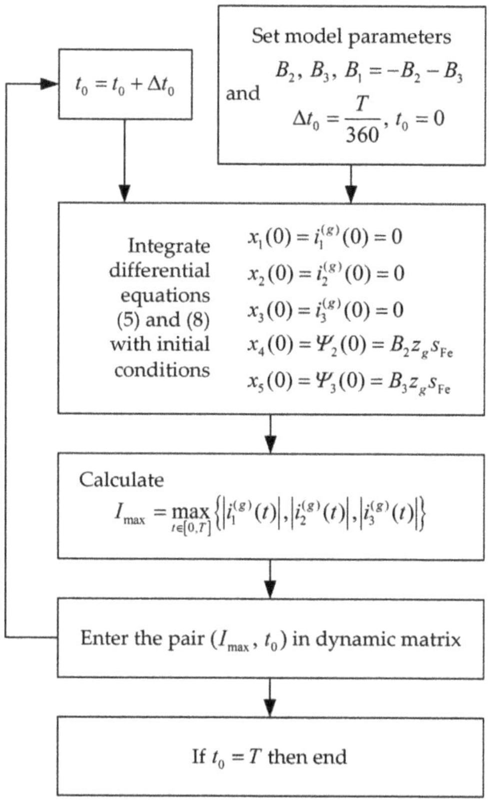

Figure 4. Structure of the computational algorithm examining the effect of the level of residual magnetism in the core and the time instant set in the supply voltage waveform on the maximum values of inrush currents of the unloaded transformer.

4. Simulation Results

Figure 5 shows the results of the analysis of the distribution of the maximum values of inrush currents of the unloaded transformer for four variants of the pre-magnetization of the core (Table 1) in a positive phase sequence system (Figure 3).

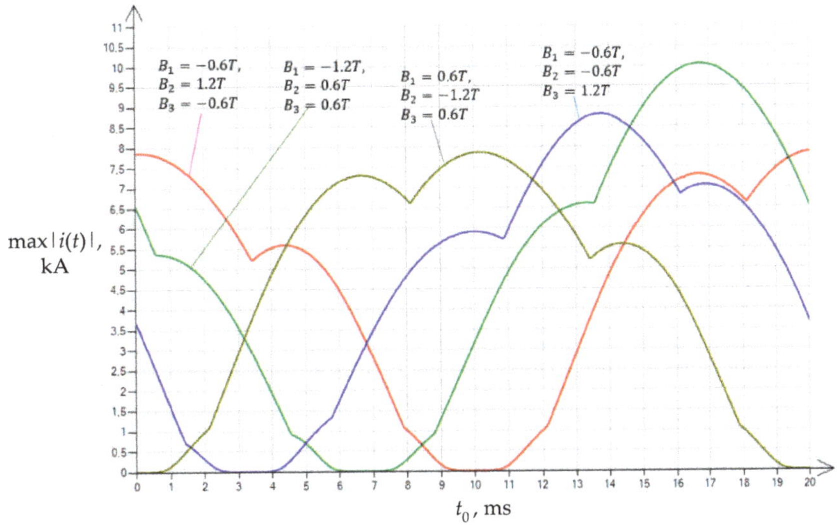

Figure 5. Distribution of the values of the maximum inrush currents of the unloaded transformer for the four variants of pre-magnetization, depending on the time instant t_0, characterizing the time instant of energizing in the course of the supply voltage for the positive phase sequence.

The conditions of the experiment in Figure 5 were used in its repeated implementation, in which the phase sequence was changed to the opposite sequence. The calculation results obtained under the conditions of such a modified power supply are shown in Figure 6.

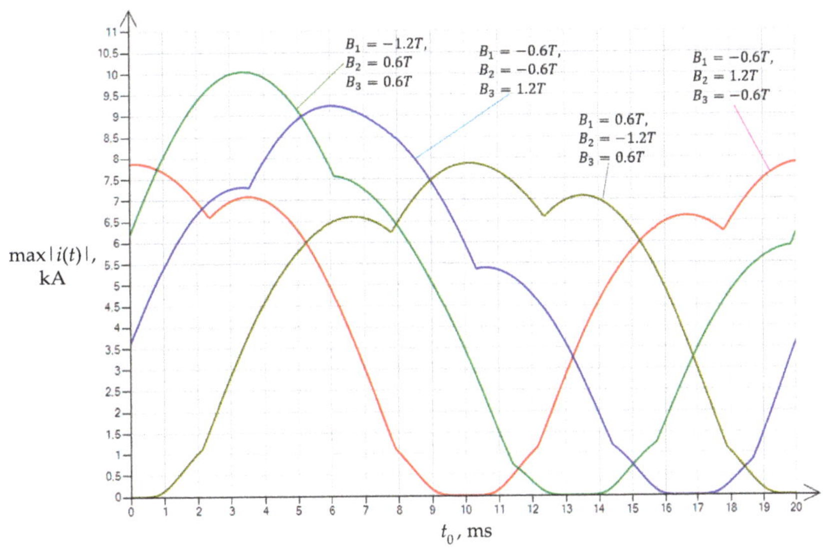

Figure 6. Distribution of the values of the maximum inrush currents of the unloaded transformer for four variants of pre-magnetization, depending on the moment t_0, identifying the time instant of energizing in the course of the supply voltage in the voltage for the negative phase sequence.

A comparison of the results of the calculations presented in Figures 5 and 6 showed that for the pre-magnetization scenario in the center column with a value twice that of the outermost columns (this refers to the absolute value), the distributions of maximum currents do not depend on the phase sequence.

For pre-magnetization scenarios 1 and 4 (Table 1), the calculations of inrush currents of the unloaded transformer led to results highly consistent with each other (similar shape and values of the obtained waveforms in Figures 5 and 6).

In a further stage of the study, it was assumed that the initial magnetization of the transformer core would be carried out according to variant 1 (Table 1).

Figure 7 shows the result of the experiment of energizing the unloaded transformer for the residual magnetism set in variant 1 and for the set time instant $t_0 = 2.77$ ms specified in the waveform of the reference supply voltage $e_{12}^{(s)}(t)$.

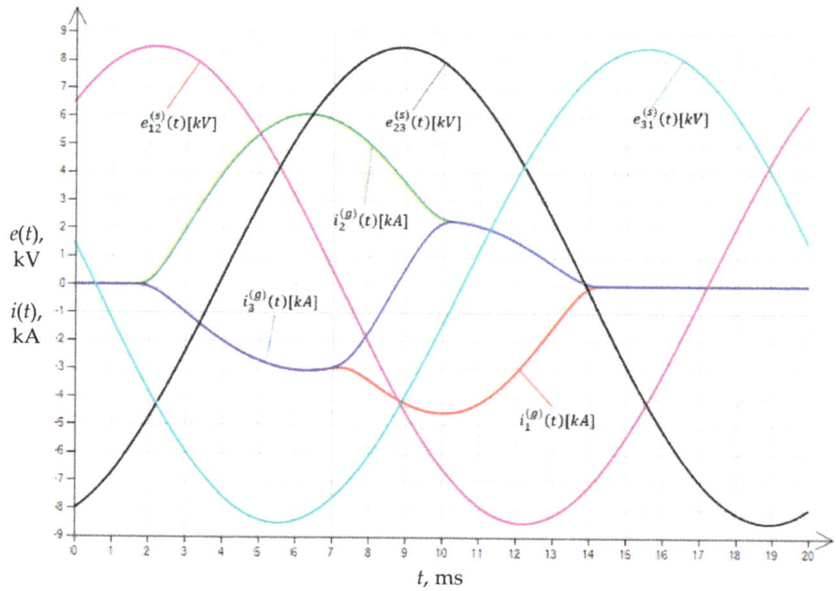

Figure 7. The waveforms of inrush currents of the unloaded transformer in the first period of the transient state; energizing occurred with residual magnetism present (variant 1) and at the time instant $t_0 = 2.77$ ms, defined in the waveform of the reference supply voltage $e_{12}^{(s)}(t)$.

The calculated maximum instantaneous value of the unloaded transformer inrush current during the first transient period was $I_{\max} = \max\limits_{t\in[0,\mathrm{T}]}\left\{\left|i_1^{(g)}(t)\right|, \left|i_2^{(g)}(t)\right|, \left|i_3^{(g)}(t)\right|\right\} = 6$ kA.

In the case of energizing the transformer after a time $t_0 = 10$ ms from the transition of voltage $e_{12}^{(s)}(t)$ through zero, the maximum instantaneous value of the inrush current of the unloaded transformer during the first transient period was reduced to only $I_{\max} = \max\limits_{t\in[0,\mathrm{T}]}\left\{\left|i_1^{(g)}(t)\right|, \left|i_2^{(g)}(t)\right|, \left|i_3^{(g)}(t)\right|\right\} = 2.5$ A (Figure 8).

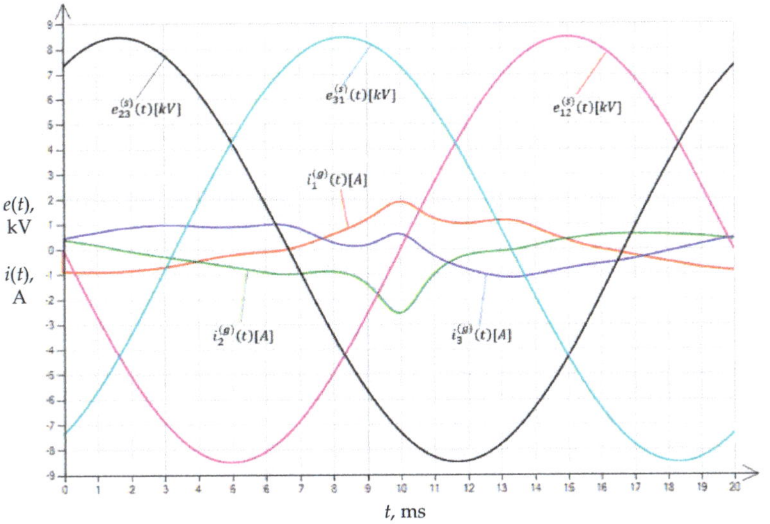

Figure 8. The waveforms of the inrush currents of the unloaded transformer in the first period of the transient state; energizing occurred with residual magnetism present (variant 1) and at the time instant $t_0 = 10$ ms, defined in the waveform of the reference supply voltage $e_{12}^{(s)}(t)$.

5. Algorithm for Selecting the Time Instant Specified in the Waveform of the Switched Supply Voltage

The results of the performed experiments and the diagnosed regularities of the model were used to develop an algorithm for selecting the time instant t_0 calculated from the reference point of the voltage $e_{12}^{(s)}(t)$ transition through zero.

With regard to the selection of the supply voltage and the terminals of the primary side of the transformer to which this voltage is to be applied, it was determined that the most optimal configuration was provided by the computationally verified voltage $e_{12}^{(s)}(t)$ and the terminals of the primary winding wound on the center column of the transformer (Figure 1), as seen in scenario 1.

5.1. Method of Implementation of Pre-Magnetization of the Core Columns of the Tested Transformer

Due to the difficulties associated with measuring the residual magnetism of three-phase transformers [12,18], it was decided that the pre-magnetization technique of the core [27] instead of its measurement should be used. The schematic diagram of the solution involving the pre-magnetization of the core of an unloaded transformer with delta-connected primary side windings is presented in Figure 9a.

The concept of the pre-magnetization of the core of the transformer under study envisages the generation of a magnetization of twice the value and opposite polarity in the center column than is the case in the outermost columns (variant 1) (Figure 9b).

As shown in the conducted numerical experiments, such as the proportion and symmetry of the initial magnetization guarantees the distribution of the maximum values of the inrush current of the unloaded transformer as a function of the switching time calculated with respect to the transition of the voltage $e_{12}^{(s)}(t)$ through zero and independent of the phase sequence of the supply voltages $e_{12}^{(s)}(t), e_{23}^{(s)}(t), e_{31}^{(s)}(t)$.

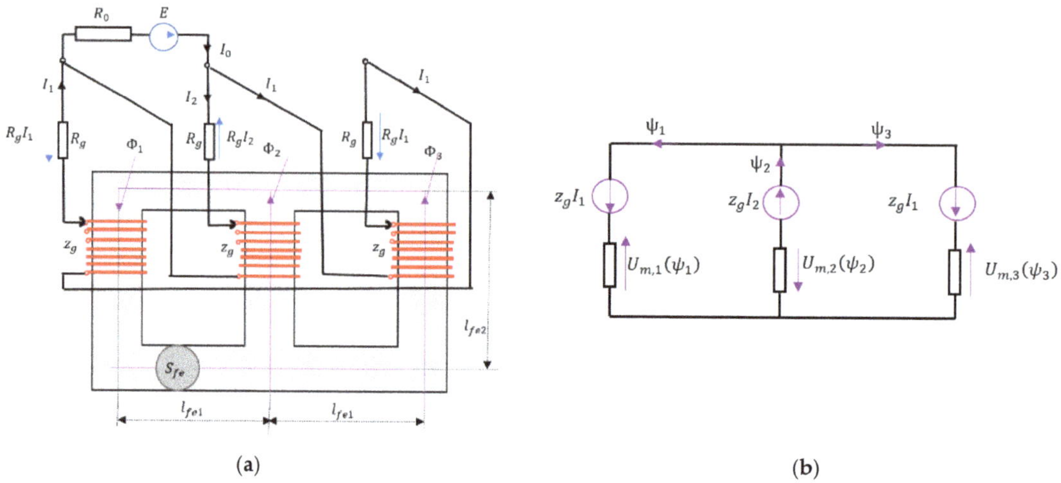

Figure 9. Schematic diagram of the method of the pre-magnetization of the core of an unloaded transformer with delta-connected primary side windings (**a**) and an equivalent diagram of the magnetic circuit exhibiting forced flux distribution (**b**).

5.2. Selection of the Currents Required for Core Pre-Magnetization

The next step was to estimate what values of current are necessary for pre-magnetization in the system shown in Figure 9.

The starting point in the calculations was the value of the transformer winding resistance R_g. Since the value of this resistance is relatively small, the condition for selecting the resistance R_0 in the circuit, as shown in Figure 9a, was formulated as [34]:

$$R_0 >> R_g \tag{15}$$

It was possible to determine the DC currents flowing through the transformer windings without any problem via the following:

$$I_0 = \frac{E_0}{R_0} \quad , \quad I_1 = \frac{1}{3}\frac{E_0}{R_0} \quad , \quad I_0 = \frac{2}{3}\frac{E_0}{R_0} \tag{16}$$

The currents determined in this way for the different windings of the three-phase transformer allowed the formulation of a system of nonlinear equations based on magnetic circuit theory (Figure 9b). In the equations, magnetic voltage drops were taken into account with the omission of the dynamic term (i.e., the Formula [27]).

$$U_{\mathrm{m},k}(\Psi_k) = h_{\mathrm{Fe},k} H \frac{\Psi_k(t)}{z_{\mathrm{g}} s_{\mathrm{Fe}}} \quad , \quad k = 1,2,3 \tag{17}$$

The system of nonlinear equations determining the distribution of flux linkages took the following form:

$$\begin{bmatrix} f_1(\boldsymbol{X}) \\ f_2(\boldsymbol{X}) \\ f_3(\boldsymbol{X}) \end{bmatrix} = \begin{bmatrix} I_0 z_{\mathrm{g}} - U_{\mathrm{m},1}(x_1) - U_{\mathrm{m},2}(x_2) \\ I_0 z_{\mathrm{g}} - U_{\mathrm{m},2}(x_2) - U_{\mathrm{m},3}(x_3) \\ -x_1 + x_2 - x_3 \end{bmatrix} = 0 \tag{18}$$

where $\boldsymbol{X} = [x_1, x_2, x_3]^{\mathrm{T}} = [\Psi_1, \Psi_2, \Psi_3]^{\mathrm{T}}$.

The formulated system of equations was solved for the given values of current I_0 by relating the solutions of $\Psi_i (i = 1, 2, 3)$ to the corresponding average values of magnetic flux density:

$$B_i = \frac{\Psi_i}{z_g s_{Fe}} \quad (19)$$

The solution of the system of Equation (18), as shown in Equation (19), for different values of current I_0 is presented in Figure 10.

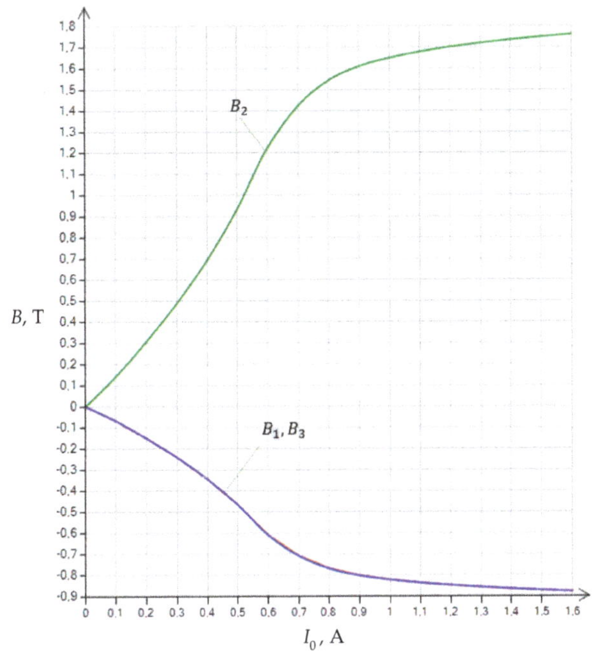

Figure 10. Dependence of magnetic flux density in different columns on magnetization current I_0.

The calculated characteristic unambiguously proves that the proposed method of the pre-magnetization of the transformer core results in flux density levels in the different columns of the core according to variant 1:

$$B_2 = B_0 \; , \; B_1 = -\frac{1}{2} B_0 \; , \; B_3 = -\frac{1}{2} B_0 \quad (20)$$

where B_0 depends on the magnetization current I_0.

For the calculated magnetization configuration (20), the opposite magnetization can also be considered. To obtain this configuration, it is enough to change the direction of current flow I_0 in the circuit shown in Figure 9a.

Turning off the flow of current I_0 in the pre-magnetization system will reduce the value of flux density in the different columns of the core. For transformer sheets (soft magnetic materials), the value of flux density B_0 will correspond to the value of remanence B_r, which is significantly lower than saturation flux density (by as much as 50%).

In the test data adapted for the simulation, it was assumed that the transformer sheet has a saturation flux density of 1.9 T. Hence, the value of the flux density B_0, representing the initial magnetization of the core in the calculations in variant (20), will vary up to a value of 1.2 T.

Results of the calculations of the maximum values of inrush currents

$$I_{\max}(t_0) = \max_{t\in[0,\mathrm{T}]}\left\{\left|i_1^{(g)}(t)\right|, \left|i_2^{(g)}(t)\right|, \left|i_3^{(g)}(t)\right|\right\} \tag{21}$$

of the unloaded transformer in the first period of the transient state from the energization, depending on the adopted value of B_0, are presented in Figure 11.

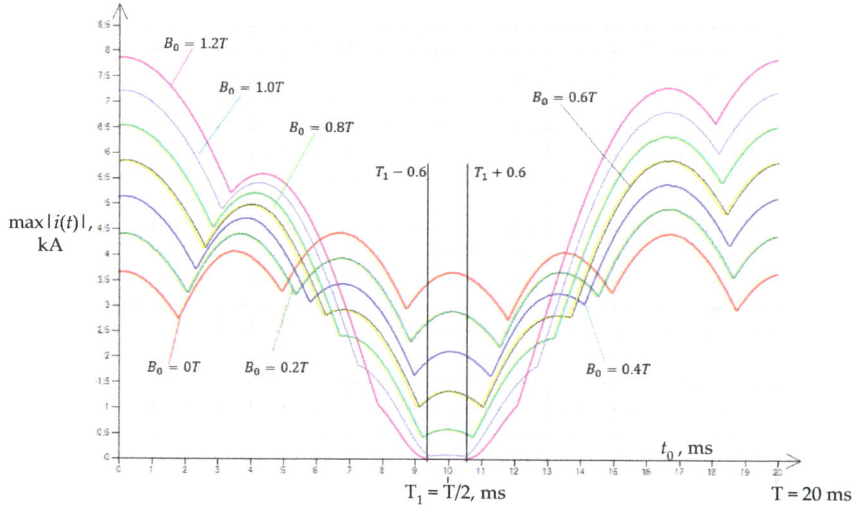

Figure 11. Distribution of maximum inrush currents of the unloaded transformer for seven variants of pre-magnetization B_0, depending on the time instant t_0, defined in the voltage waveform $e_{12}^{(s)}(t)$ (as in Figure 12), and on the pre-magnetization of the core in configuration (20) for a positive phase sequence of supply voltage.

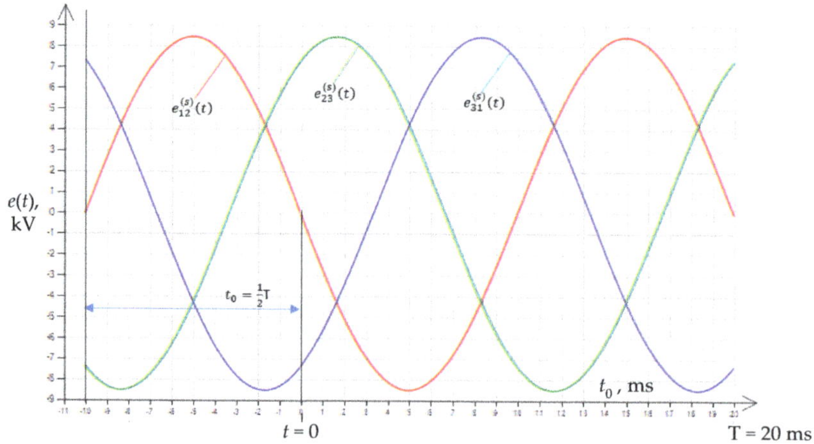

Figure 12. Visualization of the time instant of voltage switching on the transformer measured relative to the time instant when the voltage $e_{12}^{(s)}(t)$ passes through zero, for a positive phase sequence of supply voltage.

The calculated inrush currents are functions dependent on time t_0 (the time instant of switching on) calculated with respect to the transition of voltage $e_{12}^{(s)}(t)$ through zero with positive derivative at the point $\frac{\mathrm{d}}{\mathrm{d}t}e_{12}^{(s)}(t) > 0$ (Figure 12).

The results were obtained for the initial magnetization, as seen in (20), enforced according to the scheme in Figure 9 with a positive phase sequence of supply voltage.

The analysis of the effect of the value of B_0 on the distribution of the maximum values of inrush currents showed the existence of an interval of the variability of the maximum values of the currents, in which the maximum values of the currents decrease as the flux density B_0 increases.

It was found that in the interval $t_0 \in \left[\frac{1}{2}\mathrm{T} - 0.6\ \mathrm{ms}, \frac{1}{2}\mathrm{T} + 0.6\,\mathrm{ms}\right]$ for $B_0 = 1.0$ T, the maximum inrush currents reached values in the order of tens of amperes, while for $B_0 = 1.2$ T, the values decreased to a few amperes (Figure 13).

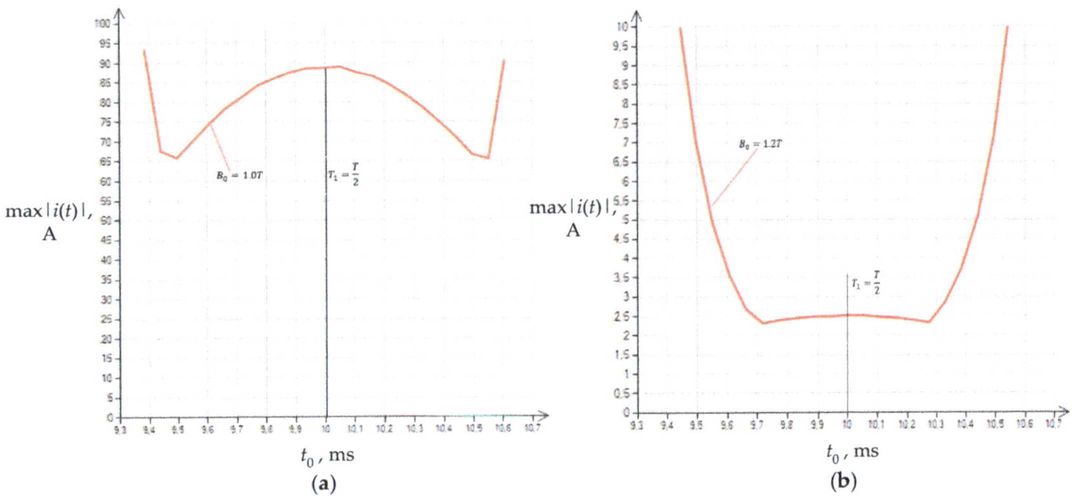

Figure 13. Distribution of the maximum inrush currents of the unloaded transformer for two values of pre-magnetization $B_0 = 1.0$ T (**a**) and $B_0 = 1.2$ T (**b**), depending on the time instant $t_0 \in \left[\frac{1}{2}\mathrm{T} - 0.6\ \mathrm{ms}, \frac{1}{2}\mathrm{T} + 0.6\,\mathrm{ms}\right]$, defined in the waveform of the switched voltage $e_{12}^{(s)}(t)$, and on the pre-magnetization of the core in configuration (20).

Outside the indicated time interval, the opposite phenomenon was observed: the increase in B_0 was accompanied by a sharp increase in the maximum values of inrush currents. For $t_0 = 0$, the currents reached the highest possible values (kiloamperes) (Figure 11).

This means that in order to eliminate the large inrush currents accompanying the energizing of an unloaded three-phase transformer, its switching should be performed after a time $t_0 = \frac{\mathrm{T}}{2}$, calculated with respect to the reference point determined by the transition of the reference voltage (here: $e_{12}^{(s)}(t)$) through zero with a positive derivative at the point $\frac{\mathrm{d}}{\mathrm{d}t}e_{12}^{(s)}(t) > 0$ (Figure 12).

At the energizing time instant $t_0 = \frac{\mathrm{T}}{2}$, the instantaneous voltage $e_{12}^{(s)}(t) = 0$, while its derivative is negative ($\frac{\mathrm{d}}{\mathrm{d}t}e_{12}^{(s)}(t) < 0$) (Figure 12).

The obtained solution is general and can be applied to any transformer, where primary winding is delta-connected and where columns are pre-magnetized according to (20) for $B_0 > 1$ T. The supply voltages $e_{12}^{(s)}(t), e_{23}^{(s)}(t), e_{31}^{(s)}(t)$ can be arranged into a positive or negative phase sequence (without affecting the outcome).

The collected data can be used to construct a special device that will execute energizing the transformer at specific time instants in the supply voltage waveforms (Table 2).

Table 2. Switching conditions of the unloaded transformer for which the minimum values of inrush currents are obtained, taking into account the pre-magnetization in configuration (20).

Type	k_{phsq}	$e_{12}^{(s)}(-t_0)$	$\frac{d}{dt}e_{12}^{(s)}(t)\mid_{t=-t_0}$	t_0	$e_{12}^{(s)}(0)$	$\frac{d}{dt}e_{12}^{(s)}(t)\mid_{t=0}$
1	−1	0	>0	$\frac{1}{2}$T	0	<0
2	1	0	>0	$\frac{1}{2}$T	0	<0

5.3. The Strategy of Energizing a Three-Phase, Three-Column Transformer Where Primary Winding Is Delta-Connected and Where the Columns Are Pre-Magnetized According to (20) at Specific Time Instants in the Supply Voltage Waveforms

The developed strategy is illustrated by the flowchart in Figure 14.

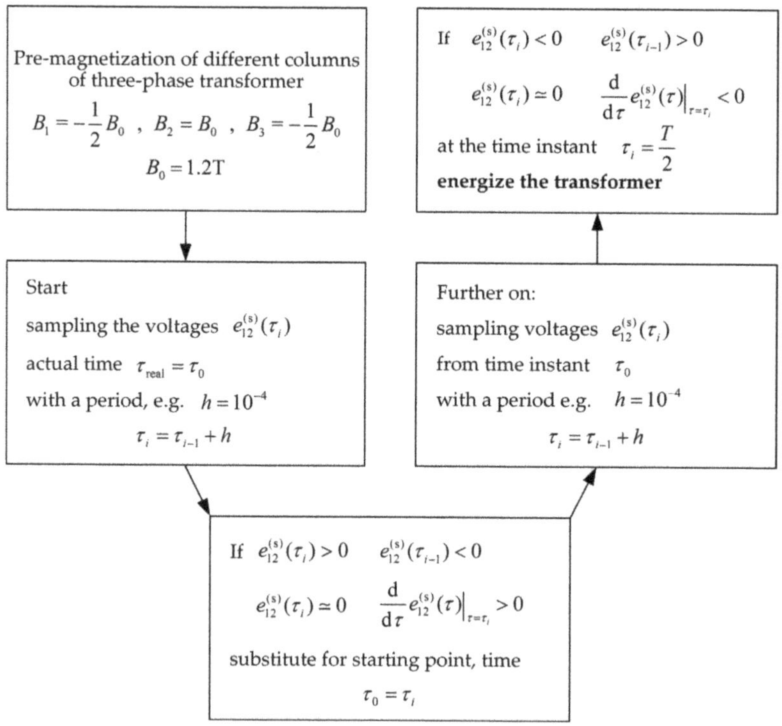

Figure 14. Inrush current reduction strategy when energizing a three-phase transformer with delta-connected primary side windings requiring the pre-magnetization of the core columns.

6. The Simplifications Implied by Using the Model Formulated in this Paper

- The initial magnetization state of the transformer core can be taken into account as residual magnetism in the form of (assignable) initial conditions for the state variables $\Psi_2(t_0)$, $\Psi_3(t_0)$, which are essentially flux linkages $\Psi_2(t_0) = B_2 z_g s_{\text{Fe}}$ and $\Psi_3(t_0) = B_3 z_g s_{\text{Fe}}$, where B_2, B_3 is the residual flux density in the different columns of the transformer core.
- The integration of the equations with the initial conditions formulated in this way leads in the first instance to a solution $(I_0 - \Psi)$ located deep in the saturation region of the ferromagnetic core curve; the maximum inrush currents of the unloaded transformer determined from this solution are calculated accurately.
- Taking any initial conditions (e.g., zero) for the systems of Equations (5) and (8) as the starting point leads to a steady-state solution, which is the limit cycle of the transient state solution; the trajectories of the transition to this limit cycle have no physical

interpretation in this case, since the assumed mathematical model does not take into account magnetic hysteresis (it does not track the history of core magnetization).
- The model is suitable for the study of steady-state conditions of the transformer when the resistances $R_{Fe,k}$ are determined based on the idling losses, in which eddy currents and magnetic hysteresis are taken into account.
- The completed tests have shown that there is no significant influence* of the parameter $R_{Fe,k}$ of the model on the maximum values of the inrush currents of the unloaded transformer in the first period of the transient state caused by energizing the transformer (*—lossiness values provided by the manufacturer of the transformer plates vary within wide limits).

Therefore, the developed model can be successfully used to study both the steady-state operation of a three-phase transformer (Dy) and to calculate the inrush currents that characterize it when it is energized under unloaded conditions.

7. Conclusions

The research task of modeling the dynamics of an unloaded three-phase transformer with a Dy connection group during the first transient period has been achieved.

The methodology and simulation results given in the article provide practical tools for studying the inrush currents of these devices.

What is particularly interesting, the developed model makes it possible to take into account the influence of residual magnetism in the different columns of the core and the time instant determined in the voltage waveform where the indicated voltage is supplied to the given winding of the transformer.

The completed calculations proved that for each scenario of core pre-magnetization of the modeled device, there is a time interval in which the energizing of an unloaded transformer is not accompanied by the occurrence of inrush current surges.

The tests showed that the width of such a time interval is more than 1 ms, and the currents occurring are several times higher than transformer's steady-state idle current.

The remaining time intervals of the first period of the transient state caused by energizing an unloaded transformer should be associated with the occurrence of large current pulses.

For a three-phase transformer with connection group Dy, the initial magnetization of the core in the form of the variant specified in (20) is the easiest to implement in the circuit shown in Figure 9.

Such an arrangement allows the symmetric and proportional magnetization of the core of a three-column transformer. One of these columns, e.g., the center one, is twice as highly magnetized as the remaining columns; moreover, the magnetization is of opposite polarity.

The symmetry and adequate distribution of core magnetization, which are demonstrated in this paper, result in the fact that the positioning of the time interval (measured relative to a reference point determined by the transition of voltage $e_{12}^{(s)}(t)$ through zero), during which the energizing of transformer is accompanied by a strong limitation of current pulses, does not depend on the phase sequence of phase-to-phase (line) voltages $e_{12}^{(s)}(t), e_{23}^{(s)}(t), e_{31}^{(s)}(t)$.

If the guidelines set out in this paper for a three-phase, three-column transformer are met, i.e., its core is pre-magnetized in a variant of (20) and it is switched on after time $t_0 = \frac{T}{2}$ is calculated from the instant when voltage $e_{12}^{(s)}(t)$ passes through zero, the values of inrush currents in the first transient period will be minimal.

The next stage of the authors' work will cover the experimental verification of the developed mathematical model, as well as the construction and testing of a device for the controlled switching of the supply voltage with the pre-magnetization of the transformer's core.

Author Contributions: Conceptualization, M.Ł., B.B. and Ł.M.; Methodology, B.B. and Ł.M.; Software, B.B. and J.K.-P.; Validation, J.K.-P.; Formal analysis, M.Ł., J.K.-P. and Ł.M.; Investigation, B.B., J.K.-P. and Ł.M.; Data curation, J.K.-P.; Writing—original draft, B.B.; Writing—review & editing, M.Ł. and Ł.M.; Visualization, B.B. and Ł.M.; Supervision, M.Ł. and Ł.M.; Project administration, M.Ł.; Funding acquisition, M.Ł. All authors have read and agreed to the published version of the manuscript.

Funding: This research received no external funding.

Data Availability Statement: Data available on request from the authors.

Conflicts of Interest: The authors declare no conflict of interest.

Nomenclature

The list of symbols and notations used in this paper:

e	sinusoidal supply voltage
Ψ_k	flux linkage associated with the transformer's primary coil, computed as $\Psi_k(t) = z_g \Phi_k(t)$, in the $k-$ th column of the transformer
$i_k^{(g)}$	transformer primary current
$R_{\mathrm{Fe},k}$	equivalent resistance representing the iron core losses
$i_{\mathrm{Fe},k}$	active component of the transformer's idle current
R_s	equivalent resistance of the power grid (power source)
L_s	equivalent inductance (reactance) of the power grid (power source)
R_g	primary winding resistance
L_g	leakage inductance of the primary winding
H	magnetic field strength
Φ	main flux leakage (effective value of the flux)
B_k	flux density
s_{Fe}	cross-sectional area of the core of the transformer
z_g	number of turns in the primary windings
$\varphi 0$	phase of the initial supplied voltage
ω	pulse

Appendix A

Table A1. Assumed data of the modeled transformer.

Quantity, Unit	Value
Primary voltage PRI/phase to phase, V	6000
Secondary voltage SEC/phase to phase, V	230
Frequency	50
Rated primary current, A	800
Number of primary winding turns, N	160
Iron power loss (for 1.7 T), W/kg	1.05
Copper power loss, kW	50
Short-circuit voltage, %	5
Core cross-sections, m^2	0.110565
Column length, m	2
Length of the yoke, m	0.76
Equivalent resistance of the network, Ω	0.1
Equivalent reactance of the network, Ω	0.2

Table A2. Function $H(B)$ coefficients obtained from solving the estimation problem based on the manufacturer's data.

k	a_k
1	29.9624271037522
2	−76.4912078883278
3	420.867774746849
4	−1274.37623231261
5	2196.27285444425
6	−2301.93260519503
7	1516.77585894405
8	−630.577358829122
9	160.397849549712
10	−22.7840824031901
11	1.38472183966324

References

1. CIGRÉ Working Group. *C4.307. Transformer Energization in Power Systems: A Study Guide*, 568th ed.; CIGRÉ Publication: Paris, France, 2014; pp. 8–73.
2. Kulkarn, S.V.; Khaparde, S.A. *Transformer Engineering: Design, Technology, and Diagnostics*, 2nd ed.; Taylor & Francis Inc.: Boca Raton, FL, USA, 2012; pp. 1–750.
3. Majka, Ł.; Baron, B.; Zydroń, P. Measurement-based stiff equation methodology for single phase transformer inrush current computations. *Energies* **2022**, *15*, 7651. [CrossRef]
4. Barros, R.M.R.; da Costa, E.G.; Araujo, J.F.; de Andrade, F.L.M.; Ferreira, T.V. Contribution of inrush current to mechanical failure of power transformers windings. *High Volt.* **2019**, *4*, 300–307. [CrossRef]
5. Steurer, M.; Frohlich, K. The impact of inrush currents on the mechanical stress of high voltage power transformer coils. *IEEE Trans. Power Deliv.* **2002**, *17*, 155–160. [CrossRef]
6. Pandey, S.B.; Lin, C. Estimation for a life model of transformer insulation under combined electrical and thermal stress. *IEEE Trans. Reliab.* **1992**, *41*, 466–468. [CrossRef]
7. Alassi, A.; Ahmed, K.H.; Egea-Alvarez, A.; Foote, C. Transformer Inrush Current Mitigation Techniques for Grid-Forming Inverters Dominated Grids. *IEEE Trans. Power Deliv.* **2023**, *38*, 1610–1620. [CrossRef]
8. Hamilton, R. Analysis of transformer inrush current and comparison of harmonic restraint methods in transformer protection. *IEEE Trans. Ind. Appl.* **2013**, *49*, 1890–1899. [CrossRef]
9. Ni, R.; Wang, P.; Moxley, R.; Flemming, S.; Schneider, S. False Trips on Transformer Inrush-Avoiding the Unavoidable. In Proceedings of the 68th Annual Conference for Protective Relay Engineers, College Station, TX, USA, 30 March–2 April 2015.
10. Nagpal, M.; Martinich, T.G.; Moshref, A.; Morison, K.; Kundur, P.P. Assessing and limiting impact of transformer inrush current on power quality. *IEEE Trans. Power Deliv.* **2006**, *21*, 890–896. [CrossRef]
11. Blume, L.F.; Camilli, G.; Farnham, S.B.; Peterson, H.A. Transformer magnetizing inrush currents and influence on system operation. *Trans. Am. Inst. Electr. Eng.* **1944**, *63*, 366–375. [CrossRef]
12. Horiszny, J. *Analysis and Reduction of Transformer Inrush Current*; Gdansk University of Technology: Gdańsk, Poland, 2016; p. monograph 159. (In Polish)
13. Chiesa, N.; Mork, B.A.; Høidalen, H.K. Transformer model for inrush current calculations: Simulations, measurements and sensitivity analysis. *IEEE Trans. Power Deliv.* **2010**, *25*, 2599–2608. [CrossRef]
14. Brunke, J.H.; Frohlich, K.J. Elimination of transformer inrush currents by controlled switching. Part I. Theoretical considerations. *IEEE Trans. Power Deliv.* **2001**, *16*, 276–280. [CrossRef]
15. Leite, J.V.; Benabou, A.; Sadowski, N. Transformer Inrush Currents Taking into Account Vector Hysteresis. *IEEE Trans. Magn.* **2010**, *46*, 3237–3240. [CrossRef]
16. Cui, Y.; Abdulsalam, S.G.; Chen, S.; Xu, W. A sequential phase energization technique for transformer inrush current reduction. Part I: Simulation and experimental results. *IEEE Trans. Power Del.* **2005**, *20*, 943–949. [CrossRef]
17. Chraygane, M.; El Ghazal, N.; Fadel, M.; Bahani, B.; Belhaiba, A.; Ferfra, M.; Bassoui, M. Improved modeling of new three-phase high voltage transformer with magnetic shunts. *Arch. Electr. Eng.* **2015**, *64*, 157–172. [CrossRef]
18. Horiszny, J. Research of leakage magnetic field in deenergized transformer. *Compel-Int. J. Comput. Math. Electr. Electron. Eng.* **2018**, *37*, 1657–1667. [CrossRef]
19. Mitra, J.; Xu, X.; Benidris, M. Reduction of three-phase transformer inrush currents using controlled switching. *IEEE Trans. Ind. Appl.* **2020**, *56*, 890–897. [CrossRef]
20. Ni, H.; Fang, S.; Lin, H. A simplified phase-controlled switching strategy for inrush current reduction. *IEEE Trans. Power Del.* **2021**, *36*, 215–222. [CrossRef]
21. Cano-González, R.; Bachiller-Soler, A.; Rosendo-Macías, J.A.; Álvarez-Cordero, G. Inrush current mitigation in three-phase transformers with isolated neutral. *Electr. Power Syst. Res.* **2015**, *121*, 14–19. [CrossRef]

22. Nicolet, A.; Delince, F. Implicit Runge–Kutta methods for transient magnetic field computation. *IEEE Trans. Magn.* **1996**, *32*, 1405–1408. [CrossRef]
23. Noda, T.; Takenaka, K.; Inoue, T. Numerical integration by the 2-stage diagonally implicit Runge–Kutta method for electromagnetic transient simulations. *IEEE Trans. Power Del.* **2009**, *24*, 390–399. [CrossRef]
24. Pries, J.; Hoffmann, H. State Algorithms for Nonlinear Time-Periodic Magnetic Diffusion Problems Using Diagonally Implicit Runge–Kutta Methods. *Magn. IEEE Trans.* **2015**, *51*, 7208612. [CrossRef]
25. Sowa, M. Ferromagnetic coil frequency response and dynamics modeling with fractional elements. *Electr. Eng.* **2021**, *103*, 1737–1752. [CrossRef]
26. Sowa, M.; Majka, Ł.; Wajda, K. Excitation system voltage regulator modeling with the use of fractional calculus. *AEU-Int. J. Electron. Commun.* **2023**, *159*, 154471. [CrossRef]
27. Baron, B.; Kolańska-Płuska, J.; Lukaniszyn, M.; Spałek, D.; Kraszewski, T. Solution of nonlinear stiff differential equations for a three-phase no-load transformer using a Runge-Kutta implicit method. *Arch. Electr. Eng.* **2022**, *71*, 1081–1106.
28. Baron, B.; Kolańska-Płuska, J.; Waindok, A.; Kraszewski, T.; Kawala-Sterniuk, A. Application of Runge–Kutta implicit methods for solving stiff non-linear differential equations of a single-phase transformer model in the no-load state. Innovation Management and information Technology impact on Global Economy in the Era of Pandemic. In Proceedings of the 37th International Business Information Management Association Conference (IBIMA), Cordoba, Spain, 1–2 April 2021; pp. 8071–8087.
29. Baron, B.; Kolańska-Płuska, J.; Kraszewski, T. Application of Runge–Kutta implicit methods to solve the rigid differential equations of a single-phase idle transformer model. *Pozn. Univ. Technol. Acad. J. Electr. Eng.* **2019**, *100*, 75–86. (In Polish)
30. Baron, B.; Kolańska-Płuska, J. *Numerical Methods of Solving Ordinary Differential Equations in C#*; Politechnika Opolska Publisher: Opole, Poland, 2015. (In Polish)
31. Hairer, E.; Wanner, G. *Solving Ordinary Differential Equations II: Stiff and Differential-Algebraic Problems*, 2nd ed.; Springer: Berlin/Heidelberg, Germany, 2010.
32. Dekker, K.; Verwer, J.G. *Stability of Runge–Kutta Methods for Stiff Nonlinear Differential Equations*; North-Holland: Amsterdam, The Netherlands; New York, NY, USA; Oxford, UK, 1984.
33. Dos Passos, W. *Numerical Methods Algorithms and Tools in C#*; CRC Press, Taylor & Francis Group LLC: Boca Raton, FL, USA; London, UK; New York, NY, USA, 2010; ebook-PDF. Available online: https://vdocuments.mx/numerical-methods-algorithms-and-tools-in-c-586e04e116cbc.html?page=1 (accessed on 11 April 2023).
34. Majka, Ł.; Szuster, D. Application of the stationary DC decay test to industrial turbogenerator model parameter estimation. *Przegląd Elektrotechniczny* **2014**, *90*, 242–245.

energies

MDPI

Article

Analysis of Core Losses in Transformer Working at Static Var Compensator

Piotr Osinski [1,*] and Pawel Witczak [2]

[1] Interdisciplinary Doctoral School, Lodz University of Technology, 90-537 Lodz, Poland
[2] Institute of Mechatronics and Information Systems, Lodz University of Technology, 90-537 Lodz, Poland; pawel.witczak@p.lodz.pl
* Correspondence: piotr.osinski@dokt.p.lodz.pl

Abstract: This article presents the comparison of 3D and 2D finite element models of a power transformer designed for reactive power compensation stations. There is a lack of studies in the literature on internal electromagnetic phenomena in the active part of a transformer operated in these conditions. The results of numerical 2D and 3D calculations of no-load current and losses in the transformer core were obtained by using various methods and models. The impact of considering the hysteresis loop phenomenon on the calculation of core losses was investigated by using the Jiles–Atherton core losses model. The results obtained in the paper show that the model of the core must contain the areas representing the influence of overlappings on the no-load current and also on the flux density field in the core. The capacitive load of the transformer increases the flux density in the core limbs by several percent, so the power losses there must also increase accordingly. As a summary of the research, differences in the values of losses in each core element between the capacitive load and no-load conditions are presented. The results presented in this paper indicate that considering nonlinearity related to the magnetic hysteresis loop has a significant impact on the calculation of the core losses of power transformers.

Keywords: power transformer; finite element analysis; SVC; core losses; capacitive load

Citation: Osinski, P.; Witczak, P. Analysis of Core Losses in Transformer Working at Static Var Compensator. *Energies* **2023**, *16*, 4584. https://doi.org/10.3390/en16124584

Academic Editor: Sérgio Cruz

Received: 8 May 2023
Revised: 5 June 2023
Accepted: 6 June 2023
Published: 8 June 2023

1. Introduction

The development of the world power supply makes it necessary to search for better methods to regulate the current flow and voltage level. Static var compensation (SVC), which belongs to the group of flexible alternating current transmission systems (FACTS), has gained great importance.

The technical solutions of SVC, which can be found in the literature [1], allow one to present the basic elements of this system, which are a load for the power transformer. The thyristor-switched capacitor (TSC) is a system in which each section contains, in addition to capacitors, thyristor switches that are turned on or off depending on the required reactive power. Thyristor-switched reactors (TSR) or thyristor-controlled reactors (TCR) consist of several three-phase sections in which thyristor switches control the inductive elements by turning them on or off depending on the reactive power that the entire compensation system must draw from the power grid. Fixed capacitors (FC) are also used in SVC stations as filters for higher harmonics. The presence of FC is necessary to eliminate interference caused by semiconductor switches.

The element that connects the reactive power compensation system to the high-voltage network (HV) is a power transformer. Its design must ensure the ability to transform reactive power. In addition, due to the operation of semiconductor switches, the power transformer must be able to handle harmonic currents and maintain voltage levels associated with all operating conditions without loss of life [2]. The basic structure scheme of the SVC reactive power compensation station is presented in Figure 1.

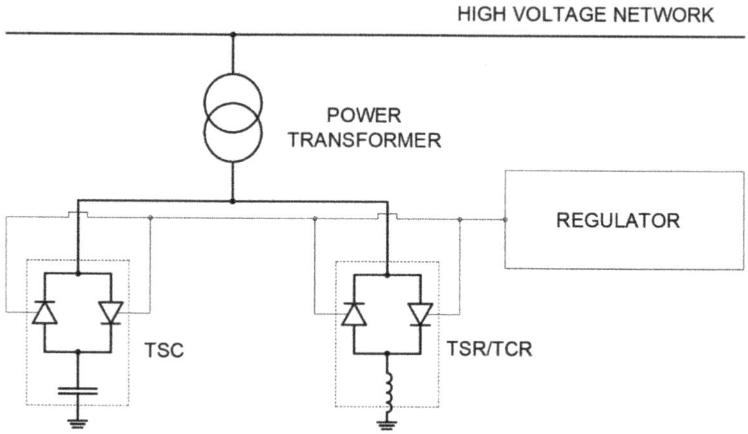

Figure 1. Basic structure scheme of the SVC station.

One of the most important aspects to consider, when analyzing the operation of a power transformer in a SVC station is the capacitive load. In the presence of a capacitive load, the voltage on the secondary side and the magnetic flux density may increase [3]. The explanation of this phenomenon is presented in Figure 2.

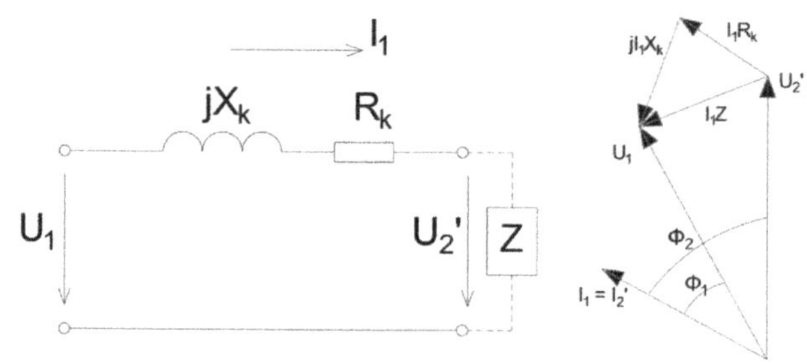

Figure 2. Simplified equivalent circuit diagram of a power transformer loaded with impedance Z and the resulting vector diagram of the transformer operation with a capacitive load.

An important point in the design of the core, in connection with the determination of its cooling parameters, is the analysis of the worst-case operating conditions of a power transformer, i.e., overexcitation and loads that lead to an increase in the core hot-spot temperature. These conditions are specified in the applicable industry standards. In a power transformer with three wounded limbs, the core hot spot is located in the geometric center of the overlap area between the top yoke and the center limb [4].

The inrush current is a form of overload that occurs when the power transformer is energized. The phenomenon occurs mainly when the voltage exceeds the value zero and when the transformer has not been switched off after its previous operation by a smooth reduction to a voltage of 0 V, leaving residual magnetic flux in the core. The inrush current is transient, and the time constant of the transient state after connecting the power transformer to the HV network depends on the characteristics of the magnetic circuit. It can last up to 1 s. Details of the power transformer inrush current calculations can be found in [5].

2. Analysis of Magnetic Field and Core Losses

The analysis of magnetic field distribution and the core losses is an essential part of the power transformer design process. Therefore, the selection of an appropriate calculation method that takes into account the hysteresis phenomenon is necessary for their correct determinations [6]. The most popular models for the calculation of core losses are comprehensively presented in [7,8]. A large group of empirical models is based on the equation that was proposed in 1982 by C. Steinmetz in [9]. In this method, the time-averaged core losses per volume P are expressed as an exponential function of peak magnetic flux density and frequency, which has the following form:

$$P = k \cdot f^{\alpha} \cdot {B_m}^{\beta} \tag{1}$$

The symbol k denotes a material parameter, f is the frequency, B_m is the peak magnetic flux density, and α and β are empirical coefficients, which are determined by fitting the model to the loss measurement data.

One of the most common models for calculating power losses in transformer cores is the Jiles–Atherton (J-A) hysteresis loop model presented in [10]. It is a mathematical nonlinear model that represents the relationship between the magnetic field strength and the magnetization at a given time. The J-A model is based on the theory that the magnetization M in ferromagnets depends not only on the strength of the magnetic field but also on the magnetization history of the material. In the J-A model, five coefficients are used to characterize the ferromagnetic material [11]. In practice, the generation of the hysteresis loop consists in the selection of the coefficients based on the power loss of the modeled core sheet for the frequencies of 50 Hz and 60 Hz and its magnetization curve. An example of the result of generating the hysteresis loop of the J-A model is shown in Figure 3.

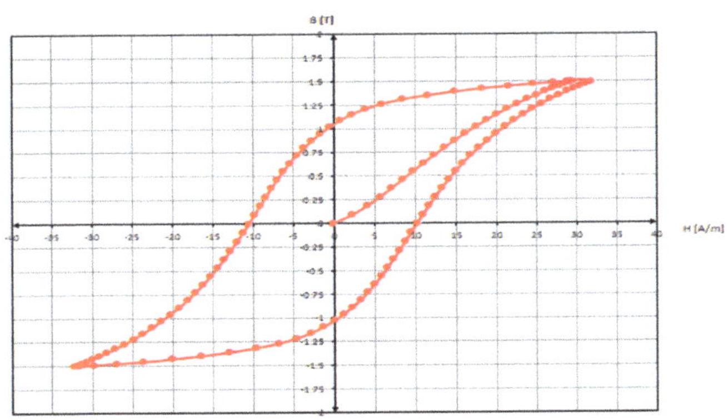

Figure 3. An example of J-A model hysteresis loop of 23ZDKH85 core steel in Simcenter MAGNET 2022.1. Red dot and line: Data points and approximation of hysteresis.

In the scientific literature [12–15], there are many examples of the application of a numerical analysis of power transformers using commercial packages that employ the finite element method (FEM). In this paper, Simcenter MAGNET 2022.1 was selected to simulate phenomena in the power transformer. This software uses the FEM to solve the 2D form of equation 2 for the magnetic potential.

$$rot\left(\frac{1}{\mu}\,rot\ A\right) = J \tag{2}$$

where μ—magnetic permeability, A—magnetic potential vector, and J—current density vector.

In this method, the problem domain is divided into a mesh of triangular elements, and the potential in each element is approximated by a simple function of x and y coordinates. Eddy currents are not considered in these calculations. The 3D formulation is based on the T-Ω method presented in [16]. The magnetic field is represented as the sum of two parts: the gradient of a scalar potential magnetic field Ω and, in conductors, an additional vector field T represented with vector-edge elements. In this method, Equation (3) incorporating Faraday's law and Equation (4), which takes into account source lessness of the H field, are solved.

$$rot\left(\frac{1}{\gamma}\,rot\,T\right) = -\frac{d}{dt}\,[\mu\,(T - grad\,\Omega)] \tag{3}$$

where γ—conductivity.

$$div\,\mu\,(T - grad\,\Omega) = 0 \tag{4}$$

Two available software solvers were used for the power transformer simulations: Time-Harmonic, in which the analysis is performed at a given frequency and non-linear properties of the ferromagnetic elements are approximated, and Transient, which is based on the time-stepping method, in which Maxwell's equations are solved. The electromagnetic field is decomposed into discrete time steps, and Maxwell's equations are solved iteratively at each step [17]. In this way, it is possible to analyze phenomena at specific time points. Both the Steinmetz and J-A models are used to calculate the core losses, although the J-A model can only be used with the Transient solver and 2D models.

3. Numerical Model of Transformer Core and Winding

Electromagnetic calculations were performed for a 300 MVA power transformer, whose main electrical parameters are HV 230 kV (wye connection) and LV 45.5 kV (delta connection). After analyzing the dimensions of the transformer, a three-dimensional finite element model was created. Considering the fact that the objective of the analysis was to determine the distribution of magnetic flux and calculate the core losses, the finite element mesh must be dense enough to achieve satisfactory accuracy. The 3D mesh of the power transformer is shown in Figure 4b. Figure 5 presents a circuit diagram of the analyzed transformer under a capacitive condition.

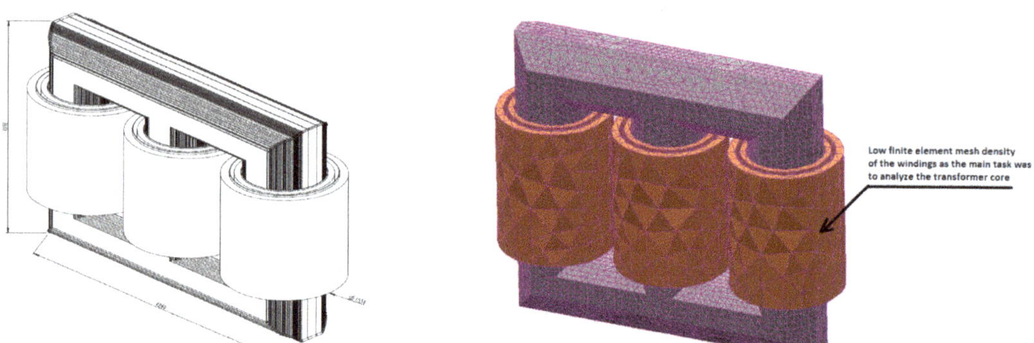

(a) (b)

Figure 4. Core and windings of a 300 MVA power transformer: (**a**) basic dimensions in mm; (**b**) 3D model with finite element mesh.

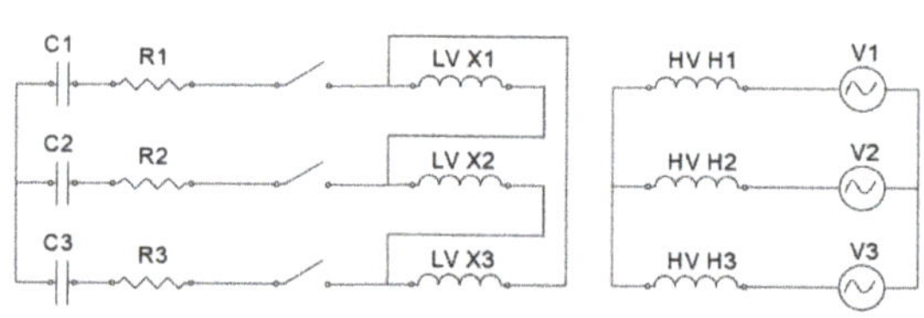

Figure 5. Circuit diagram of 300 MVA power transformer model under capacitive load condition.

4. Results of 3D and 2D Analysis of Power Transformer

The analysis of the no-load state of the 3D model of the power transformer was performed using the Time-Harmonic solver. In the three-dimensional model, the region of the core sheet overlap was not considered due to the high degree of complexity of this issue. Performing a 3D analysis using the time-stepping method in the case of a 3D core model required an enormous amount of computing time.

Figure 6 presents the calculation results showing the distribution of flux density in the core of the power transformer at no-load. Attention is drawn to the region where the root mean square value (RMS) of magnetic flux densities exceeds 6T. It was found that such values occur only near the vertices of the individual mesh elements in the corners of the core window. This is a numerical calculation error that has a minimum impact on the overall parameters, such as no-load current or power losses in the core.

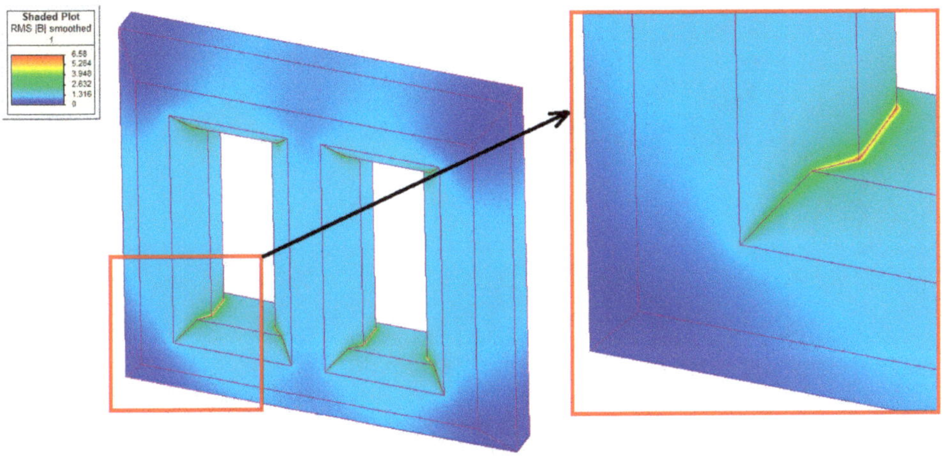

Figure 6. Flux density field in 3D model of power transformer under no-load condition; calculation method: Time-Harmonic; core losses calculation model: Steinmetz.

The analysis of the limb cross-section shows that the magnetic flux in the transformer core is distributed evenly. The RMS value of the magnetic flux densities in the middle point of the limb is B_{RMS} = 1.15 T which gives the maximum value B_{max} = 1.63 T.

In the case of a 2D analysis of the load condition of a power transformer, it is necessary to scale the model, calculate the corresponding depth of the geometry and determine the scaling factor of the magnetic permeability. The criterion for similarity between the 2D model and the 3D model is the magnitude of the stored energy of the magnetic field in the whole object. This is possible if the magnetic field strength is identical in both the 2D and 3D models. Under load conditions of the power transformer, most of the magnetic field energy is concentrated in the winding zone, which is in the duct between the HV and LV windings. Therefore, it is required in the 2D model to resolve the value of the magnetic permeability of the winding zone area by the scaling factor expressed in Equation (5):

$$\frac{\mu_{2Dk}}{\mu_{3Dk}} = \frac{d_k}{d} \tag{5}$$

where μ_{2Dk}—magnetic permeability of the k area of the 2D model, μ_{3Dk}—magnetic permeability of the k area of the 3D model, d_k—limb cross-section divided by its diameter, and d—depth of the 2D model. A detailed explanation of this issue can be found in [18,19].

In order to determine the scaling factor for the magnetic permeability in the area of the winding zone, the depth of the 2D model was chosen to correspond to the real cross-section of the core. The rescaling of the magnetic permeability increases the value of the leakage flux, resulting in a good agreement between the magnetic fluxes entering the core calculated by two and three-dimensional models. The value of the magnetic permeability in the overlapping area was calculated as accurately as possible using measurements of the real no-load current. Figure 7 presents an example of the core sheet overlapping area. The 2D model of the power transformer and the finite element mesh are shown in Figure 8.

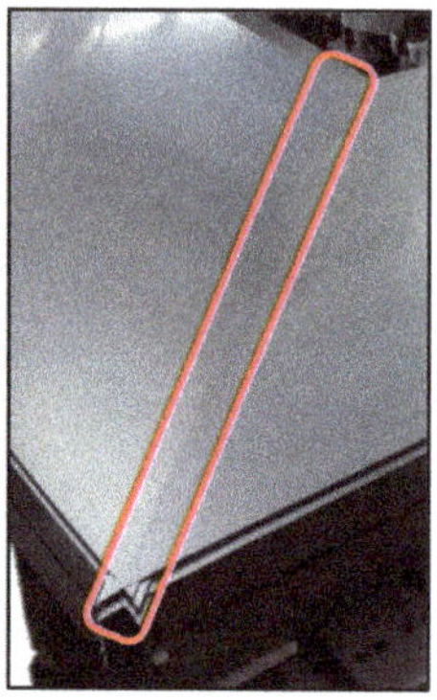

Figure 7. Example of core sheet overlapping area.

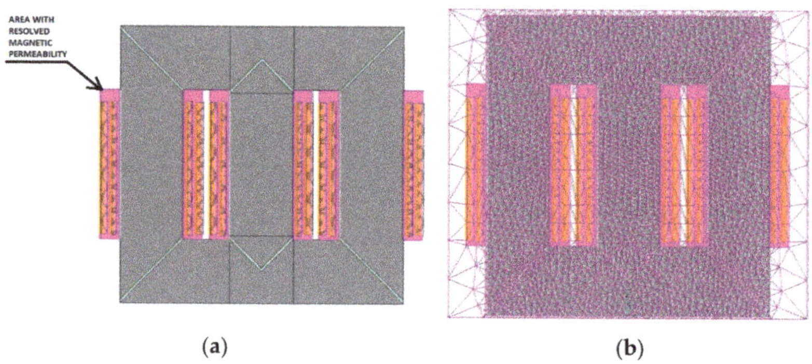

Figure 8. Two-dimensional model of 300 MVA power transformer: (**a**) numerical model; (**b**) finite element mesh.

In addition to the periodic component, the inrush current also contains an aperiodic component whose value depends on the phase of the voltage at the time the transformer is turned on. The sum of the instantaneous aperiodic components of the three phases causes the magnetization of the transformer core with a random vector in space. To minimize this phenomenon, it was assumed that the supply voltage increases exponentially to the nominal value during transformer turn on. Function (6) was used for this purpose.

$$u(t) = \left(1 - e^{-\frac{t}{\tau}}\right) \cdot U_m \cdot e^{j\omega t} \quad (6)$$

where t—time instant (ms), τ—time constant, assumed τ = 20 ms, ω—pulsation, and U_m—maximum voltage value (V).

Due to the relatively rapid increase in voltage values in successive periods, there is an aperiodic component present in the calculations. A comparison was made between the values of the phase currents in the no-load state obtained in the test field and the values calculated in the software using the time-stepping method with time steps equal to 0.1 ms. The current waveform in the 2D model of the power transformer for the time-stepping method, using the Steinmetz and J-A core losses models, is presented in Figure 9.

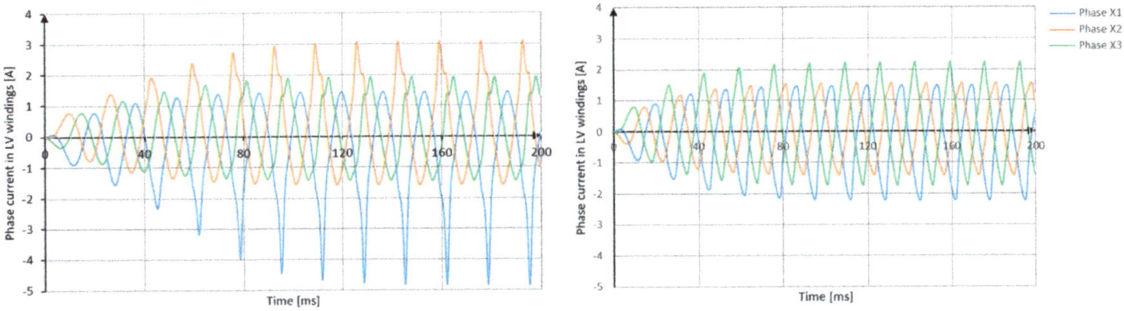

(a) (b)

Figure 9. The current waveforms in the 2D model of the power transformer: (**a**) calculation method: Transient 2D, core losses calculation model: Steinmetz; (**b**) calculation method: Transient 2D, core losses calculation model: Jiles–Atherton.

To check the accuracy of the numerical model of the 300 MVA transformer, the no-load inrush current obtained from the test field was compared with the RMS values of the current waveform calculated using Time-Harmonic 2D and Transient 2D models with Steinmetz and Jiles–Atherton core losses models. The results are displayed in Figure 10.

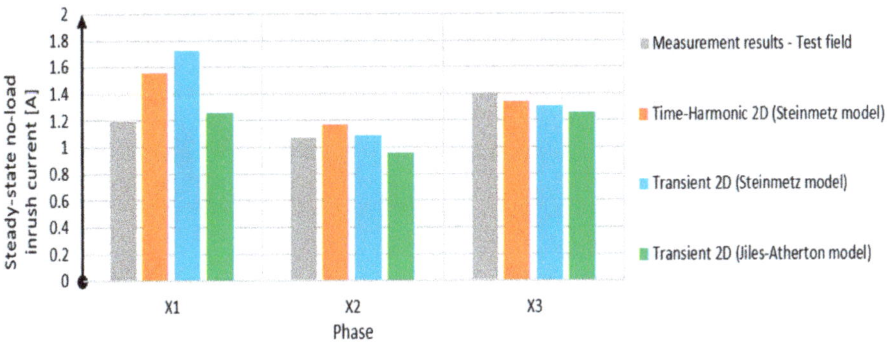

Figure 10. Steady-state no-load inrush current of transformer for Time-Harmonic 2D and Transient 2D calculation methods with Steinmetz and J-A core losses models, divided into phases X1, X2 and X3.

The best convergence with the measured no-load currents in the test field was obtained for the Transient 2D method, using the J-A core losses model with a range of 5 to 21% for individual phases. The results of the RMS values of the transformer no-load currents differ between the windings on the outer limbs. The main factors affecting this are differences in the properties of the core laminations and unavoidable production tolerances that determine the allowable differences in the dimensions of air gaps in the overlap region. The waveform of the current in LV windings under capacitive load conditions was also analyzed. The current harmonic inserted by the additional current sources on the secondary side had no effect on the flux density inside the core.

The differences between the values of the no-load losses of the real transformer measured in the test field and the calculated values in the software are presented in Figure 11. The difference expressed as a percentage is placed on the ordinate axis, while the successive calculation methods with core losses models are placed on the axis of abscissa.

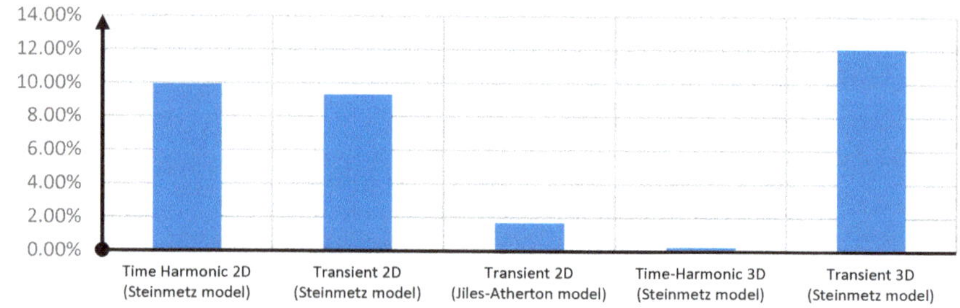

Figure 11. Difference in no-load losses between the test field measurement values and the values calculated in software expressed as a percentage against measurements.

Based on the differences, the Transient 2D method using the J-A core losses model and the Time-Harmonic 3D method with the Steinmetz model show the best convergence with the measurements of the real unit. However, the good accuracy of the 3D Time-Harmonic results seems to be accidental because the model does not contain overlapping.

To compare the differences between the core losses in the no-load state and in the capacitive load state, Figures 12 and 13 are displayed, showing the flux density distribution in the 2D model core. The core hot spot was divided into two parts: the part located in the upper yoke (point A) and the part located in the middle limb (point B). In addition, the central part of the middle limb (point C) was also verified. For these points, the maximum flux density at a given moment in time was calculated. Determination of the magnetic flux density distribution was performed using the time-stepping method.

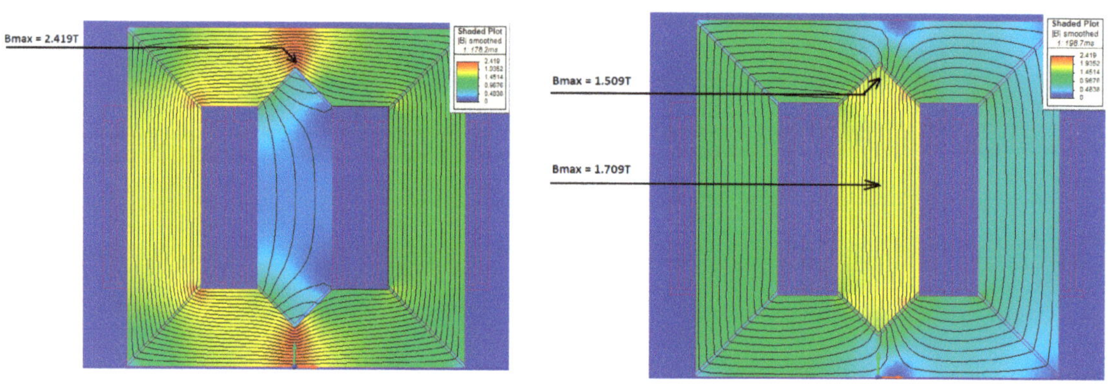

(a) (b)

Figure 12. Instantaneous flux density distribution in 2D model under no-load condition. Calculation method: Transient 2D, core losses model: Jiles–Atherton. (**a**) The highest instantaneous values of flux density in the core hot spot; (**b**) the highest instantaneous values of flux density in the core limb.

The value of flux density at point C increases significantly in the case of the capacitive loading of the transformer compared to the no-load condition. The flux density also increases in the limb part of the core hot spot (point B), while in the yoke part (point A), the flux density values remain similar. The explanation of this phenomenon is that the leakage flux that occurs under load enters the limbs of the transformer core, resulting in an increase in the magnitude of magnetic flux density in this area. In the yoke section, on the other hand, the additional leakage flux does not occur. For both investigated operating states of the power transformer, the magnitude of flux density in the core hot spot is significantly higher compared to the other parts of the core.

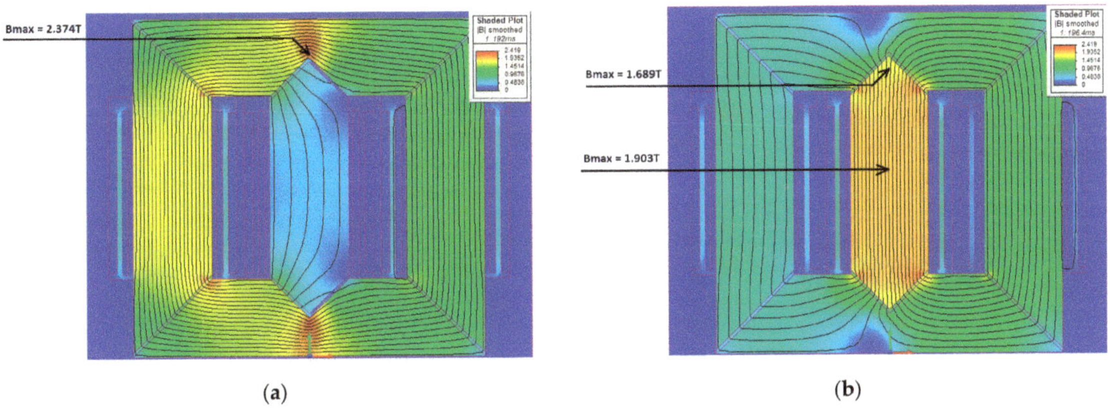

Figure 13. Instantaneous flux density distribution in 2D model under capacitive load condition. Calculation method: Transient 2D, core losses model: Jiles–Atherton. (**a**) The highest instantaneous values of flux density in the core hot spot; (**b**) the highest instantaneous values of flux density in the core limb.

The analysis of the differences in core losses between the capacitive load and no-load conditions was also performed, examining individual elements of the transformer core. The 2D model of the core was divided into smaller elements numbered from 1 to 11. The results obtained from the Transient 2D method using the J-A core losses model were selected for comparison. The obtained differences were related to the values of core losses in the no-load state. The results, showing the changes expressed as a percentage are presented in Figure 14, maintaining the division into hysteresis and eddy current losses.

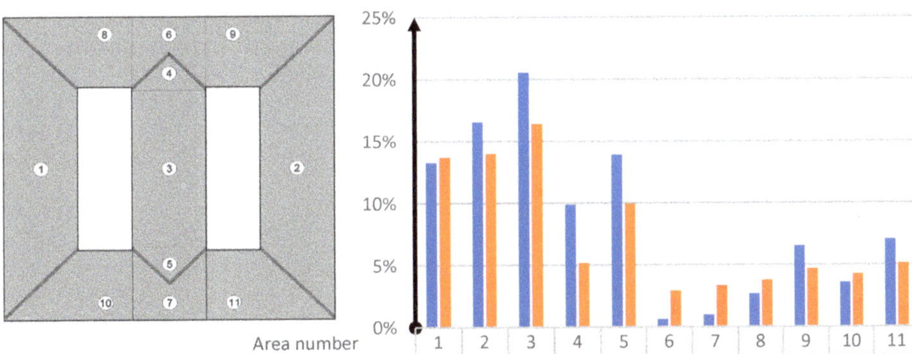

Area number	1	2	3	4	5	6	7	8	9	10	11
Hysteresis losses	13%	17%	21%	10%	14%	1%	1%	3%	7%	4%	7%
Eddy current losses	14%	14%	16%	5%	10%	3%	3%	4%	5%	4%	5%

Figure 14. Percentage change between the values of core losses in the capacitive load condition and in the no-load condition, referred to the no-load losses.

The core losses of the transformer in the capacitive load condition increased compared to those calculated for the no-load condition. The largest change in the value of hysteresis losses, amounting to 21%, was observed in element 3, the middle part of the center limb. The percentage change values of more than 10% were also observed in the outermost limbs and in elements 4 and 5, which represent the core hot-spot area in the middle limb.

5. Conclusions

1. The obtained results of core losses calculated using various methods showed an accuracy within a few percent of the results obtained from measurements of the real power transformer in the test field. The Transient 2D calculation method with the J-A core losses model and the Time-Harmonic 3D method with the Steinmetz model proved to be the best. It is supposed that in the case of the second method, such a good result is accidental because the 3D model did not consider the region of overlap; moreover, the hysteresis loop was also not included in the calculations. The important issue is that core loss values calculated using the Steinmetz model are subject to an error resulting from the need to extrapolate the magnetic permeability and iron losses characteristic in the saturation area.
2. The presented two-dimensional model of the power transformer was prepared considering the areas where the core sheets overlap. Thanks to this approach, it was possible to obtain a more realistic distribution of flux density in the core as well as the values of phase currents.
3. The analysis of the core hot-spot area and the central point of the core limb showed that the capacitive load causes a significant increase in the value of magnetic flux density in the core limb compared to the value obtained in the no-load condition. The change in flux density was particularly noticeable in the limb part of the core. Considering the preparation of the power transformer design for operation in reactive power compensation stations, an increase in the magnetic flux density value results in the core entering the saturation state more quickly. In areas with the worst oil circulation, local overheating may occur, leading to a weakening of the insulation system. The consequence of this may be a significantly accelerated aging of the power transformer, resulting in increased operation costs for the entire power grid. Furthermore, the determination of the core area with the highest density of power losses makes it possible to accurately determine the installation of sensors monitoring the condition of the transformer.
4. The paper discusses aspects related only to the core losses in the power transformer, including the effects of magnetic hysteresis, different calculation methods and the impact of capacitive load. The analysis of windings losses was not included. The operation of thyristors in SVC stations causes the appearance of higher current harmonics and increased load losses. The results presented in this paper can serve as a basis for further research on the topic of power transformer total losses in reactive power compensation systems.

Author Contributions: Conceptualization, P.O. and P.W.; methodology, P.O.; software, P.O.; validation, P.O. and P.W.; formal analysis, P.O.; investigation, P.O.; resources, P.O.; data curation, P.O.; writing—original draft preparation, P.O.; writing—review and editing, P.O. and P.W.; visualization, P.O.; supervision, P.W.; project administration, P.W. All authors have read and agreed to the published version of the manuscript.

Funding: This research received no external funding.

Data Availability Statement: Not applicable.

Conflicts of Interest: The authors declare no conflict of interest.

References

1. Dixon, J.; Rodriguez, J.; Moran, L.; Domke, R. Reactive Power Compensation Technologies: State-of-the-Art Review. *Proc. IEEE* **2005**, *93*, 2144–2164. [CrossRef]
2. Xu, W.; Martinich, T.G.; Sawada, J.H.; Mansour, Y. Harmonics from SVC transformer saturation with direct current offset. *IEEE Trans. Power Deliv.* **1994**, *9*, 1502–1509. [CrossRef]
3. Arslan, E.; Sakar, S.; Balci, M.E. On the No-Load Loss of Power Transformers under Voltages with Sub-harmonics. In Proceedings of the 2014 IEEE International Energy Conference (ENERGYCON), Dubrovnik, Croatia, 13–16 May 2014. [CrossRef]
4. Girgis, R.; teNyenhuis, G.E. Hydrogen Gas Generation Due to Moderately Overheated Transformer Cores. In Proceedings of the IEEE Power & Energy Society General Meeting, Calgary, AB, Canada, 26–30 July 2009. [CrossRef]

5. Girgis, R.; teNyenhuis, G.E. Characteristics of Inrush Current of Present Designs of Power Transformers. In Proceedings of the IEEE Power Engineering Society General Meeting, Tampa, FL, USA, 24–28 June 2007. [CrossRef]
6. Jain, S.A.; Pandya, A.A. Three Phase Power Transformer Modeling Using FEM for Accurate Prediction of Core and Winding Loss. *Kalpa Publ. Eng.* **2017**, *1*, 75–80.
7. Krings, A.; Soulard, J. Overview and comparison of iron loss models for electrical machines. *J. Electr. Eng. Elektrotech. Cas.* **2010**, *10*, 162–169.
8. Tekgun, B. Analysis, Measurement and Estimation of the Core Losses in Electrical Machines. Master's Thesis, The University of Akron, Akron, OH, USA, 2016.
9. Steinmetz, C.P. On the law of hysteresis. *Trans. Am. Inst. Electr. Eng.* **1984**, *72*, 197–221. [CrossRef]
10. Jiles, D.; Atherton, D. Theory of ferromagnetic hysteresis. *J. Magn. Magn. Mater.* **1986**, *61*, 48–60. [CrossRef]
11. Guerin, C.; Jacques, K.; Sabariego, R.V.; Dular, P.; Geuzaine, C.; Gyselinck, J. Using a vector Jiles-Atherton hysteresis model for isotropic magnetic materials with the FEM, Newton-Raphson method and relaxation procedure. *Int. J. Numer. Model. Electron. Netw. Devices Fields* **2016**, *30*, e2189. [CrossRef]
12. Zakrzewski, K.; Tomczuk, B. Magnetic Field Analysis and Leakage Inductance Calculation in Current Transformers by Means of 3-D Integral Methods. *IEEE Trans. Magn.* **1996**, *32*, 1637–1640. [CrossRef]
13. Constantin, D.; Nicolae, P.M.; Nitu, C.M. 3D Finite Element Analysis of a Three Phase Power Transformer. In Proceedings of the EuroCon 2013, Zagreb, Croatia, 1–4 July 2013. [CrossRef]
14. Xing, J.; Dai, Z.; Song, Y.; Wang, D.; Yu, W.; Zhang, Y.F. Finite Element Calculation Loss Analysis of Large Transformers under the Influence of GIC. In Proceedings of the 10th International Conference on Power Science and Engineering, Istanbul, Turkey, 21–23 October 2021. [CrossRef]
15. Mousavi, S.; Shamei, M.; Siadatan, A.; Nabizadeh, F.; Mirimani, S.H. Calculation of Power Transformer Losses by Finite Element Method. In Proceedings of the IEEE Electrical Power and Energy Conference (EPEC), Toronto, ON, Canada, 10–11 October 2018. [CrossRef]
16. Nakata, T.; Takahasi, N.; Fujiwara, K.; Okada, Y. Improvements of the T-Omega method for 3-D Eddy current analysis. *IEEE Trans. Magn.* **1988**, *24*, 94–97. [CrossRef]
17. Hussain, S.; Chang, K. Effects of Incorporating Hysteresis in the Simulation of Electromagnetic Devices, Siemens. Technical Report. 2019. Available online: https://www.researchgate.net/publication/325943830_Effects_of_Incorporating_Hysteresis_in_the_Simulation_of_Electromagnetic_Devices (accessed on 7 May 2023).
18. Ciesielski, M.; Witczak, P. The Equivalence Conditions of 2D and 3D Models of Phase Shifting Transformer. In Proceedings of the 7th International Advanced Research Workshop on Transformers (ARWtr), Baiona, Spain, 2–26 October 2022. [CrossRef]
19. Ciesielski, M.; Witczak, P. The use if FEM modeling to analyze phase shifting transformer in steady-state service conditions. *COMPEL Int. J. Comput. Math. Electr. Electron. Eng.* **2022**, *41*, 1214–1222. [CrossRef]

energies

MDPI

Article

Numerical and Analytical Determination of Steady-State Forces Acting on Cleats and Leads Conductor of the Power Transformer

Michał Smoliński [1,2,*] and Paweł Witczak [2]

[1] Hitachi Energy, 91-205 Lodz, Poland
[2] Institute of Mechatronics and Information Systems, Lodz University of Technology, 90-924 Lodz, Poland; pawel.witczak@p.lodz.pl
* Correspondence: michal.smolinski@hitachienergy.com

Abstract: Electromagnetic forces acting on conductors of the cleats and lead of a power transformer can cause permanent damage to the insulation of conductors. Determining the force acting on the conductor of cleats and leads cannot be performed using the standard analytical formula because those conductors are in close proximity to the construction part of the active part made of ferromagnetic material. To calculate those forces in a steady state of a three-phase AC current, a parametric numerical simulation was conducted. Based on the simulation, a new analytical formula for forces acting on cables near the ferromagnetic plate was proposed by the authors. It was also noted that the presence of the ferromagnetic plate can increase the forces up to 60% compared to the same geometry without the plate. This publication also discusses how eddy currents and the proximity effect influence forces acting on conductors.

Keywords: electromagnetic forces; cleats and leads; power transformer; numerical simulation

check for updates

Citation: Smoliński, M.; Witczak, P. Numerical and Analytical Determination of Steady-State Forces Acting on Cleats and Leads Conductor of the Power Transformer. *Energies* **2023**, *16*, 3600. https://doi.org/10.3390/en16083600

Academic Editors: José Antonio Domínguez-Navarro and Marco Merlo

Received: 13 March 2023
Revised: 10 April 2023
Accepted: 12 April 2023
Published: 21 April 2023

1. Introduction

In the three-phase power transformer, the electromagnetic forces acting on cleats and leads (C&L) conductors can be enormous. Especially in the case of an operational short circuit of power transformers with tertiary winding with a reduced power, the current density in C&L may reach the value of 280 A/mm^2, which gives a peak phase current of approximately 100 kA, as in the case of the power transformer shown in Figure 1. The damage of the insulation of the cables in this transformer after the short circuit test is shown in Figure 1. This figure depicts three cables from three different phases near the core press plate. The insulation of those cables was destroyed near the supporting ribs due to the oscillation of cables during the short circuit test. This kind of failure is very difficult to predict during the design phase of a power transformer. To prevent the failure of insulation, the electromagnetic forces acting on the cables must first be known. Because cables are close to the core press plate, which is made of ferromagnetic material, forces acting on the cables cannot be determined by using an analytical formula. These forces can be determined by using magnetic stimulation. After knowing the force acting on cables, dynamic mechanical simulation must be conducted to simulate the oscillation of cables, and some failure criteria for insulation must be proposed. Building such a complex multiphysics model is time-consuming and much knowledge is needed. As each power transformer is designed individually according to customer needs, the cleats and leads designer does not have enough time to perform such a time-consuming algorithm. The idea of this publication was to develop analytical formulas that will approximate forces acting on conductors in the steady state of AC current. Having such analytical formulas helps the designer to calculate forces very quickly without using magnetic numerical simulation. This paper will not discuss the mechanical part of the problem. The mechanical model of the cable is awaited in further research.

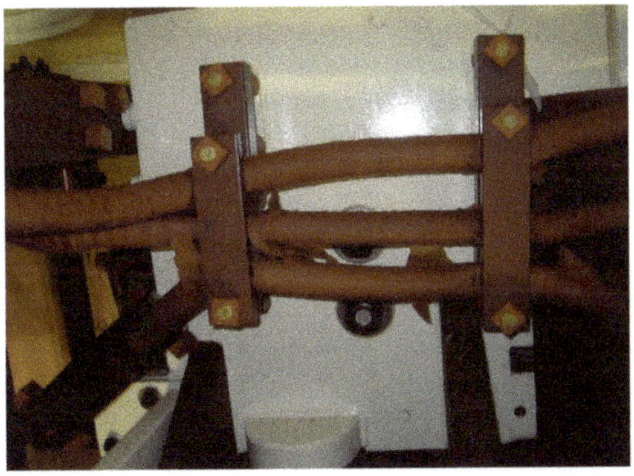

Figure 1. Damaged insulation during short circuit test.

There is a very limited number of publications about forces acting on C&L conductors. However, there is a great number of papers focusing on short-circuit forces acting on the windings, and some of them will be described here. To determine short-circuit forces in windings, three main methods were used: numerical simulation, analytical formulas, and experiments. In ref. [1], a finite element method (FEM) simulation was used to calculate the electromagnetic field in the core and windings and determine forces during short-circuit and inrush conditions. The authors prepared two-dimensional and three-dimensional models of windings and compared axial and radial forces obtained from both simulations. The main conclusion was that there was a significant difference between forces calculated by 3D and 2D simulations. The radial force in the case of the 2D simulation was underestimated. In ref. [2], a magnetostatic analysis of the winding conductor was used to calculate the peak value of the force acting on windings. The peak forces calculated by the magnetostatic analysis were transferred to a static mechanical simulation to calculate the maximum deformation of the winding and determine its mechanical stiffness. Based on this coupled simulation, the critical buckling load for transformer winding was determined. The authors also prepared an experimental stand and measured the force in the short-circuit state and confirmed that the critical buckling load from the simulation differs by only 2.3% from the critical value obtained from the experiment. This publication shows how effective a good FEM method for the multiphysics system can be. Publication [3] presents a full dynamic multiphysics model of windings. The authors used a 2D transient simulation to calculate the dynamic electromagnetic forces acting on windings during the short circuit. The authors developed a mechanical model of windings and applied calculated forces from electromagnetic simulation to those windings. Based on a transient mechanical simulation, the deformation in winding and stresses in conductors as a function of time was determined. A similar approach of a multiphysics dynamic simulation was used in ref. [4] and ref. [5]; however, here, both the electromagnetic simulation and mechanical simulation were made using 3D transient solvers in COMSOL software.

Based on the presented publications, it can be concluded that using a multiphysics model to calculate forces acting on winding conductors and the mechanical response is widely used. However, building such a model is time-consuming. Therefore, an analytical approach to calculating forces is needed. Analytical formulas for radial and axial short-circuit forces were developed in ref. [6] and compared to results from numerical simulations. The results of the numerical simulation are in good agreement with the analytical formulas presented by the authors.

This paper concentrates on forces acting on the C&L of the power transformer. The authors found some selected publications discussing the geometry of straight conductors, which are similar to the problem in this paper. In ref. [7], a geometry consisting of three parallel busbars was analyzed. The forces acting on the busbars were calculated using the analytical formula. A mechanical transient simulation was conducted to determine the deformation and mechanical stresses in the busbar. It should be mentioned that the use of analytical formulas to determine the forces was possible because the geometry was simple and there was no ferromagnetic material near the conductor, which is not the case with the system described in Figure 1. In ref. [8], a full multiphysics numerical model concerning electromagnetic, thermal, and mechanical phenomena was discussed. Based on this model, the dynamic deformation and equivalent stresses in busbars caused by electromagnetic forces and thermal effects were determined.

Comparing ref. [9], which presents the classical approach to modeling of short circuit state of power transformers to the more up-to-date refs. [10,11], it can be concluded that the requirements for modeling of the short circuit state operation of power transformers have significantly increased and currently numerical simulations are used more frequently to better model physics of short circuit state.

The discussed problem of the failure of the insulation is very complex because it involves both electromagnetic and mechanical phenomena. In order to fully solve the discussed problem, a multiphysics simulation is needed, consisting of an electromagnetic simulation and mechanical simulation. This paper focuses only on determining forces acting on the conductors of the cleats and leads of a power transformer. The mechanical modeling of a cleats and leads conductor is in progress and will be discussed in the next paper. As the conductor of cleats and leads consists of hundreds of small copper wires wounded helically and insulation wrapped around the copper part, the mechanical modeling of such a system by the use of a numerical simulation is very difficult. Due to the complex geometry of the cables, the system is highly nonlinear, and force cannot be applied by one step, but by gradually increasing the force, which makes a mechanical simulation very difficult and time-consuming. Moreover, in this publication, the authors did not concentrate on the modeling of one case of a power transformer or a few cases, but rather developed a general approach on how to determine forces acting on cleats and leads conductors, analyzing the results from thousands of design points from parametric simulations and analytical formulas.

In Section 2, the analytical formula for forces acting on the cleats and leads conductors in a standard three-phase system without a ferromagnetic plate is determined. In Section 3, the authors describe the methods and results of the parametric numerical simulation of a magnetic field, concentrating on calculating forces acting on mentioned conductors. Firstly, the case without a ferromagnetic plate is discussed, and the results of the simulation are compared to the analytical equation from Section 2. As the next step, a number of parametric simulations with geometry consisting of three cables and a magnetic plate are presented. The forces determined by the numerical simulation were compared to the standard analytical formula from Section 2. In Section 3, the influence of geometrical parameters, as well as eddy currents in plate and cables, on forces acting on conductors is discussed. Based on parametric numerical simulations, the authors checked which parameters have a significant influence on the forces acting on the conductors and developed new improved formulas for forces acting on conductors near a ferromagnetic conductive plate, which are discussed in Section 4. In Section 5, all conclusions from the whole publication are presented.

2. Analytical Formulas for AC Forces in Steady State without the Ferromagnetic Plate

Before discussing a more advanced analysis, the theoretical formula for forces acting on cables in the symmetric three-phase in-plane system shown in Figure 2a will be discussed. The geometry mentioned in Figure 2a does not contain ferromagnetic plates, so simple analytical formulas can be discussed. The following assumptions were made for this analysis:

- The conductor is very long, so it can be assumed that the magnetic field is constant across the whole length of the conductor.
- The diameter of the cable is small with respect to the distance between cables.

- The currents flowing through cables are sinusoidal changing with angular frequency ω. The steady state is analyzed here. The phase of currents in cables 1, 2, and 3 are, respectively, 0, 120, and 240 degrees.

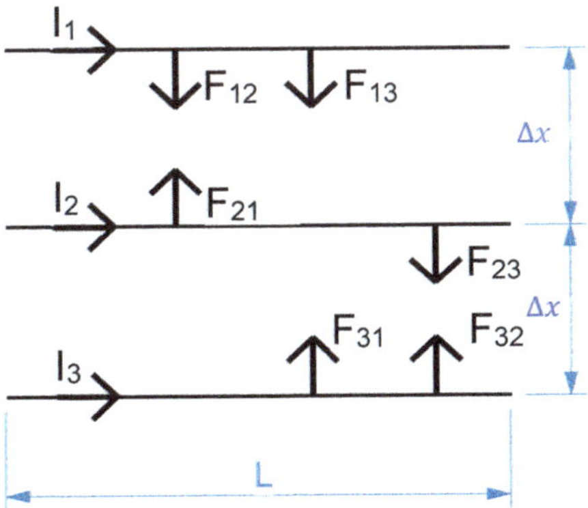

(a)

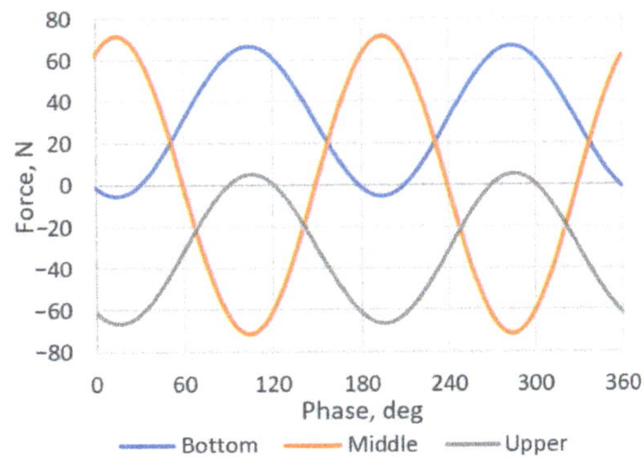

(b)

Figure 2. (a) Forces acting on 3 parallel conductors; (b) instantaneous forces acting on plane, 200 mm long system of three-phase conductors in air shifting each other by 60 mm and carrying 10 kA peak current.

Combining the Lorentz equation and Ampere law, the classic form of the instantaneous force $f_{km}(t)$ per unit length between two parallel conductors shifted by distance Δx and carrying the currents $i_m(t)$ and $i_k(t)$ holds [12]:

$$f_{km}(t) = \frac{\mu_0}{2\pi\Delta x} i_m(t) i_k(t) \quad (1)$$

The currents are symmetrically shifted in time but the set of conductors is not symmetric in space. We may distinguish the "middle" and two "side" conductors. The resultant force acting on the k-th conductor $f_k(t)$ is the sum of interactions with the remaining two.

$$f_k(t) = f_{km}(t) + f_{kn}(t) \quad (2)$$

It can be decomposed into two components: the constant in time F_{kDC} and alternating in time with the doubled network frequency 2ω and amplitude F_{kAC}.

$$f_k(t) = F_{kDC} + F_{kAC} e^{j(2\omega t + \varphi_k)} \quad (3)$$

The phase angles φ_k amount to 0 and $\pm 2\pi/3$. These amplitudes are obtained after simple trigonometric manipulations inserting sinusoidal relationships for phase currents. Their values are presented in Table 1 following the compact formula using the factor k. I stands for the amplitude of the AC current and the peak force F amounts to

$$F = k \frac{\mu_0}{\pi \Delta x} I^2 \quad (4)$$

Table 1. Values of factor k for set of three conductors supplied from three-phase network; subscripts are as in Figure 2a.

Force Component	F_{1DC}	F_{1AC}	F_{2DC}	F_{2AC}	F_{3DC}	F_{3AC}
k	$\frac{3}{16}$	$\frac{\sqrt{3}}{8}$	0	$\frac{\sqrt{3}}{4}$	$\frac{3}{16}$	$\frac{\sqrt{3}}{8}$

3. Numerical Simulations

The main goal of this publication was to study the forces acting on a three-phase system of cables situated close to the active part of the power transformer and carrying sinusoidal currents. This system is usually located near the ferromagnetic plate pressing the core as is shown in Figure 1. The said plate is made of a conductive, ferromagnetic material and, as will be described later, its presence has a great influence on the forces acting on the cables. Therefore, the analytical Equation (4) cannot be used directly. In order to calculate forces acting on the conductors, a numerical simulation must be applied.

The numerical simulation was conducted in ANSYS Maxwell 2021 R1 software in the eddy current module. This solver assumes that all quantities, such as the current density and magnetic field density, are sinusoidal changing with a defined frequency, so it only calculates the steady state of the *AC* current. The mentioned solver numerically solves the Maxwell equations in the form presented below [12].

$$\nabla \cdot D = \frac{\rho_e}{\epsilon_0} \tag{5}$$

$$\nabla \times E = -j\omega B \tag{6}$$

$$\nabla \cdot B = 0 \tag{7}$$

$$\nabla \times H = J + j\omega D \tag{8}$$

As a next step, the instantaneous force acting on each conductor was calculated by integrating Equation (9) over the volume of the cable [12].

$$F(t) = \int J(t) \times B(t) dV \tag{9}$$

Firstly, a parametric model consisting of three parallel cables supported by wooden brackets near the ferromagnetic plate was created. (Figure 3a) The presented parametric geometry is very close to the real geometry shown in Figure 1. By modifying the parameters (d, Δx, Δy, L, H, h), the geometry of the model can be changed and the influence of geometrical parameters on the force acting on the conductors can be investigated. The boundary condition assigned to all external surfaces of the model is a magnetic field tangent. As the excitation, the total current flowing through the cross-section of the cable was assigned to be the AC with amplitude I.

In the first simulation, the plate was omitted and eddy currents in cables were neglected. The instantaneous forces acting on cables obtained from the simulation are shown in Figure 2b. The force acting on the upper and bottom conductor consists of an *AC* component of 2ω frequency and a *DC* component, whereas, for the middle conductor, only the *AC* component is present. This is exactly as discussed in Section 2. Moreover, the numerical values of the *DC* force and *AC* force amplitude from theoretical Equation (4) are the same as those calculated by Maxwell software. This consistency in results between the simulation and theoretical formulas confirms that the simulation model is correct.

In order to investigate the influence of the plate on forces, a simulation with a plate made of ferromagnetic material with constant relative permeability $\mu_r = 290$ was conducted. Linear B-H characteristics were used to improve the speed of the simulation. The selected value of relative permeability was chosen according to the authors' experience. The results of selected linear simulations were compared to the simulation with nonlinear characteristics, and the difference between those results was considerably low. Four different simulations were carried out for the same geometry to investigate the influence of eddy currents in the plate and conductors on forces acting on the cables. In the first simulation, the plate was assumed to be nonconductive, and the conductors were stranded (eddy currents were neglected). The forces acting on the cable in the x-direction are shown in

Figure 4a. Comparing forces acting on cables near a non-conductive plate (Figure 4a) to forces acting on the conductor when the plate is not present (Figure 2b), it can be observed that the amplitude of forces is 25% higher when the plate is present. The force acting on the cables close to the nonconductive plate in the y-direction is shown in Figure 4b. In the case of simulation without the plate, the force in the y-direction is zero. From those simulations, two main conclusions can be drawn. Firstly, adding a ferromagnetic plate increases the magnitude of the force in the x-direction. Secondly, when the plate is present, a new component of force in the y-direction is present. This finally confirms that using the standard analytical Equation (4) is not enough because it will underestimate the forces.

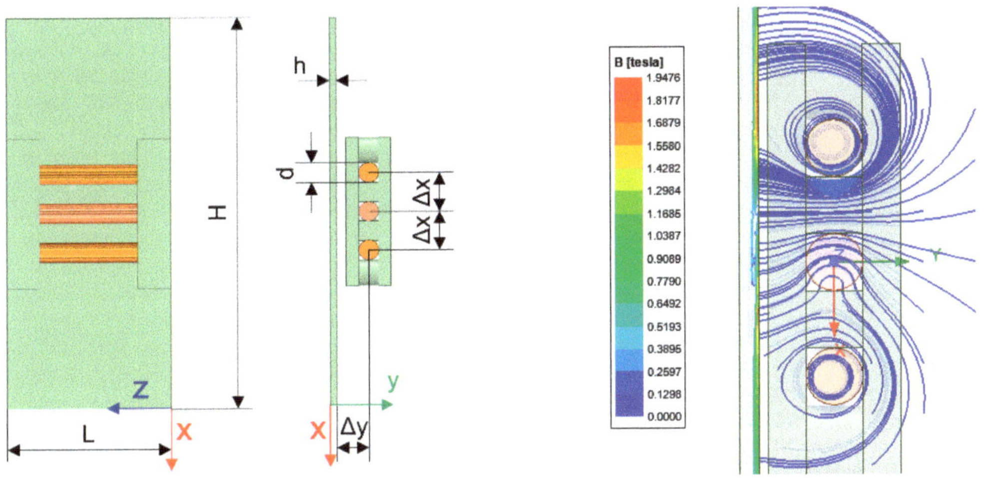

(**a**) (**b**)

Figure 3. (**a**) Model of 3 cables near magnetic plate in ANSYS Maxwell; (**b**) magnetic flux density for simulation with conductive plate.

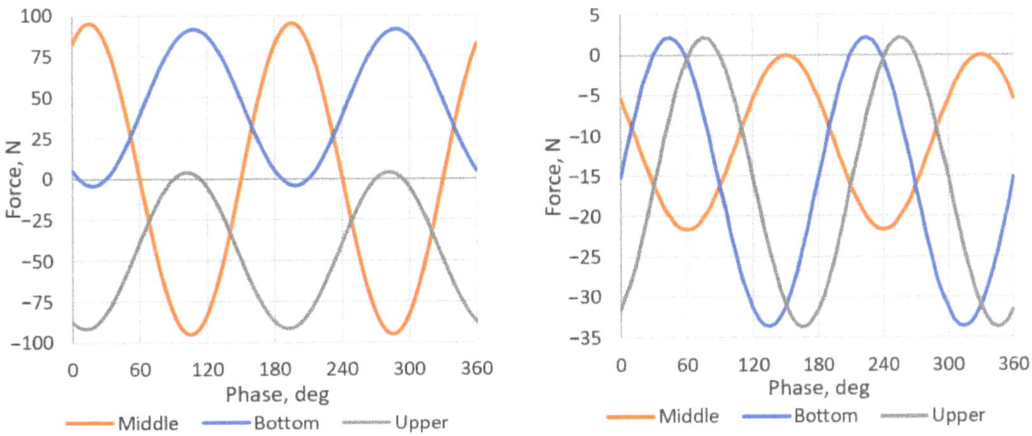

(**a**) (**b**)

Figure 4. Instantaneous forces acting on stranded cables near a non-conductive core plate in (**a**) x direction; (**b**) y direction (d = 30 mm, L = 200 mm, Δx = 60 mm, Δy = 40, H = 600 mm, I = 10 kA).

In the second simulation, the eddy currents in plates were added, but eddy currents in cables were neglected. The assumed conductivity of the plate was 6.7×10^6 S/m. The graphs of instantaneous forces are very similar to those shown in Figure 4, so they will be omitted. After adding eddy currents to the plate, the force in the x-direction is lower than when eddy currents were not included in the model. Moreover, the induced eddy currents in the plate caused a nonzero DC force acting on the middle conductor, which was not the case for simulations without eddy currents.

In the third simulation, only eddy currents in the cable were activated and the plate was non-conductive. In this case, the *DC* force for the middle conductor was again nonzero. This clearly shows that adding the eddy currents either in cables or in the plate causes the *DC* force in the middle conductor and that the system is unsymmetric (the amplitudes of forces acting on the upper and lower conductor are also not the same). Moreover, the *AC* component of force is bigger if eddy currents are present in cables. In the last simulation, both currents in the cable and plate were taken into consideration, and the superposition of effects discussed earlier was present in this case.

In order to investigate the influence of geometrical parameters such as the distance between cables Δx or distance between cables and plate Δy on forces, a parametric simulation for each of the previously mentioned simulations was conducted. In order to better analyze the results, the forces acting on cables without a plate calculated by analytical Formula (4) were treated as the reference and the correction factors were defined to better visualize the influence of the ferromagnetic plate and eddy currents in cables and in the plate. The correction factors for the *AC* and *DC* force for the bottom and upper conductor and AC force for the middle conductor, given in Equations (10) and (11), can be used. The mentioned correction factors define the ratio of the force when the plate is present to the force obtained from the analytical formula. The same concept of the correction factor cannot be used for the *DC* force of the middle conductor, so the correction factor for the *DC* force for the middle conductor is defined, according to (12), as the ratio of the *DC* force with a plate to the *AC* force according to the analytical formula. As the forces in the y direction are significantly lower than in the x direction, they are not discussed in such detail.

$$k_{AC} = \frac{F_{AC\ wit\ cor\ plate}}{F_{A\ withou\ cor\ plate}} \tag{10}$$

$$k_{DC} = \frac{F_{DC\ wit\ cor\ plate}}{F_{D\ withou\ cor\ plate}} \tag{11}$$

$$k_{DC\ middle} = \frac{F_{D\ middle\ wit\ cor\ plate}}{F_{A\ middle\ withou\ cor\ plate}} \tag{12}$$

Now, the results of the mentioned parametric simulation will be discussed using corrections factors. Analyzing Figure 5a,b, it can be concluded that adding a plate always increases the force acting on the bottom conductor. The increase in force is bigger when the cables are very close to the ferromagnetic plate. Secondly, adding the eddy currents in plates always decrease the force in the bottom cable. However, adding the eddy current in the cables increases the force in the bottom cable. As can be seen, the increase in the force may reach 60% for the *AC* component and 40% for the *DC* component. This clearly shows that the influence of ferromagnetic plates and eddy currents in cables should not be neglected during the design process.

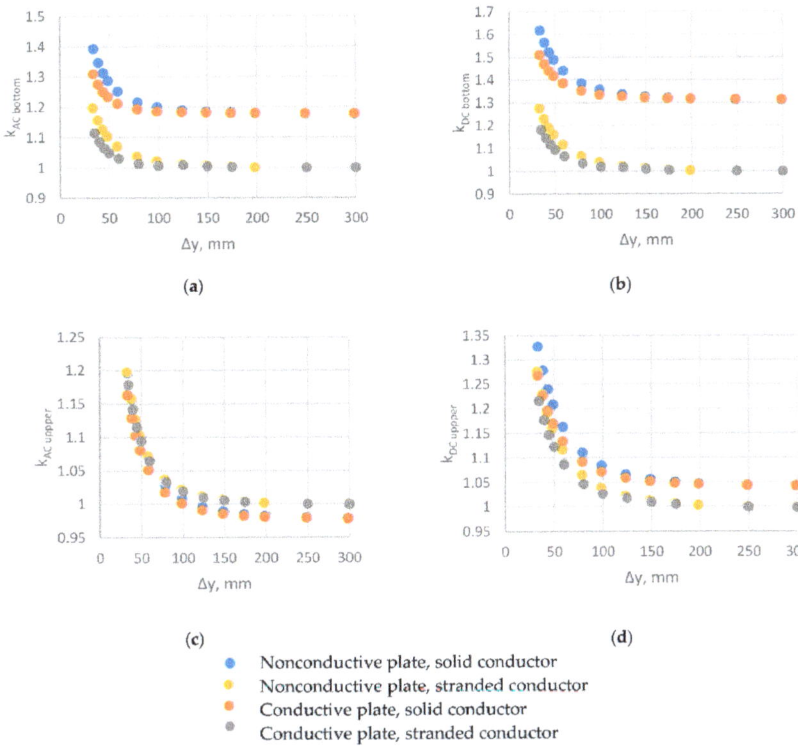

Figure 5. Correction factors: (**a**) $k_{AC\ bottom}$; (**b**) $k_{DC\ bottom}$ (**c**) $k_{AC\ upper}$; (**d**) $k_{DC\ upper}$ for $\Delta x = 35$ as a function of core plate distance.

Analyzing Figure 5c,d, it can be concluded that the forces acting on the upper conductor are lower than the forces acting on the bottom conductor (Figure 5a,b) if eddy currents are included either in cables or in the plate. This means that the eddy currents make the system unsymmetric. Similarly, as for the lower conductor, the eddy currents in the plate decrease forces acting on the upper conductor, and eddy currents in cables increase the forces acting on the upper conductor. This asymmetry depends on different phases of currents in bottom and upper conductors, and not on the geometry of the system.

The same conclusions can be drawn for the *AC* force acting on the middle conductor based on Figure 6a. Eddy currents in the plate decrease the *AC* force acting on the middle conductor, and eddy currents in conductors increase this force. Figure 6b–d show the correction factor for the DC force acting on the middle conductor for different distances between cables (35, 50, and 100 mm, respectively). In the case where eddy currents are neglected both in the plate and conductors, the DC component is zero. After adding conductivity to the plate, this force is nonzero due to eddy currents in the plate. Similarly, in the case of a solid conductor near a non-conductive plate, the force is also nonzero due to eddy currents in cables. However, the direction of force caused by eddy currents in a conductor is opposite to the direction of force caused by eddy currents in the ferromagnetic plate. This cannot be seen in the graphs below because those graphs only analyze absolute values of forces. Finally, after adding the eddy currents in both the conductor and plate, the superposition of both forces exists. As the directions of the forces are opposite, the values of the forces are subtracted. The component of the force coming from eddy currents in the conductor mainly depends on the distance between conductors and the diameter of the conductor. The closer the conductors are to each other, the more important this force is. On the other side, the force caused by eddy currents in the ferromagnetic plate is strongly dependent on the distance between cables and the plate. The closer the cables are to the plate, the higher this force is. As a result, depending on the geometry, either the force caused by eddy currents in cables or the force caused by eddy currents in plates is higher, which can be seen in Figure 6b–d.

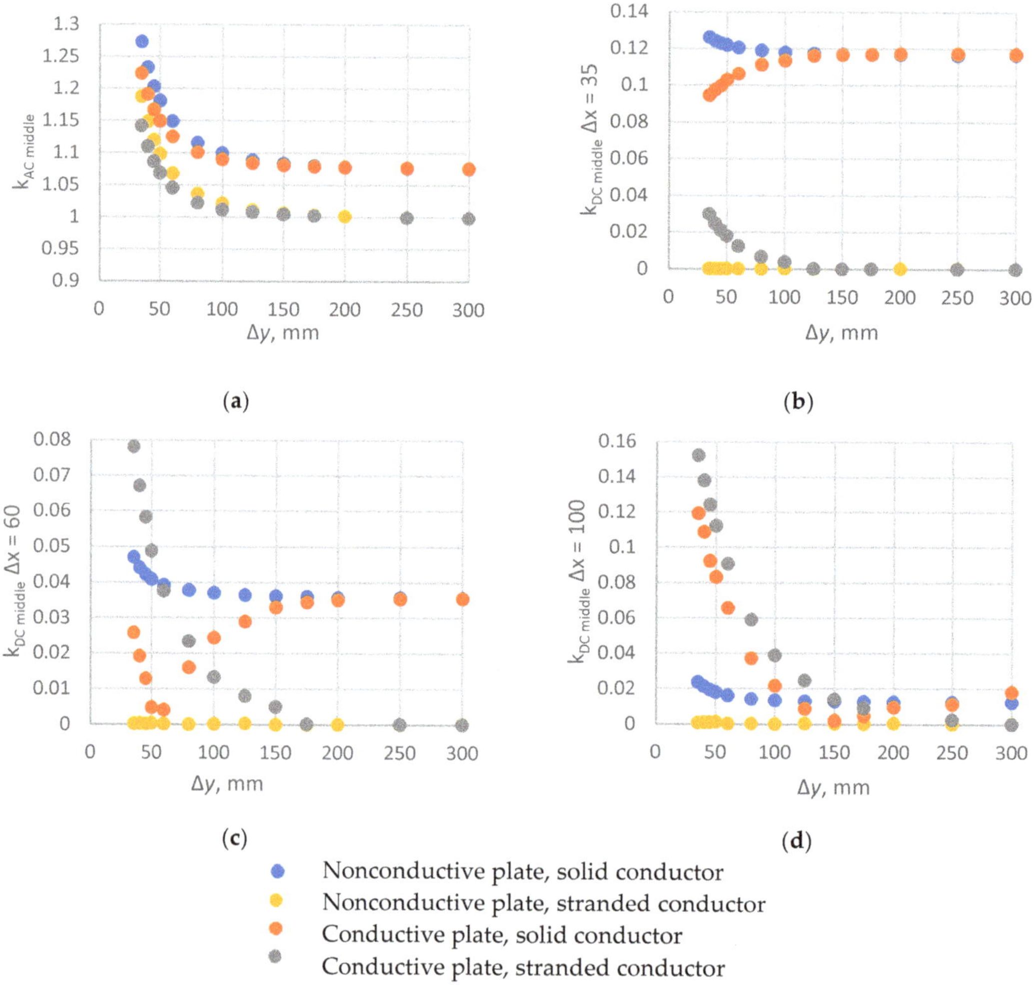

Figure 6. Correction factors: (**a**) $k_{AC\ middle}$ for Δx = 35; (**b**) $k_{DC\ middle}$ for Δx = 35: (**c**) $k_{DC\ middle}$ for Δx = 60; (**d**) $k_{DC\ middle}$ for Δx = 100; as a function of core plate distance.

In the previous simulation, the diameter of the cable was assumed to be 30 mm. However, eddy currents in cables strongly depend on the diameter of the cable and the distance between the conductors. For this reason, a parametric simulation with the diameter of the cable changing from 10 to 30 mm and the distance between cables changing from 35 mm to 300 mm was performed. Figure 7a presents the ratio of the *AC* force acting on the bottom solid conductor to the *AC* force acting on the stranded one for different values of the diameter of the cable and the distance between cables. This figure clearly shows that the influence of eddy currents is more significant for a large diameter of the cable and when cables are close to each other. This can be explained by the proximity effect. Due to the proximity effect, the magnetic field produced by one cable influences the current distribution in the second cable. The smaller the distance between the cables, the bigger the influence of the magnetic field induced by one conductor in the other one. The bigger the diameter of the cable, the smaller the resistance of the cable, and eddy currents in the cable have higher values. In Figure 7b, the current density in the cable and the ferromagnetic plate is shown for phase angle 0. It can be seen that the current distribution is not uniform because one magnetic field of a neighboring conductor influences the current distribution in another one.

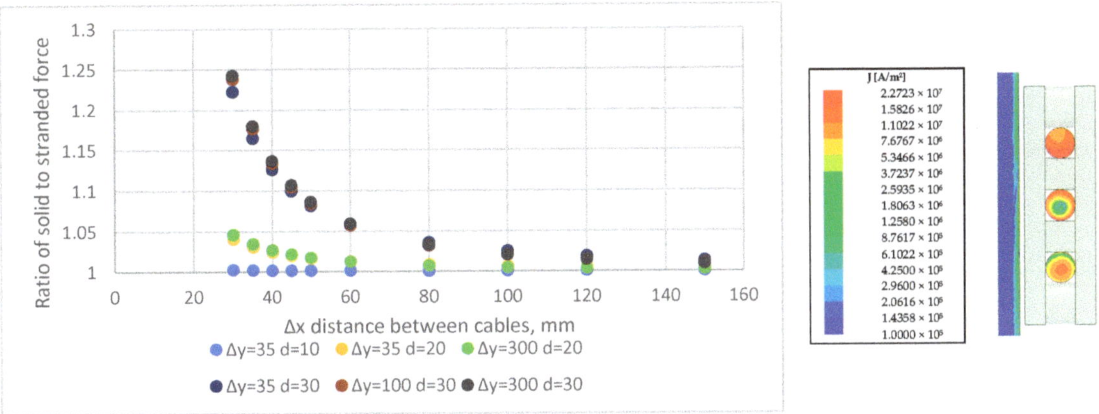

(a) (b)

Figure 7. (**a**) Ratio of AC force acting on a solid conductor to force acting on stranded one for upper cable; (**b**) proximity effect for current density distribution.

The mentioned ratio of solid to stranded *AC* force for the bottom conductor can be approximated by Equation (13). A similar analysis was conducted for the other components of the force and the approximation functions will be described in the next section.

$$\frac{F_{x\ solid}}{F_{x\ stranded}} = 1 + C\frac{d^4}{\Delta x^2} \tag{13}$$

4. Improved Analytical Formulas

Based on all parametric simulations, analytical formulas are proposed approximating the forces acting on the cables near the conductive ferromagnetic plate. At the beginning, it was checked which parameters have a significant influence on the forces and which have negligible influence. The parameters with negligible influence (for example, the thickness of plate h) were not taken into consideration while developing improved formulas. The authors studied the influence of each parameter separately and the influence of a combination of parameters, trying to find the best mathematical function approximating the results of the simulations. After defining the proper mathematical formula, the least square method was used to determine the unknown constants in the equations.

Equations (14)–(16) can be used for three stranded cables near the conductive plate. The *DC* force for the middle conductor in the *x*-direction is described by Equation (16). All other forces in the *x*-direction are expressed by Equation (14). Equation (15) can be used to determine the forces in the y-direction.

$$F_{x\ stranded} = \left(1 + Ae^{-\alpha\frac{\Delta y}{\Delta x}}\right)k\frac{\mu_0 I^2 L}{\pi \Delta x} \tag{14}$$

$$F_{y\ stranded} = B\frac{\mu_0 I^2 L}{\pi \Delta x}e^{-\beta\frac{\Delta y}{\Delta x}} \tag{15}$$

$$F_{x\ DC\ middle\ stranded} = \left(Ae^{-\alpha\frac{\Delta y}{\Delta x}}\right)k\frac{\mu_0 I^2 L}{\pi \Delta x} \quad (16)$$

Figure 8 presents the results from the simulation for different parameters and the approximation function. The maximum relative error of approximation is less than 5%, which is acceptable for engineering purposes.

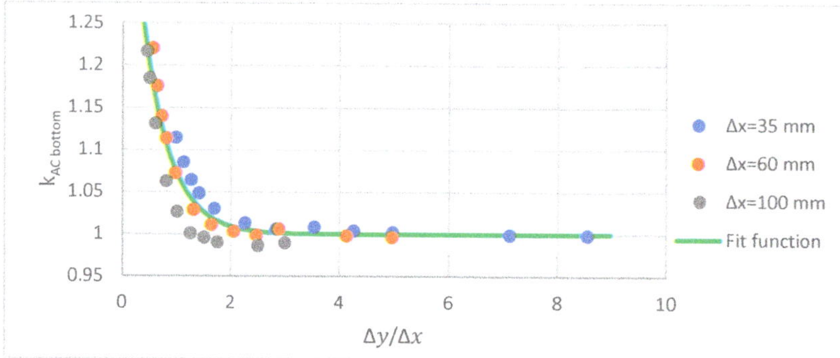

Figure 8. Simulation results and approximation function for stranded conductor.

The analytical formulas for forces acting on the solid cables near the conductive plate were determined and expressed by Equations (17)–(19). For all x-forces except for the *DC* force acting on the middle conductor, Equation (17) can be used.

$$F_{x\ solid} = \left(1 + C\frac{d^4}{\Delta x^2}\right)\left(1 + Ae^{-\alpha\frac{\Delta y}{\Delta x}}\right)k\frac{\mu_0 I^2 L}{\pi \Delta x} \quad (17)$$

For the *DC* component of the force acting on the middle conductor, Equation (18) must be used:

$$F_{x\ DC\ middle\ solid} = \left|C\frac{d^4}{\Delta x^2} - Ae^{-\alpha\frac{\Delta y}{\Delta x}}\right|k\frac{\mu_0 I^2 L}{\pi \Delta x} \quad (18)$$

For all y-forces, Equation (19) can be used.

$$F_{y\ solid} = \left(1 + D\frac{d^4}{\Delta x^2}\right)B\frac{\mu_0 I^2 L}{\pi \Delta x}e^{-\beta\frac{\Delta y}{\Delta x}} \quad (19)$$

As it is depicted in Figure 9 the approximation function are with very good agree-ment to the results of simulations.

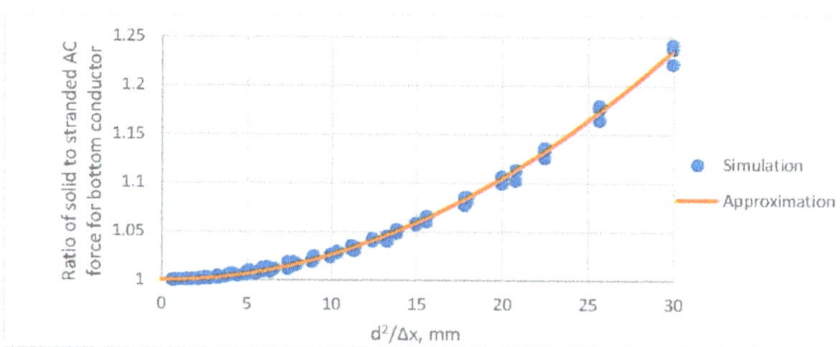

Figure 9. Simulation results and approximation function for solid conductor.

5. Conclusions

1. The electromagnetic forces acting on conductors of C&L can damage the insulation of the conductor. In order to prevent such a failure, it is necessary to know the forces acting on these conductors. Cables are very frequently in close proximity to ferromagnetic elements as core press plates, which makes the calculation of forces difficult.
2. In order to determine the forces acting on the conductor near ferromagnetic elements, the classical Equation (4) for long cables in a vacuum cannot be used because it will underestimate the forces acting on the conductors. Adding the ferromagnetic plate near to the cables causes an increase in the force acting on these cables in the direction parallel to the plate. The increase in force can be up to 60%, and it is stronger the closer the cables are to the plate.
3. Adding the ferromagnetic plate nearby the cable also causes the force in a direction perpendicular to the cables' plane. This force attracts cables toward the plate and it decreases when increasing the distance between the plate and cables. However, forces in the y-direction are always significantly smaller than forces in the x-direction.
4. The eddy currents in the ferromagnetic plate and cables significantly change the forces acting on the conductor. After adding eddy currents in the cables or the plate, the system is unsymmetric. The magnitude of the force acting on the upper and bottom conductor is different due to different phases of the current in those conductors. Furthermore, the eddy currents cause a nonzero DC force on the middle conductor.
5. Eddy currents in the plate cause a decrease in forces in the x-direction, except for the DC force acting on the middle conductor, which is nonzero, in contrast to the situation where eddy currents are neglected.
6. Eddy currents in cables increase the forces in the x-direction due to the proximity effect. The effect of eddy currents on forces increases when increasing the diameter of the cable and decreases when increasing the distance between two cables.
7. Based on the parametric simulation, new improved analytical formulas were developed to calculate forces acting on the cables near the ferromagnetic plate, which will be used in foreseen non-linear mechanical calculations of C&L set during operational short-circuit conditions.
8. Developing improved analytical equations is a big step forward for solving the problem discussed here. However, in order to use the mentioned equations in real engineering applications, a mechanical model of the conductor must be developed. Due to the complex geometry of cleats and leads conductors, performing the mechanical simulation is very difficult. Currently, the authors are in an advanced stage of developing the mechanical simulation, which will be described in the next paper.
9. In the future, experimental verification is foreseen by measuring the deflection of the realistic cable support geometry. This should be compared with the results from a complex non-linear numerical model of the structure, where the applied forces are given by the formulas given here. The finite element meshes in the electromagnetic and structural domains are completely different; therefore, the coupled magneto-mechanical approach cannot be used. A dynamic analysis of the structure seems to go beyond the scope of the presented work. The authors decided to trust the correctness of electromagnetic calculations, which were verified only for variants for which there is an analytical solution. The value added by this paper is the synthesis of thousands of finite elements solutions, showing the importance of several geometric dimensions influencing the final force value.

Author Contributions: Conceptualization, M.S. and P.W.; methodology, M.S. and P.W.; software, M.S.; validation, P.W.; formal analysis, P.W.; investigation, P.W.; resources, M.S. and P.W.; data curation, M.S.; writing—original draft preparation, M.S. and P.W.; writing: M.S. and P.W.; visualization, M.S.; project administration, P.W.; funding acquisition, P.W. All authors have read and agreed to the published version of the manuscript.

Funding: This research was funded by Technical University of Lodz.

Data Availability Statement: Not applicable.

Conflicts of Interest: The authors declare no conflict of interest.

References

1. Faiz, J.; Ebrahimi, B.M.; Noori, T. Three- and Two-Dimensional Finite-Element Computation of Inrush Current and Short-Circuit Electromagnetic Forces on Windings of a Three-Phase Core-Type Power Transformer. *IEEE Trans. Magn.* **2008**, *44*, 590–597. [CrossRef]
2. Geißler, D.; Leibfried, T. Short-Circuit Strength of Power Transformer Windings-Verification of Tests by a Finite Element Analysis-Based Model. *IEEE Trans. Power Deliv.* **2017**, *32*, 1705–1712.
3. Zhang, H.; Yang, B.; Xu, W.; Wang, S.; Wang, G.; Huangfu, Y.; Zhang, J. Dynamic Deformation Analysis of Power Transformer Windings in Short-Circuit Fault by FEM. *IEEE Trans. Appl. Supercond.* **2014**, *24*, 4. [CrossRef]
4. Zhou, D.; Li, Z.; Ke, C.; Yang, X.; Hao, Z. Simulation of transformer windings mechanical characteristics during the external short-circuit fault. In Proceedings of the 2015 5th International Conference on Electric Utility Deregulation and Restructuring and Power Technologies (DRPT), Changsha, China, 26–29 November 2015; pp. 1068–1073.
5. Yadav, S.; Mehta, R.K. FEM Based Study of Short Circuit Forces on Power Transformer Windings. In Proceedings of the 2019 3rd International Conference on Recent Developments in Control, Automation & Power Engineering (RDCAPE), Noida, India, 10–11 October 2019; pp. 540–544.
6. Ebrahimi, B.M.; Fereidunian, A.; Saffari, S.; Faiz, J. Analytical estimation of short circuit axial and radial forces on power transformers windings. *IET Gener. Transm. Distrib.* **2014**, *8*, 250–260. [CrossRef]
7. Szulborski, M.; Łapczyński, S.; Kolimas, Ł.; Kozarek, Ł.; Rasolomampionona, D.D. Calculations of Electrodynamic Forces in Three-Phase Asymmetric Busbar System with the Use of FEM. *Energies* **2020**, *13*, 5477. [CrossRef]
8. Kadkhodaei, G.; Sheshyekani, K.; Hamzeh, M.; Dadjo Tavakoli, S. Multiphysics analysis of busbars with various arrangements under short-circuit condition. *IET Electr. Syst. Transp.* **2016**, *6*, 250–260. [CrossRef]
9. Bertagnolli, G. *The ABB Approach to Short-Circuit Duty of Power Transformers*, 1st ed.; ABB Ltd.: Zurich, Switzerland, 1996.
10. Cuesto, M.; Porrero, J.; Munoz, M.; Camara, J.; Hurlet, P.; Tanguy, A.; Ryadi, M. *Short Circuit Design Conception and Validation of a 570 MVA, Single-Phase GSU-Transformer by SC-Withstand Tests on a Mock-Up Unit*; Cigre: Paris, France, 2016; A2-206.
11. Forslin, J.; Tillery, J.; Muñoz, M.; Saccone, D. *State of the Art in Short-Circuit for Transformers*; Cigre: Paris, France, 2022; A2-10506.
12. Griffiths, D.J. *Introduction to Electrodynamics*, 4th ed.; Cambridge University Press: Cambridge, UK, 2017.

energies

Article

Transmission of Vibrations from Windings to Tank in High Power Transformers

Pawel Witczak [1,*] and Michal Swiatkowski [2]

1 Institute of Mechatronics and Information Systems, Lodz University of Technology, 90-924 Lodz, Poland
2 Hitachi Energy, 91-205 Lodz, Poland; michal.swiatkowski@hitachienergy.com
* Correspondence: pawel.witczak@p.lodz.pl

Abstract: This article presents a step-by-step methodology for calculating transformer tank vibrations caused by electromagnetic forces. This approach uses 3D finite element models for both magnetic and structural calculations. Particular attention was paid to the description of momentum transfer between structural and fluid areas of the transformer. The actual geometry of the coils in the phase windings was taken into account. The dominant role of the axial component of the Lorentz force is the main conclusion of the article. The results are given in the form of three-dimensional displacement fields of the transformer tank presented together with the acoustic pressure field in the oil. The theoretical analysis is verified by laser-scanned vibration patterns on the tank wall.

Keywords: power transformer; finite elements; electromagnetic forces; structural vibration; fluid vibration

1. Introduction

The windings in power transformers have a complex and heterogeneous structure that is mechanically highly anisotropic and relatively weak, especially in high-voltage units where the amount of insulation materials is substantial. Two main points of interest for the structural analysis of the winding area are the winding's resistance to buckling under an operational short circuit and its vibrational behavior under a rated load. On first sight, they look similar: in both cases, the displacement field in the coils is enforced by the electromagnetic Lorentz forces. However, because of the completely different values and time profiles of phase currents in these two issues, the methods of analysis and introduced simplifications must also differ. When the question of possible damage to the winding is discussed, the transient electromagnetic state in the first few milliseconds is solved together with coupled structural calculations. The case of possible buckling of inner winding, usually LV, under radial stress is considered in [1]. This paper includes the evaluation of the bending stiffness of a winding disk of multiple-turn laminated construction, the effect of supporting spacers on the deformation of a winding and the plastic deformation of a winding caused by circumferential stress. A similar problem is also solved in [2], where additional moments of force caused by the wiring process are also investigated. The axial stress of electromagnetic origin may also lead to buckling. The position of windings in the core window may be non-symmetric and which leads to additional force components acting on the winding [3]. The important problem here is the solution to stability questions. The key difficulty in this area is the homogenization process of the winding volume, which means the introduction of artificial material with equivalent properties to represent the real object. The nonlinear material constants, in both electromagnetic and mechanical analyses [4], are the main source of the considerable computational effort required for these calculations.

It is easier to analyze the winding vibrations: the electromagnetic and mechanical stress values are so small that the linearization of the material properties can be applied. However, on the other hand, the solid model of the winding is usually coupled with the fluid area because the transformers immersed in oil are the main object of interest. This

check for updates

Citation: Witczak, P.; Swiatkowski, M. Transmission of Vibrations from Windings to Tank in High Power Transformers. *Energies* **2023**, *16*, 2886. https://doi.org/10.3390/en16062886

Academic Editor: Sérgio Cruz

Received: 5 February 2023
Revised: 9 March 2023
Accepted: 17 March 2023
Published: 21 March 2023

requires a different mathematical treatment with special attention paid to the smoothness of the boundary between solid and fluid volumes. A regular and possibly simple shape of solid parts of the numerical model is extremely helpful in obtaining the convergence of the solution. Nevertheless, such a calculation is quite long and reported examples of its application in the transformer domain are a good illustration of the progress in numerical tools used in the analysis. The detailed model of the transformer, including the winding structure reduced to its orthotropic equivalent, was analyzed in [5] by means of the coupled magneto-mechanical approach. A similar treatment but extended to the real coil distribution in space may be found in [6]. In both of these works, a two-stage 2D-3D transformer model was used to take into account construction details. The coupling between magnetic, structural and acoustic phenomena inside the transformer tank computed by different software can also be obtained by the external supervising algorithm [7]. The influence of resonance effects of the transformer tank on additional devices attached to it is given in [8]. Nevertheless, the full simultaneous solution of coupled electromagnetic-structural-acoustic fields in a power transformer is not available yet due to geometric and material complexity. The use of several simplified models analyzed in series is necessary, but on the other hand, it helps to find which geometric or material parameters are most responsible for the resultant vibrations.

The significance of winding vibration in power transformers also results from the progress in the manufacture of core lamination [9]. Modern sheets have a low magnetostriction factor, under 1 $\mu m/(mT^2)$ [10], and the load-controlled noise dominates in units with power over 100 MVA [11]. Knowing the path of vibration transmission from the phase windings to the surface of the tank makes it possible to detect possible damage within the winding [12–14]. Regardless of the progress in materials technology, the clear tendency of the simultaneous growth of the rated power of transformers, accompanied by the pursuit of the reduction of their costs, means that the development of accurate theoretical models of winding behavior under possible load conditions is still an important issue.

The paper presents, in a detailed way, how the winding vibrations are transmitted via oil in the form of the acoustic pressure wave to the transformer tank.

2. Analysis of Magnetic Field and Forces in Phase Windings

Electromagnetic calculations were carried out for a 120 MVA transformer using the numerical model obtained with Simcenter Magnet 2022.1 finite elements (FE) software and presented in Figure 1. It contains half of the transformer structure accompanied by the appropriate symmetry boundary conditions. It means that the small asymmetry of the position of the active part inside the tank was neglected together with the asymmetry of the magnetic field distribution on the opposite walls of the tank. However, the magnetic flux inside the axial channels of transformer windings is not sensitive to that. The case of the short circuit test in rated conditions was analyzed. Such a test is obligatory during the acceptance tests of the transformer, usually, the acoustic emission measurements are also made during it to simulate these phenomena in the conditions of rated operation.

The solution was obtained in terms of time-harmonic 3D analysis using the T-Ω approach, where the nonlinear properties of ferromagnetic parts were included in an approximated way. The Lorentz volumetric forces acting on HV and LV windings were extracted from the solution field and displayed in Figure 1. Due to symmetry, only half of the windings may be considered. This figure shows the amplitudes only, however, the directional properties of these forces in whole windings were included in further analysis.

These forces possess the following properties:

- The value of both force components does not change along the winding circumference, the small changes from this statement, occurring mostly for the side limbs, may be neglected,
- The value of the radial component of the volumetric force varies linearly from zero to maximum along the radial direction,

– The value of the axial component of the volumetric force remains almost constant along the radial direction.

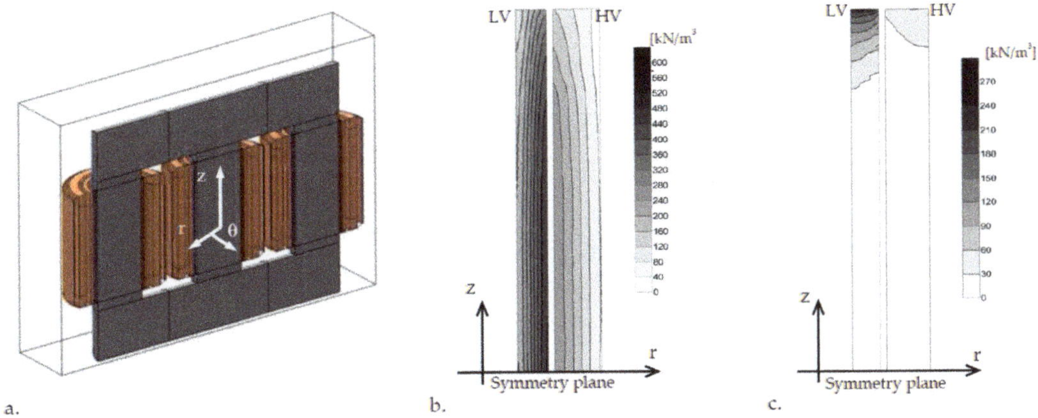

Figure 1. Results of force calculations in a rated short circuit test, the geometry of the FE model (**a**), distributions of AC components of magnitudes of volumetric forces in radial (**b**) and axial (**c**) directions.

These properties will significantly simplify further analysis. The time dependence of the Lorentz force contains two components—DC and AC, having the same amplitudes. From the point of view of transformer vibrations, the DC part has no meaning, and it may be omitted. Calculation of the volumetric force field **f** in the winding is performed simply by calculating the vector product of the constant in space current density J (small variations in this parameter close to the top and bottom of the winding were not accounted for here) by the flux density field B

$$f = J \times B \tag{1}$$

In order to allow the easy transfer of electromagnetic forces into mechanical considerations, some additional tasks are required. The static equivalence between the magnetic stress field σ_{ij} and force density f is given by

$$div\ \sigma_{ij} = -f \tag{2}$$

Expanding (2) in cylindrical coordinates and including the axisymmetric properties for the radial component, we have [15]

$$\frac{\partial \sigma_{rr}}{\partial r} + \frac{\partial \sigma_{rz}}{\partial z} + \frac{\sigma_{rr} - \sigma_{\theta\theta}}{r} = -f_r \tag{3}$$

and for the axial one

$$\frac{\partial \sigma_{zr}}{\partial r} + \frac{\sigma_{zr}}{r} + \frac{\partial \sigma_{zz}}{\partial z} = -f_z \tag{4}$$

The diagonal elements of the Maxwell stress, taking into account $B_\theta = 0$, are given by

$$[\sigma_{rr}, \sigma_{\theta\theta}, \sigma_{zz}] = \frac{1}{2\mu_0}\left[B_r^2 - B_z^2, -B_r^2 - B_z^2, -B_r^2 + B_z^2\right] \tag{5}$$

and the remaining off-diagonal element is equal to

$$\sigma_{rz} = \frac{1}{\mu_0} B_r B_z \tag{6}$$

Further simplifications may be obtained for the middle zone of the winding where $B_r = 0$ and $B_z = \text{const}$. The equations of equivalence (3) and (4) are simply

$$\begin{aligned} f_r &= -\tfrac{\partial \sigma_{rr}}{\partial r} \\ f_z &= 0 \end{aligned} \tag{7}$$

The surface AC traction f_{Sr} acting in the radial direction on the middle section of the winding bounded by the outer radii r_1 and r_2 equals to

$$f_{Sr} = \int_{r_1}^{r_2} f_r(r,z)\,dr \cong 0.5JB_z(r_m,z)(r_2 - r_1) \tag{8}$$

where r_m is the mean value of radii r_2 and r_1. The volumetric force at one end equal to zero. This traction is compensated by the hoop stress appearing in windings along their circumference—the outer winding, usually, HV is elongated, and the inner one is compressed. A different situation exists for the axial stress; in both windings, the electromagnetic axial traction is self-compensated, creating the compression of the winding structure. Assuming once again that means axial stress appears in points having the mean radius, we may calculate the axial traction applied to the winding part between coordinates (0,z) as

$$f_{Sz} = \int_{0}^{z} f_z(r_m,z)\,dz \tag{9}$$

The self-compensation of axial forces concerns the case with the symmetric placement of phase windings on the core. When there is a slight asymmetry in the position of the windings on the core limb, an additional force component appears, causing the windings to move in the form of a rigid body. That displacement is limited by the properties of the pressboard support, and this issue was not considered here. Integral (9) has no analytic expression, and it must be calculated numerically. The volumetric force fields shown in Figure 1 extracted and integrated along lines with the mean radii for both windings give the distributions of the mean traction in radial and axial directions along the winding's height, displayed in Figure 2. Some perturbations visible in the picture of radial traction were caused by a much smaller number of finite elements in the radial direction.

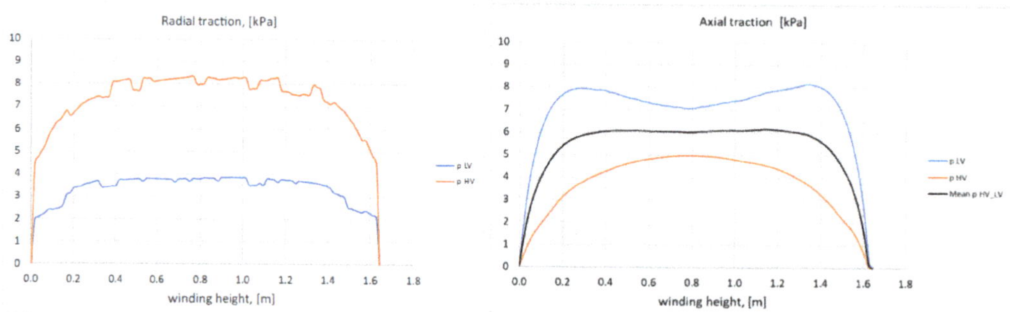

Figure 2. Distributions of traction amplitudes (AC components) along winding height in transformer windings.

3. Numerical Model of Radial Duct in Windings

The main windings in high-power transformers are usually wired with a Continuously Transposed Cable (CTC) and have a dozen cooling radial ducts. The outlook of the section of such a winding is presented in Figure 3.

The majority of radial ducts are understood as the volumes limited by symmetry planes (r,z) of subsequent spacers and symmetry planes (r,θ) of subsequent coils that have the same dimensions in the z-direction. Each of them has two additional planes of symmetry cutting out a quarter of the duct volume. Inside this sub-volume, we have different materials: copper wires (half of CTC), axial and radial spacers and transformer oil filling the empty space between them. This sub-volume was modeled by the ANSYS system as the coupled structural-fluid continuum and solved in terms of a linear harmonic problem.

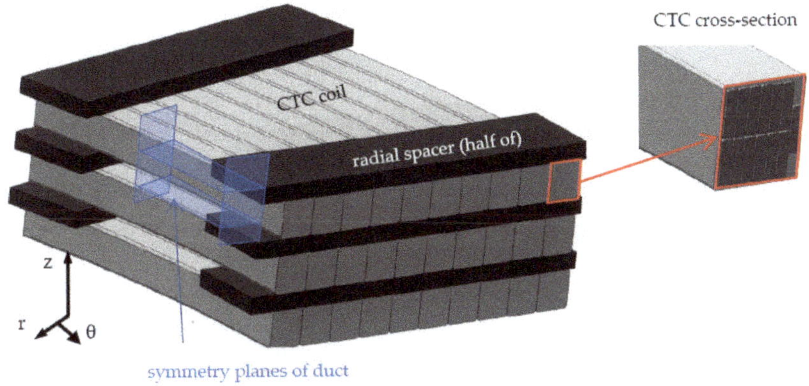

Figure 3. Outlook of the section of winding wired with CTC.

The structural part was represented by the 3D solid elements having three unknown displacements $\{u\}$ in each node of the mesh. The solved set of the equation has the general form

$$(-\omega^2[M] + j\omega[C] + [K])\{u\} = \{F\} \tag{10}$$

where $[M]$, $[C]$ and $[K]$ denote matrices of mass, damping and stiffness, $\{F\}$ is the nodal force vector and ω means the angular frequency.

The fluid mesh was created with 3D acoustic elements solved in terms of scalar acoustic (velocity) potential Φ in element nodes. The acoustic state variables pressure p_a and particle velocity $\mathbf{v}_a$ are defined as

$$\begin{aligned} p_a &= \rho \tfrac{d\Phi}{dt} \\ \mathrm{v}_a &= \nabla \Phi \end{aligned} \tag{11}$$

where ρ is the oil density. The state variables and potential Φ fulfill the wave equation

$$\nabla^2 p_a = \frac{1}{c^2}\frac{d^2 \mathrm{p_a}}{dt^2} \tag{12}$$

where c is the sound velocity. Introducing the FE approximation, it is possible [16] to convert (12) into the form in (10). The ANSYS 2022 software recognizes the fluid-structure boundary where the fluid has additional unknowns describing the fluid displacements. Therefore, definition (11) allows the preservation of continuity of normal stress and normal velocity in both continua. The geometry of an analyzed model of the radial duct is displayed in Figure 4. The model starts from the outer surface of the tertiary winding (r = 0.41 m) and ends on the outer surface of the Tap winding (r = 0.76 m). These windings are shell-type without radial ducts. Tertiary winding is placed almost directly on the core, so it may be treated by the fluid as the stiff boundary. The Tap winding is fixed by the appropriate symmetry boundary conditions on its (r,z) ending planes.

The vibration analysis was performed separately for radial and axial excitations having a frequency of 120 Hz. The pressure representing the radial forces was applied to surfaces between oil and windings in the main insulation axial channel. The axial pressure was forced on the surfaces of both windings belonging to the boundary of the whole model. The pressure values were taken from the traction distributions shown in Figure 2 at half of the winding height. The oil pressure on the inner surface of the Tap winding is the main parameter determining the transmission of vibrations to the walls of the tank, and its space profile is shown in Figure 5. The oil pressure in the axial channel between the HV and Tap windings created by the radial excitation is about two thousand times smaller than at the axial case. This effect is due to the significantly different stiffness of the materials limiting the deformation in both cases. When we have the radial excitations, they are practically fully balanced by the hoop stress in the windings and the oil area is almost not deformed. In the case of axial forces, the displacement values follow the properties of radial spacers and oil, for which the elastic moduli are many times smaller than for copper.

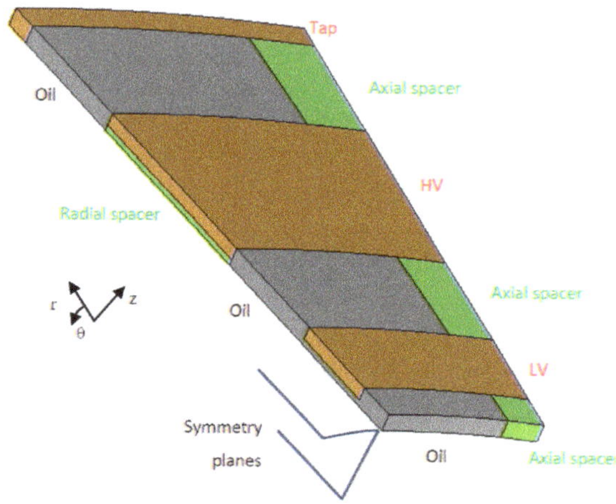

Figure 4. Outlook of the geometry of radial duct in phase windings.

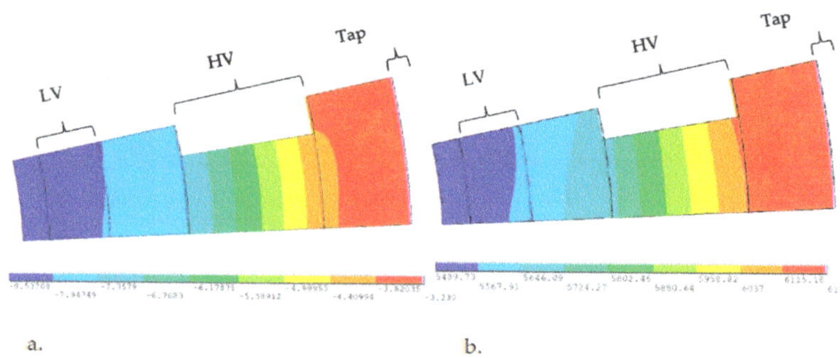

Figure 5. Distribution of oil pressure (AC magnitudes, [Pa]) in the radial channel of the phase windings zone obtained for a short circuit test at 120 Hz for (**a**) radial and (**b**) axial traction.

Additional effects appear when the frequency of traction varies. Having the same amplitude as before, the axial case was solved for three frequency values. The results are given in Figure 6.

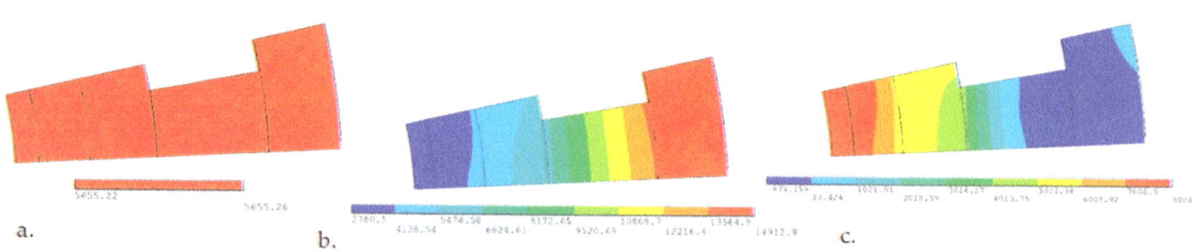

Figure 6. Distributions of oil pressure (Real component, [Pa]) in the radial channel of the winding zone at constant amplitude and varying frequency, (**a**) at 1, (**b**) at 300 and (**c**) at 900 Hz.

When increasing the frequency of excitation, a new spatial component appears, which has the form of a standing wave across the radial length of the channel and the magnitude grows with the frequency. For sufficiently high frequency, here 900 Hz, its magnitude rapidly falls down and simultaneously, the phase is reversed into an opposite value. This phenomenon was caused by a resonance form occurring inside the oil channel where the elasticities and masses of the low voltage winding (axial vibrations) and Tap winding (radial vibrations) were coupled through the oil. The maximum vibrations occur at 365 Hz. The shape of this eigenmode is presented in Figure 7, and it is in good agreement with the forced mode shown in Figure 5. This means that the contribution of the higher frequencies that are always present in the force spectrum will be enhanced.

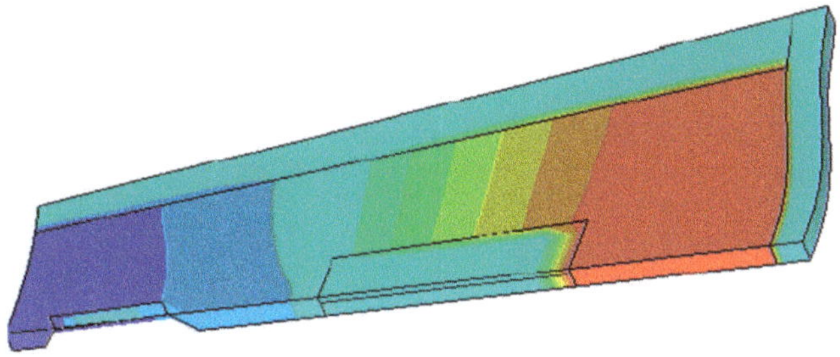

Figure 7. Eigenmode (pressure) of radial channel structure at 365 Hz, green color represents the null value.

4. Numerical Model of Transformer Tank

The FE model of the transformer tank consists of a rectangular steel tank having walls 10 mm thick equipped with vertical stiffeners of 150 × 10 mm and nine magnetic shields of 1650 × 450 × 30 mm connected to the tank in three points along each vertical edge. The shields are "transparent" for the oil interaction, increasing the stiffness and mass of the tank wall only. This simplification was dictated by the need to maintain the regularity of the FE mesh in the oil area, which is of decisive importance from the point of view of the stability of the solution to the entire task. Inside the tank exists the set of Tap windings having the null displacement conditions on surfaces inside the core windows. The outlook of the structural part of the tank is shown in Figure 8. The interior of the tank shown in Figure 9 is filled with acoustic elements modeling the oil. Its volume is limited to the area outside the HV windings and the core yokes, which means that the outer surfaces of these parts represent infinitely stiff media. The model is excited on the outer surfaces of the HV windings, where the known values of the normal oil particle velocity v_{ak} were introduced in form

$$\mathrm{v}_{ak}(t) = V_m e^{j(\omega t - \varphi_k)} \tag{13}$$

where ω is angular frequency and $\phi_k = 0, \pm 2\pi/3$, respectively. This mode of the exciting results from the movements of the oil particles inside the radial channels, as was calculated above.

The magnitude of the velocity V_m has been chosen in such a way that the mean value of the resultant pressure extracted along the winding circumference in the duct between HV and Tap is equal to the value of 6100 Pa calculated previously in the model of the radial duct. The velocity distribution is applied to oil along the cylindrical surface a bit shorter than the winding height, which gives the same value of excitation force. The comparison of these two approaches is shown in Figure 10.

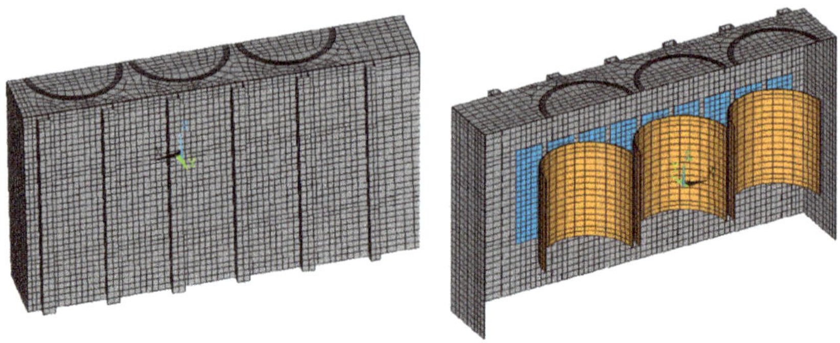

Figure 8. Structural elements of the transformer tank model.

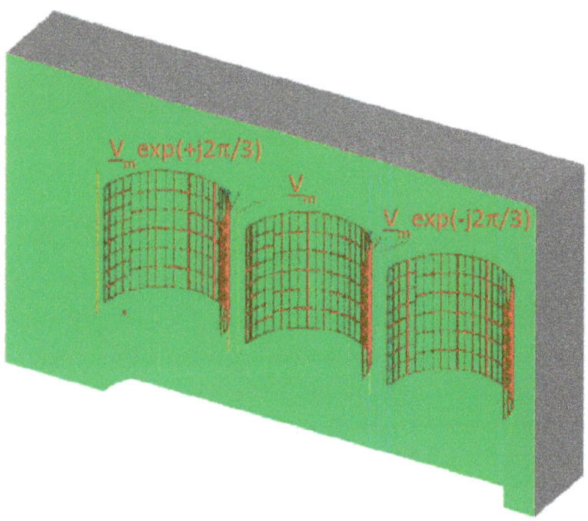

Figure 9. Surfaces with forced oil particle velocity in transformer tank model.

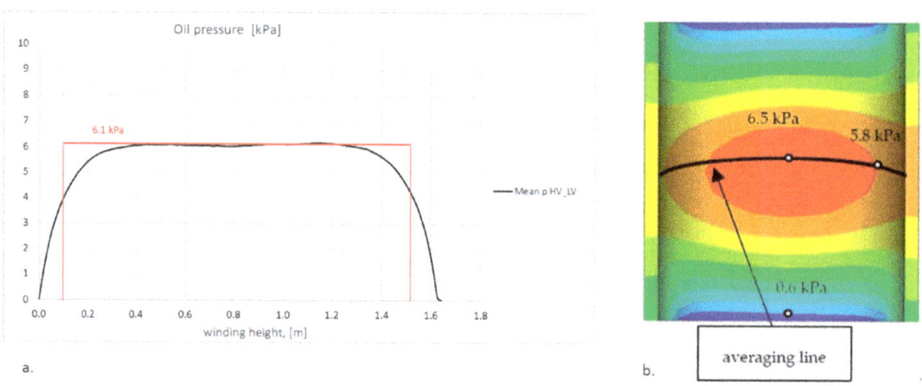

Figure 10. Comparison of oil pressure distribution on the outer surface of HV winding. (**a**) Calculated from electromagnetic analysis, uniform along winding circumference and (**b**) resultant values after insertion into the structural-fluid model of tank.

The explanation of this procedure is as follows. First, by inserting any value of the vibration velocity on the outer surface of the HV winding, say $V_{m0} = 1$, and solving the model, we have a certain sound pressure distribution. Taking the spatially averaged pressure value P_{a0} along the line shown in Figure 10b, we obtained a specific acoustic impedance Z_{a0} for the requested system of excitation

$$Z_{a0} = \frac{P_{a0}}{V_{m0}} \tag{14}$$

The desired velocity amplitude V_m giving approximately the pressure P_{am} calculated with a single radial channel is then simply

$$V_m = \frac{P_{am}}{Z_{a0}} \tag{15}$$

The next two figures show the instantaneous fields of the tank wall velocity and the sound pressure on the tank wall.

Here, we see similar shapes for the real pressure and imaginary velocity components, as well as an inverted pair. From a fluid point of view, this means that the acoustic field inside the tank does not create any propagating wave, and only near-field disturbances are observed. The critical frequency f_{cr} below which the acoustic wave cannot propagate may be estimated for the idealized case of a cylindrical source of height L vibrating in a stiff baffle toward the infinite fluid [15]

$$f_{cr} = \frac{c}{2L} \tag{16}$$

where the sound velocity in oil is c = 1480 m/s, and the winding height is L = 1.64 m. It gives f_{cr} = 451 Hz, which is well above the excitation frequency of 120 Hz. The presence of obstacles, such as the tank wall or adjacent windings in the real case, will increase the f_{cr} value.

Velocity patterns move horizontally due to the phase shift between the axial forces acting in the phase windings on adjacent core columns. They contain both traveling and standing wave elements. Having in mind the harmonic relation between the displacement u and velocity v

$$v = j\omega u \tag{17}$$

Field $Im(v_n)$, shown in Figure 11b, also represents the field $-Re(u_n)$, which, in turn, is almost identical after the sign change to the pressure field, $Re(p_a)$, displayed in Figure 12a. The lack of a significant phase shift between pressure and displacement means that resonance effects observed at the tank wall at a given frequency of excitation are of second-order in the analyzed case.

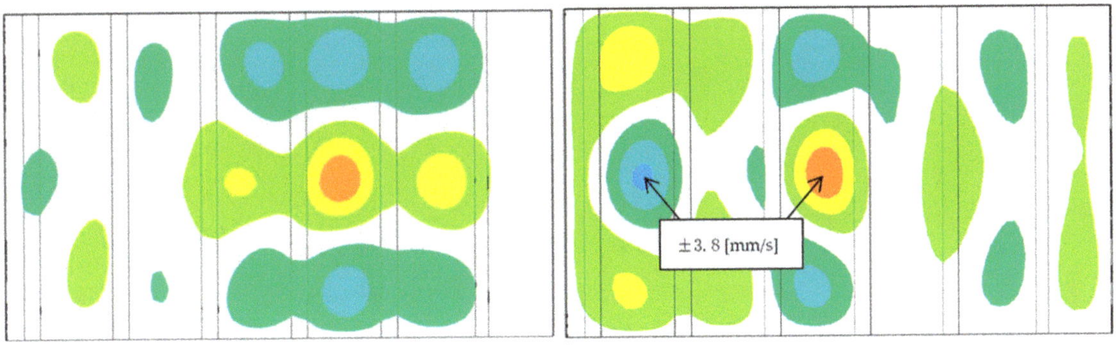

a.

b.

Figure 11. Instantaneous fields of normal velocity at the tank wall. (**a**) $Re(v_n)$, (**b**) $Im(v_n)$.

Some comments should be added to the value of pressure magnitude obtained in the calculations. The tank wall is also excited by the leakage field, which creates normal stress directly on the magnetic shields' surfaces. This stress, not accounted for in the presented analysis, is almost in-phase with the pressure transmitted via oil because the phase currents are their common origin. The places on the tank wall where magnetic stress occurs are sketched in Figure 13. The slight phase shift occurring between magnetic and acoustic components results from wave phenomena present in the oil because the pressure at a given point on the tank wall depends not only on the component produced by the nearest pair of phase windings but also on the other windings carrying currents displaced in phase and located at a different distance from this point.

a. b.

Figure 12. Instantaneous fields of in-fluid pressure on tank wall. (**a**) $Re(p_a)$, (**b**) $Im(p_a)$.

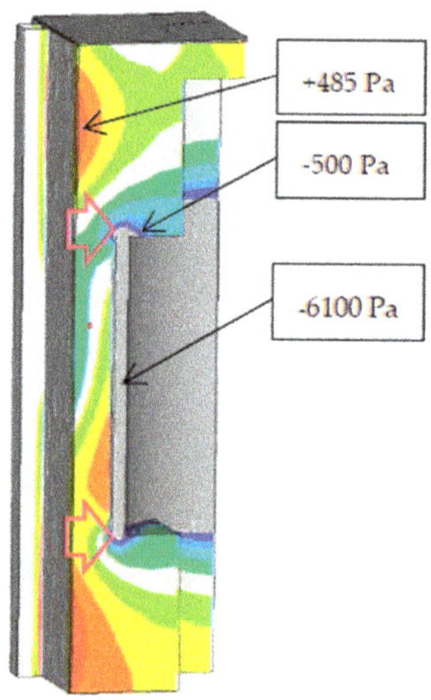

Figure 13. Cross-section of in-fluid pressure inside tank, shortened color map, block arrows show places of magnetic stress application.

The amplitude (AC component) of magnetic stress σ_m is about 150 Pa, which results [17] from

$$\sigma_m = \frac{B_m^2}{4\mu_0} \tag{18}$$

where B_m is the amplitude of the magnetic flux density close to the shield surface. Magnetic stresses occur slightly further from the top and bottom of the wall than acoustic stresses, thus creating larger moments of force bending the tank wall. Comparing it with amplitudes equal to 485 Pa presented in Figures 12 and 13 means that both effects are of similar significance.

5. Experimental Validation

The vibration velocity was measured with the PSV400 laser scanning vibrometer during the short-circuit test of analyzed transformer. The results are shown in Figure 14. It should be mentioned that the calculated and measured Realis time points are not exactly the same, as no special triggering was used during the experiments.

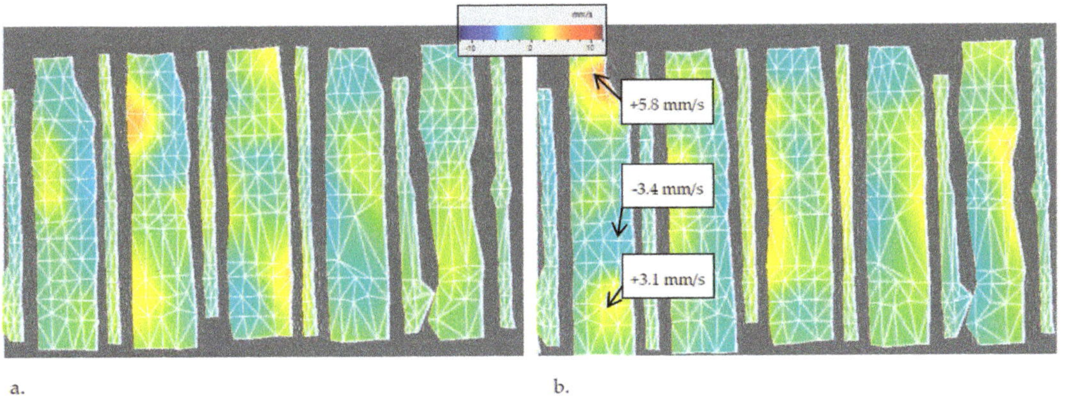

Figure 14. Instantaneous fields of normal velocity at the tank wall measured at rated short circuit test. (**a**) $Re(v_n)$, (**b**) $Im(v_n)$.

The measured patterns of vibrations are not as clear as the calculated ones. It is probably caused by the connection between the shielding system and the tank wall, which, in reality, is not as precise as in the theoretical model. The welding technology applied in the real transformer for the assembly of shields may create additional local prestress, and also, the resultant number of connection points between the shield box and the wall remains unknown. The lack of calculations of the magnetic forces applied directly to the tank, which is shifted in the phase in relation to the acoustic excitation, also affects the shape of the vibration isolines, which was discussed above. The values and shape of measured velocities of the tank surface are in acceptable agreement with calculations because the dominated vibrations on the tank wall appear close to the bottom and top of the wall, as is also indicated by the calculations. Local differences in the magnitude of the measured and calculated velocity fields are about 30%, which is almost nothing on the dB scale commonly used in vibration analysis. However, the high value of the local maximum shown in Figure 14b, which is almost twice as large as the calculated amplitudes, has not been confirmed by theoretical considerations.

6. Conclusions

This article describes, in detail, the method of transferring winding vibrations caused by electromagnetic forces to the transformer tank. It was found that axial forces are responsible for most of the vibration energy appearing on the tank surface. The pressure of acoustic origin on the tank wall results mainly from the compression of the oil inside the radial channels in the phase windings. Therefore, the forced vibration analysis requires the use of 3D numerical models of the transformer in both the magnetic, structural and acoustic domains. The basic problem is the matching of finite element grids modeling acoustic phenomena in the oil area close to the structural elements of the transformer. Due to the numerical complexity of the analyzed problem, a direct approach is still not possible, and it is necessary to take several intermediate steps. This approach, however, allows for a deeper insight into the physics of the considered phenomena, which, in turn, can explain the origin of excessive vibrations and emitted acoustic noise in real transformers.

Author Contributions: Conceptualization, P.W. and M.S.; methodology, P.W.; validation, P.W. and M.S.; investigation, P.W.; resources, M.S.; data curation, P.W.; writing—original draft preparation, P.W.; writing—review and editing, M.S.; project administration, M.S.; funding acquisition, M.S. All authors have read and agreed to the published version of the manuscript.

Funding: This research was funded by Hitachi Energy Lodz, grant Theoretical analysis, numerical simulations, measurements of vibration, 2018–2022.

Data Availability Statement: Not applicable.

Acknowledgments: The authors would like to thank Christoph Ploetner for many inspiring discussions during the implementation of this project and also Grzegorz Kmita and Michal Kozupa for their contribution during measurements.

Conflicts of Interest: The authors declare no conflict of interest. The funders had no role in the design of the study; in the collection, analyses, or interpretation of data; in the writing of the manuscript; or in the decision to publish the results.

References

1. Kojima, H.; Miyata, H.; Shida, S.; Okuyama, K. Buckling Strength Analysis of Large Power Transformer Windings Subjected to Electromagnetic Force under Short Circuit. *IEEE Trans. Power Appar. Syst.* **1980**, *PAS-99*, 1288–1297. [CrossRef]
2. Bakshi, A.; Kulkarni, S.V. Analysis of Buckling Strength of Inner Windings in Transformers under Radial Short-Circuit Forces. *IEEE Trans. Power Deliv.* **2013**, *29*, 241–245. [CrossRef]
3. Ahmad, A.; Javed, I.; Nazar, W. Short circuit stress calculation in power transformer using finite element method on high voltage winding displaced vertically. *Int. J. Emerg. Technol. Adv. Eng.* **2013**, *3*, 301–308.
4. Minhas, M.S.A. Dynamic Behavior of Transformer Winding under Short-Circuits. Ph.D. Thesis, Johannesburg University, Johannesburg, South Africa, 2007.
5. Ertl, M.; Landes, H. Investigation of load noise generation of large power transformer by means of coupled 3D FEM analysis. *COMPEL-Int. J. Comput. Math. Electr. Electron. Eng.* **2007**, *26*, 788–799. [CrossRef]
6. Rausch, M.; Kaltenbacher, M.; Landes, H.; Lerch, R.; Anger, J.; Gerth, J.; Boss, P. Combination of finite and boundary element methods in investigation and prediction of load-controlled noise of power transformers. *J. Sound Vib.* **2002**, *250*, 323–338. [CrossRef]
7. Reyne, C.; Magnin, H.; Berliat, G.; Clerc, C. A supervisor for the successive 3D computations of magnetic, mechanical and acoustic quantities in power oil inductors and transformers. *IEEE Trans. Magn.* **1994**, *30*, 3292–3295. [CrossRef]
8. Kim, Y.; Shim, J.; Park, W.; Kim, S.; Hyun, D.; Lee, D. Structure-vibration analysis of a power transformer (154 kV/60 MVA/single phase). *Int. J. Electr. Power Energy Syst. Eng.* **2009**, *2*, 249–253.
9. Wakabayashi, D.; Enokizono, M. Analysis of vector magnetic property and two-dimensional magnetostriction in electrical steel sheets. *Int. J. Appl. Electromagn. Mech.* **2014**, *44*, 317–330. [CrossRef]
10. Qiang, H.; Jingkai, N.; Songyang, Z.; Weimin, X.; Shengchang, J.; Xin, C. Study of Transformer Core Vibration and Noise Generation Mechanism Induced by Magnetostriction of Grain-Oriented Silicon Steel Sheet. *Shock Vib.* **2021**, *2021*, 8850780. [CrossRef]
11. Ertl, M.; Voss, S. The role of load harmonics in audible noise of electrical transformers. *J. Sound Vib.* **2014**, *333*, 2253–2270. [CrossRef]
12. Hu, Y.; Zheng, J.; Huang, H. Experimental Research on Power Transformer Vibration Distribution under Different Winding Defect Conditions. *Electronics* **2019**, *8*, 842. [CrossRef]
13. Liu, J.; Zhao, Z.; Tang, C.; Yao, C.; Li, C.; Islam, S. Classifying Transformer Winding Deformation Fault Types and Degrees Using FRA Based on Support Vector Machine. *IEEE Access* **2019**, *7*, 112494–112504. [CrossRef]
14. Jin, M.; Pan, J.; Huang, H.; Zhou, J. Transmission of vibration of a power transformer from the internal structures to the tank. In Proceedings of the Acoustics, Nantes, France, 23–27 April 2012; pp. 1–7.
15. Cremer, L.; Heckl, M.; Petersson, B.A.T. *Structure-Borne Sound. Structural Vibrations and Sound Radiation at Audio Frequencies*; Springer: Berlin/Heidelberg, Germany, 2004.
16. *Acoustic Analysis Guide*; ANSYS, Inc.: Canonsburg, PA, USA, 2022.
17. Witczak, P.Z.; Swiatkowski, M. Magnetic forces applied to the tank walls of a large power transformer. *COMPEL-Int. J. Comput. Math. Electr. Electron. Eng.* **2016**, *35*, 2087–2094. [CrossRef]

energies

MDPI

Article

A New Wall Current Transformer for Accurate Beam Intensity Measurements in the Large Hadron Collider

Michal Krupa * and Marek Gasior *

Beam Instrumentation Group, Accelerator Systems Department, the European Organization for Nuclear Research (CERN), 1211 Geneva, Switzerland
* Correspondence: michal.krupa@cern.ch (M.K.); marek.gasior@cern.ch (M.G.)

Abstract: The Large Hadron Collider (LHC) stores two high-energy counter-rotating particle beams consisting of multiple bunches of a nanosecond length. Precise knowledge of the number of particles within each bunch, known as the bunch intensity, is crucial for physicists and accelerator operators. From the very beginning of the LHC operation, bunch intensity was measured by four commercial fast beam current transformers (FBCTs) coupling to the beam current. However, the FBCTs exhibited several shortcomings which degraded the measurement accuracy below the required level. A new sensor, the wall current transformer (WCT), has been developed to overcome the FBCT limitations. The WCT consists of eight small radio frequency (RF) current transformers distributed radially around the accelerator's vacuum chamber. Each transformer couples to a fraction of the image current induced on the vacuum chamber by the passing particle beam. A network of RF combiners sums the outputs of all transformers to produce a single signal which, after integration, is proportional to the bunch intensity. In laboratory tests and during beam measurements, the WCT performance was demonstrated to convincingly exceed that of the FBCT. All originally installed FBCTs were replaced by four WCTs, which have been serving as the LHC reference bunch intensity sensors since 2016.

Keywords: beam instrumentation; beam intensity measurements; current transformers

Citation: Krupa, M.; Gasior, M. A New Wall Current Transformer for Accurate Beam Intensity Measurements in the Large Hadron Collider. *Energies* **2023**, *16*, 7442. https://doi.org/10.3390/en16217442

Academic Editors: Elzbieta Lesniewska, Xose Lopez-Fernandez and Pawel Witczak

Received: 26 September 2023
Revised: 25 October 2023
Accepted: 2 November 2023
Published: 4 November 2023

1. Introduction

The Large Hadron Collider (LHC) at CERN is the largest and highest-energy particle accelerator in the world [1]. It has a circumference of twenty seven kilometres and accelerates two counter-rotating particle beams to energies as high as 6.8 TeV. The beams travel for the most part in two separate vacuum chambers which are joint only close to the four large physics experiments where the beams cross each other, the collisions take place, and the paths and momenta of newly created particles are determined.

The LHC accelerates the beams of protons or positively charged ions of heavier elements. The beams are ultra-relativistic (the Lorentz factor exceeding 7200 for proton beams) and they travel very close to the speed of light (over 99.999999% *c* for proton beams).

As the particles flow around the accelerator, this motion of electric charge constitutes an electric current referred to as the beam current. Consequently, the sum of electric charges contained within a beam is called the beam charge. Another closely related beam parameter is its intensity, which is defined as the number of particles contained within the beam and can be simply calculated by dividing the beam charge by the elementary charge.

The beam particles do not form a continuous stream but are longitudinally grouped into discrete packets called bunches. The distribution of particles within a bunch is approximately Gaussian with a standard deviation ranging from 200 to 400 ps. A quantity which is typically used to describe the longitudinal bunch shape is the bunch length, assumed to be four standard deviations of the distribution.

The LHC revolution period of 88.925 µs is divided into 3564 bunch slots, each of which is approximately 25 ns long. However, the beam production mechanism and equipment safety aspects require that approximately 20% of the available bunch slots are free of particles. The LHC beams consist of up to some 2800 bunches which are interleaved with empty bunch slots following complicated filling patterns. One complete revolution of the beam around the LHC is often referred to as a "turn".

Each bunch is characterised by its own current, charge and intensity. In most cases, the LHC uses two kinds of bunches: pilot bunches with an intensity of 5–10 $\times$ 10^9 ppb (protons per bunch) and nominal bunches with an intensity of 1–2.2 $\times$ 10^{11} ppb. The corresponding peak bunch currents are approximately 1–2 A and 20–48 A, respectively. Figure 1 illustrates a hypothetical LHC bunch pattern and the turn numbering convention. Continuous and accurate measurements of the intensity of each bunch on a turn-by-turn basis are required to optimise the operation of the LHC (e.g., intensities of individual bunches should be similar), to ensure LHC equipment safety (e.g., maximal bunch intensity must be controlled), and to properly analyse the outcomes of high-energy collisions (e.g., individual intensities of the colliding and not colliding bunches must be known).

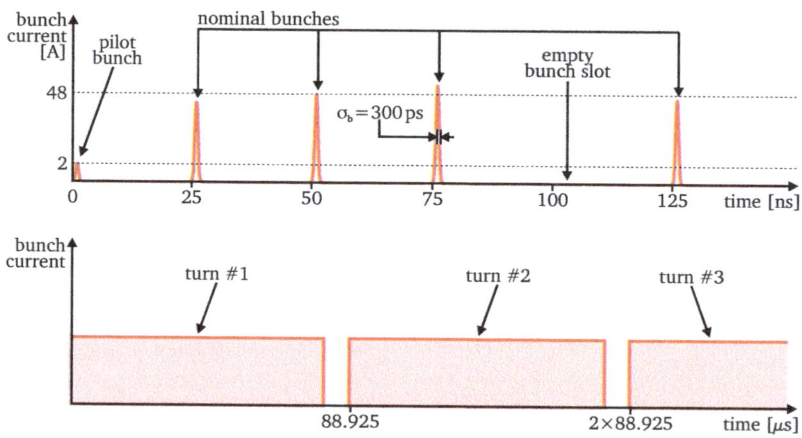

Figure 1. Temporal structure of the LHC beam.

Precise measurements of beam and bunch intensities are particularly crucial for calculating the luminosity $\mathcal{L}$ at the beam–beam collision points. The knowledge of this fundamental parameter is necessary to determine whether the outcomes of high-energy collisions are within the limits defined by the existing models or a new physics phenomenon has been discovered. For colliders, such as the LHC, the luminosity is given by:

$$\mathcal{L} = \frac{N_1 N_2 f_{\mathrm{rev}}}{4\pi\sigma_{\mathrm{x}}\sigma_{\mathrm{y}}} \quad (1)$$

where N_1 and N_2 are the intensities of the two respective beams (or bunches), f_{rev} is the frequency at which the beams (or bunches) collide, and σ_{x} and σ_{y} are the width and height of the effective overlap region of the two beams (or bunches) in the transverse plane. The beam and bunch intensity measurement error directly propagates to the luminosity calculation error, and therefore, the performance of a collider is strictly linked to the quality of its beam intensity monitoring system.

Beam and bunch intensity can be measured using one of the many techniques developed and applied in accelerators over the years [2]. In the LHC, both those quantities are measured with current transformers which couple to the electromagnetic field carried by the beam. In the first years of the LHC operation, the bunch intensity was monitored with the fast beam current transformers (FBCT) [3], which are the most widely used sensor for this purpose. A total of four FBCTs were in service, two redundant sensors for each LHC beam. However, the FBCTs exhibited an undesired sensitivity to the transverse beam position and an excessively long output pulse which prevented measurements with sufficient accuracy [4]. Both these shortcomings limited the LHC FBCT performance to such an extent that studies were launched to find an alternative solution. This manuscript describes a new sensor resulting from these studies carried out within the framework of a Ph.D. thesis [5], the wall current transformer (WCT), optimised for accurate measurements of the bunch intensity in the LHC. This manuscript summarises the work and shows the most important measurements to disseminate this emerging technology within a wider community. The presented results are supported by seven years of successful and reliable operation in the LHC. The developed technology can find applications in other accelerators and be further optimised.

2. LHC Bunch Intensity Measurements with the FBCT

The instantaneous bunch current $i_B(t)$ and the bunch charge Q_B are closely related to the bunch intensity N_B and can be used for its indirect measurement. Integrating $i_B(t)$ over the duration of a bunch slot T_B results in Q_B which, after dividing by the elementary charge e_0, gives N_B:

$$N_B = \frac{1}{e_0} Q_B = \frac{1}{e_0} \int_{t_0}^{t_0+T_B} i_B(t)\, dt \quad (2)$$

Therefore, the LHC bunch intensity can be calculated by continuously measuring the instantaneous bunch current and integrating it over windows of approximately 25 ns, corresponding to the distance between two consecutive bunches.

Moreover, it is not needed to measure the full frequency spectrum of the bunch current. A low-pass filter does not alter the signal's integral if its insertion loss at DC is negligible [5]. A bunch current pulse stretched by a low-pass filter can still serve as a basis for bunch intensity measurements as long as the pulse does not extend beyond the 25 ns window.

From the very first days of the LHC operation, the bunch intensity was measured with four commercial fast beam current transformers (FBCTs) [3], two per LHC ring. Similar devices are widely used in other particle accelerators but are more generally referred to as AC current transformers (ACCTs). Figure 2 illustrates their principle of operation.

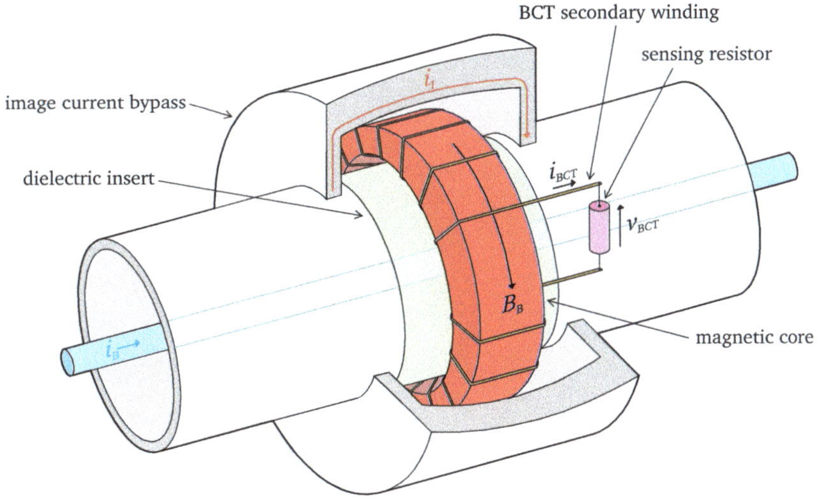

Figure 2. Principle of operation of the fast beam current transformer (FBCT) [5].

The FBCT employs a high-permeability toroidal magnetic core, which is installed over a dielectric insert. The conducting walls of the accelerator's vacuum chamber enclose the electromagnetic field carried by the charged beam in motion. The dielectric insert serves as a "window" through which the core can couple to the beam's field. Due to electromagnetic induction, this coupling results in a current flow through the secondary wire wound around the core. For ultra-relativistic beams travelling close to the speed of light, such as the LHC beams, the induced current i_{BCT} is proportional to the passing beam current i_B. This is a consequence of the relativistic Lorentz contraction phenomenon, illustrated in Figure 3, which shortens the longitudinal component of the field lines as the field's source approaches the speed of light. Consequently, the secondary current i_{BCT} has the same temporal structure as the instantaneous beam current i_B.

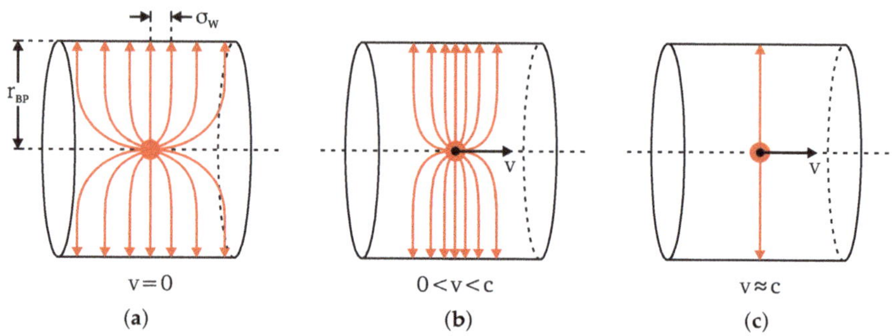

Figure 3. Illustration of the Lorentz contraction of the electric field lines of a single charge [5]. (**a**) Particle at rest. (**b**) Particle travelling below the speed of light. (**c**) Particle travelling close to the speed of light.

The proportionality factor between i_{BCT} and i_B for an FBCT in which the secondary wire makes N turns around the core is equal to N^{-1}. The current i_{BCT} is typically measured as a voltage drop v_{BCT} across a load impedance Z_{BCT}. Hence, the bunch intensity N_B can be measured with the FBCT as:

$$N_B = \frac{N}{Z_{BCT}\, e_0} \int_{t_0}^{t_0+T_B} v_{BCT}(t)\, dt \tag{3}$$

In practice, the proportionality constant between the bunch intensity and the integral of the FBCT signal is established via cross-calibration with other beam-sensing instruments. In the LHC, the reference values for total beam intensity are obtained from DC beam current transformers (DC BCTs) [6].

Figure 4 shows a simple electrical model of the FBCT, which can be used to calculate the expected signal levels and analyse the low-frequency behaviour of the monitor.

As the sensor is essentially a current transformer in which the beam represents the primary winding, it has no response at DC and its low cutoff frequency is given by:

$$f_{l,BCT} = 2\pi \frac{R_{BCT} \parallel Z_L}{L_{BCT}} \tag{4}$$

where R_{BCT} is the internal load resistance built into the FBCT, Z_L represents the input impedance of the acquisition system connected in parallel (||) to R_{BCT}, and L_{BCT} is the inductance of the secondary winding.

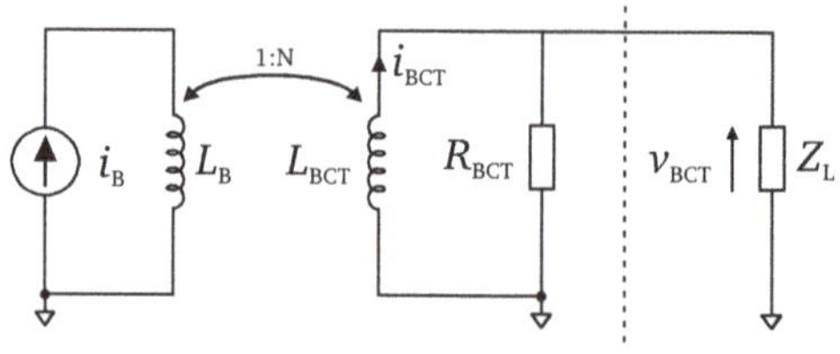

Figure 4. Low-frequency electrical model of the FBCT.

Due to the lack of low-frequency components, the FBCT signal is affected by the so-called baseline droop, as shown in Figure 5, which becomes more and more severe for increasing values of $f_{l,BCT}$. As the signal's baseline decreases, the droop lowers the value of the signal's integral. This leads to an error when using FBCT signals for bunch intensity monitoring. However, this error can be significantly reduced by setting $f_{l,BCT}$ such that the amount of the baseline droop over the effective integration window is negligible. For accurate bunch intensity measurements in the LHC, the low cutoff frequency of the FBCT cannot exceed 400 Hz [3]. As for the LHC FBCT $R_{BCT} = Z_L = 50\,\Omega$, which is typical for high-frequency systems, achieving a low $f_{l,BCT}$ requires sufficiently high L_{BCT}. For the LHC FBCT, $f_{l,BCT} \approx 200$ Hz was obtained by winding forty secondary turns around a high-permeability nanocrystalline magnetic core [3].

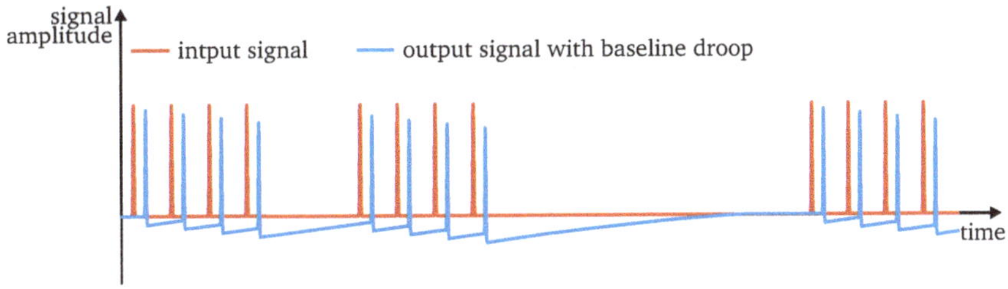

Figure 5. Baseline droop in the output signal of a sensor with no DC response.

In the first years of the LHC operation, the FBCT signals were acquired with a system based on custom analogue integrators [7]. It was later replaced by a fast-sampling system, still used operationally, implementing the numerical integration of the digitised signal [8].

Forty secondary turns terminated with $R_{BCT} \parallel Z_L = 25\,\Omega$ translate into an FBCT transimpedance v_{BCT}/i_B of 625 mΩ which results in a very large output pulse amplitude of 12.5 V for even a modest nominal bunch with a peak current of 20 A. Therefore, the LHC FBCT signal has to be strongly attenuated before it can be measured by a high-speed data acquisition system.

Already in the first years of the LHC operation, some limitations of the FBCTs were observed. Most importantly, the sensors exhibited an undesired sensitivity to the transverse beam position and the bunch length. The bunch intensity measured by the FBCT would change when the beam was transversely displaced inside the vacuum chamber even though the true bunch intensity remained constant. Similar behaviour was seen when the bunch length changed. These two effects contributed to a measurement error of a few percent exceeding the original specification of $\pm 1\,\%$ for measurements averaged over 1 s [9]. The FBCTs outputs were fitted with analogue 80 MHz low-pass filters which reduced the sensitivity to the beam position and bunch length [10]. However, the effects remained measurable and were deemed a significant limitation of the LHC bunch intensity monitoring accuracy. Any modelling of these effects seemed very difficult, if at all possible, so they were studied only empirically.

The most severe limitation of the FBCT, the dependence of its output signal on the transverse beam position, was traced to the fact that the distribution of the beam's electromagnetic field was changing with respect to the beginning and the end of the core's secondary winding as the beam changed its position. Then, due to core losses, the field induced further from the winding end had a smaller contribution to the output signal than the field induced closer to the end of the winding connected to the output terminal.

The second crucial limitation of the FBCT was the size of its core, with an external diameter of 130 mm, an internal diameter of 90 mm, and a thickness of 25 mm, which required some 2.4 m of wire to make the forty-turn winding. Such a long winding is prone to parasitic capacitance between turns, which gives rise to resonances at frequencies as low as 30 MHz. This makes the FBCT inadequate to fully separate the consecutive LHC bunches spaced by 25 ns. Moreover, the forty-turn secondary winding of the large high-permeability magnetic core forms a lossy delay line. The resulting frequency dispersion of the signals originating from different parts of the winding significantly deteriorates the FBCT's frequency response.

These two effects were convincingly demonstrated in a laboratory with a setup consisting of a one-turn loop, acting as the primary winding of the FBCT toroid, connected to a pulse generator (for time-domain measurements) or a network analyser output (for frequency-domain measurements). The output of the forty-turn secondary winding was connected to an oscilloscope or an input of the network analyser. When the one-turn primary loop was moved along the FBCT winding, large signal changes were observed in both the time and frequency domains.

3. Desired Characteristics of a New LHC Bunch Intensity Sensor

The FBCT imperfections were understood as coming from the monitor itself, rendering them very improbable to effectively overcome. Therefore, CERN decided to launch a fully in-house development aiming to find a solution that would eventually overcome the limitations and replace the LHC FBCTs.

In order to limit the changes to other LHC components, it was decided that the new sensor must ensure mechanical compatibility with the existing vacuum chamber and the dielectric insert over which the FBCTs were installed. These requirements constrained the mechanical dimensions of the new sensor to a minimum internal diameter of 84 mm, an external diameter smaller than 300 mm, and a total length of 40–290 mm.

The performance specifications have also been revised compared to the original LHC requirement of $\pm 1\,\%$ accuracy. For the new monitor, the combined bunch intensity measurement error should not exceed 0.1% for measurements averaged over 1 s. The sensor's bandwidth should span from the low cutoff frequency $f_{\mathrm{l}} < 640$ Hz up to the high cutoff frequency $f_{\mathrm{h}} > 59$ MHz.

The output signal of the new sensor must be adapted to acquisition electronics based on both analogue integrators and fast sampling. The signal amplitude at the input of the data acquisition electronics should not exceed ± 1.2 V and the duration of a pulse generated by a single LHC bunch should be less than 22 ns.

4. LHC Wall Current Transformer

To overcome the inherent FBCT design flaws, its replacement cannot employ a large magnetic core. Therefore, the standard solution with a single toroid around the beam vacuum chamber was immediately rejected. It is then evident that the new sensor cannot directly sense the beam current itself but should rather rely on the beam image current. To remain insensitive to the beam position changes, the image current must be probed at several radial positions. The corresponding individual signals must be summed with a power combiner to ensure the linearity of the signal superposition and the independence of the beam position. Finally, to maximise the sensor's bandwidth, its magnetic cores should be as small as possible and contain only a few secondary turns wound with a short wire.

With the above considerations in mind, the new sensor, named by the authors the wall current transformer (WCT), was conceived. It does not couple directly to the bunch current but rather to the current induced by it on the conductive walls of the vacuum chamber [5]. Following Gauss's law, the beam charge Q_{B} induces an equal charge of the opposite sign Q_{W} on the inner walls of the accelerator's vacuum chamber [11]. This phenomenon, as illustrated in Figure 6, is typically referred to as the image or wall charge.

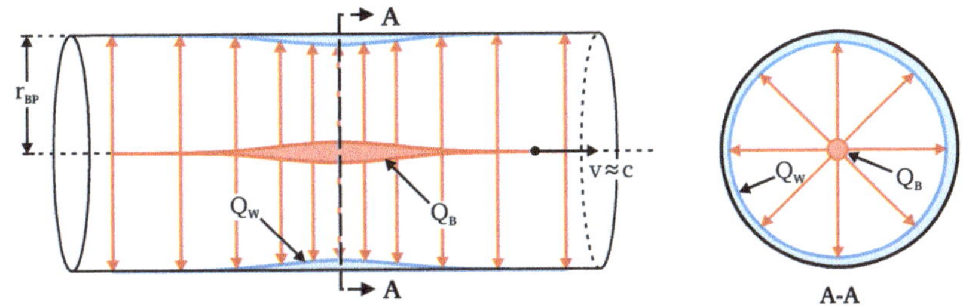

Figure 6. Image charge induced on the wall of a vacuum chamber by an ultra-relativistic beam [5].

Due to the Lorentz contraction of the field lines, the longitudinal distribution of the image charge for ultra-relativistic particle beams travelling very close to the speed of light is the same as that of the bunch charge. Both charges move together, which gives rise to an electric current flowing on the walls of the vacuum chamber. This current is interchangeably referred to as the image current or the wall current i_W. However, as the induced image charge is of the opposite sign to the bunch charge, both currents also have the same magnitude but the opposite polarity:

$$i_W(t) = -i_B(t) \tag{5}$$

Instead of using one large high-permeability toroidal magnetic core, the WCT uses eight small toroidal cores made of a nanocrystalline magnetic material. The cores are attached to a printed circuit board (PCB) and are evenly distributed around the dielectric insert embedded into the vacuum chamber. Since, for the wall current, the dielectric insert constitutes a high-impedance discontinuity, the current is forced to flow via conductive screws going through the centre of each toroid. The cores serve as RF current transformers with their primary windings formed by the screws. A thin wire is wound around each core as the secondary winding. An additional wire making a single turn around each core acts as a calibration winding allowing for a reference calibration current to be sent to the sensor.

Devices using a similar beam-coupling method were realised in the past [12–14] but they were all designed for measuring the transverse beam position at relatively low frequencies. The WCT developed by the authors is optimised for the precise measurement of the intensity of the LHC short bunches.

Figure 7 illustrates the WCT principle of operation. When the wall current reaches the monitor, all the components above the low cutoff frequency flow through the conductive screws and are sensed by the RF transformers. To avoid radiating the beam's electromagnetic field towards other accelerator components, the sensor is enclosed in a conductive housing through which low-frequency image current can flow. The WCT housing is filled with high-permeability ferrite cores which magnetically load the housing to increase its inductance. This lowers the frequency at which the image current starts flowing through the conductive screws. Both sides of the dielectric insert are connected with an additional RF bypass circuit composed of capacitors and resistors to provide a well-defined path for the high-frequency image current components. Such high frequencies are beyond the operational bandwidth of the WCT and should therefore be bypassed to avoid an excitation of the parasitic RF cavity formed by the WCT housing, which would deteriorate the longitudinal beam-coupling impedance of the WCT.

An equivalent electric model of the WCT is shown in Figure 8, which illustrates its principle of operation and can be used to calculate the most important parameters.

The image current i_I has an equal magnitude but the opposite polarity to the beam current i_B. Inside the WCT, i_I is divided into three constituents:

- The low-frequency i_{LF} flowing through the housing with inductance L_{LF} defined by the housing geometry and the ferrite permeability;
- The very-high-frequency i_{RF} flowing through the RF bypass with capacitance C_{RF}, resistance R_{RF}, and some parasitic inductance L_{RF};
- The intermediate-frequency i_{W} flowing through the screws with inductance L_{W} and resistance R_{W} defined by the RF transformer.

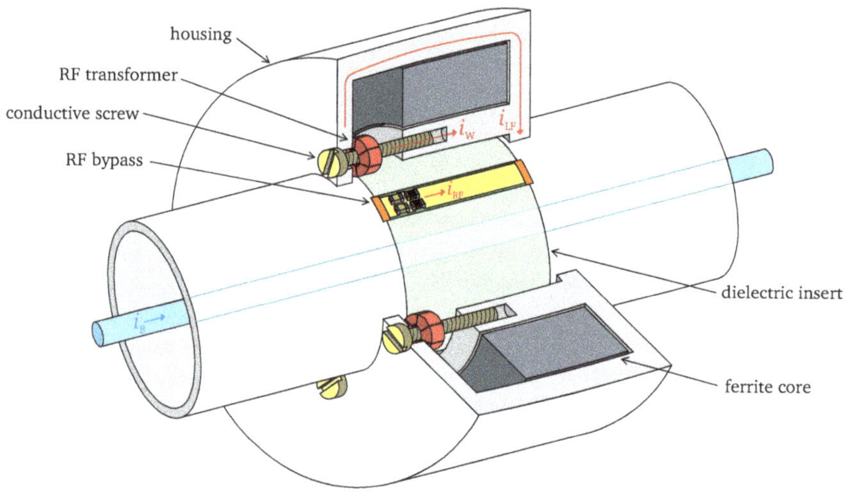

Figure 7. Principle of operation of the wall current transformer (WCT) [5].

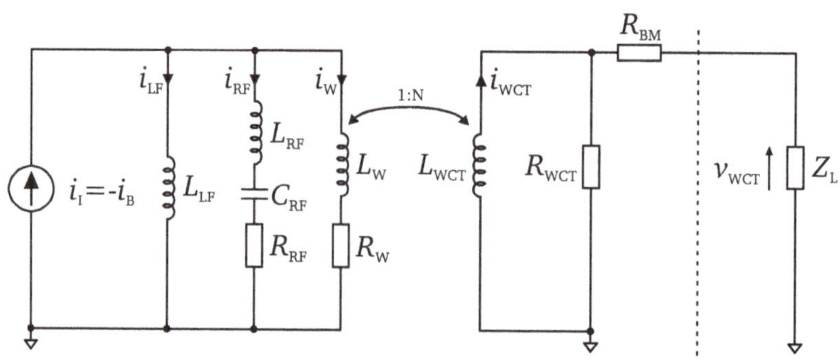

Figure 8. Electrical model of the WCT [5].

As in any current transformer, i_{W} induces a current i_{WCT} on the RF transformer's secondary side represented by L_{WCT}. This rises a voltage v_{WCT} across the load resistor R_{WCT}. To match the WCT output to the standard 50 Ω characteristic impedance of high-frequency transmission lines, the sensor includes a back-matching resistance R_{BM} which sets the monitor's output impedance to 50 Ω. The WCT output signal can then be measured by an acquisition system, represented by Z_{L}.

For frequencies at which $i_{\mathrm{W}} \approx -i_{\mathrm{B}}$, the WCT output voltage v_{WCT} is given by:

$$v_{\mathrm{WCT}} = -\frac{1}{N}\frac{R_{\mathrm{WCT}}Z_{\mathrm{L}}}{R_{\mathrm{WCT}} + R_{\mathrm{BM}} + Z_{\mathrm{L}}}i_{\mathrm{B}} \tag{6}$$

After matching the source and load impedances by selecting resistors such that $R_{WCT} + R_{BM} = Z_L$, the above equation becomes:

$$v_{WCT} = -\frac{R_{WCT}}{2N} i_B = -Z_{WCT}\, i_B \tag{7}$$

As v_{WCT} is proportional to the instantaneous beam current, the WCT can measure bunch intensity. The factor Z_{WCT} is called the transimpedance and it is helpful for comparing various bunch intensity monitors. In practice, Z_{WCT} is a function of frequency.

The frequency range at which Equation (7) is valid is determined by the sensor's bandwidth defined by its low and high cutoff frequencies f_L and f_H, respectively.

Analysis of the circuit shown in Figure 8, assuming that $L_{LF} \gg L_W$ on the primary side and that $R_{WCT} \ll R_{BW} + Z_L$ on the secondary side, leads to a simple equation for an approximate value of f_L:

$$f_L \approx \frac{1}{2\pi}\left(\frac{R_W}{L_{LF}} + \frac{R_{WCT}}{L_{WCT}}\right) \tag{8}$$

As $R_W \approx R_{WCT}/N^2$ and $L_{WCT} \propto N^2$, increasing the number of secondary turns N quickly decreases f_L. However, as shown before in Equation (7), a higher N also reduces the WCT output voltage, and therefore a compromise between the cutoff frequency and the sensitivity must be found.

The precise modelling of the WCT on the high-frequency side is difficult, mostly due to many parasitic capacitive effects that can be collectively referred to as the interwinding capacitance. Nevertheless, from the circuit shown in Figure 8, assuming that, at high frequencies, the screw's impedance is dominated by L_W and that the RF bypass capacitance C_{RF} can be neglected, the following equation can be drawn up:

$$f_H \approx \frac{R_{RF}}{2\pi\sqrt{L_W^2 - L_{RF}^2}} \tag{9}$$

Deriving a corresponding formula for the FBCT is not feasible, mostly due to the fact that its high frequency behaviour is not well defined as it is not evident how the image current traverses the sensor. The FBCT could also be equipped with an RF bypass, similar to the one used in the WCT, which would limit the sensor's longitudinal impedance presented to the beam. Nevertheless, such an addition to the FBCT would not solve its other fundamental limitations addressed by the WCT design.

The circuit shown in Figure 8 does not implicitly account for the eight RF transformers installed in parallel inside the WCT. The components only represent the effective values of a simplified equivalent circuit rather than physical components installed in the sensor.

At the heart of the WCT, there are eight RF transformers based on Vacuumschmelze T60006-L2009-W914 toroids made of nanocrystalline iron-based VITROPERM 500 F material. They were selected due to their high inductance factor $A_L = 25.5\,\mu H$ at 10 kHz and a sufficiently small size. The cores, after stripping them of their protective plastic casing, have dimensions of only 6.5 mm (inner diameter) by 9.9 mm (outer diameter) by 4.8 mm (height).

Each core is wound with $N = 10$ secondary turns and loaded with 5 Ω. With eight transformers in parallel, this translates into an effective $R_{WCT} = 625\,m\Omega$. From Equation (7), the WCT transimpedance is $Z_{WCT} = 31.25\,m\Omega$ which is 20 times smaller than that of the FBCT. With the typical LHC bunch currents, before any filtering or attenuation, the WCT output would have an amplitude of 31 mV for a pilot bunch and 625 mV for a nominal bunch. Such levels are perfectly appropriate for typical front–end signal conditioning electronics.

To improve the low-frequency behaviour of the WCT, the housing was filled with Ceramic Magnetics CMD5005 nickel–zinc machined ferrite cores with relative permeability $\mu_r = 2100$ up to 600 kHz and overall dimensions of 176.8 mm (outer diameter), 117.2 mm (inner diameter), and 60 mm (length). With such a high-permeability core inside, the WCT housing has an inductance of $L_{LF} = 10\,\mu H$.

The effective secondary-side inductance $L_{WCT} = 320\,\mu H$ is formed by eight parallel cores with a ten-turn winding each. Together with the aforementioned housing inductance L_{LF} and the secondary-side resistance R_{WCT}, it is possible to use Equation (8) to calculate the low cutoff frequency of the WCT as $f_L = 410\,Hz$.

An internal RF bypass made from two flexible PCBs controls the WCT's high-frequency behaviour. The bypass consists of series resistors and capacitors with effective values of $R_{RF} = 1.67\,\Omega$ and $C_{RF} = 60\,nF$. The capacitance decouples the bypass at low frequencies, forcing the current to flow through the conductive screws instead. At very high frequencies, the bypass and the screws form a current divider. The screws' impedance is dominated by their self-inductance of approximately $L_W = 250\,pH$. The RF bypass' parasitic inductance was conservatively estimated as $L_{RF} < 50\,pH$. Therefore, the theoretical high cutoff frequency of the WCT calculated from Equation (9) is at least 1.1 GHz.

To produce a single WCT output, the signals generated by the eight RF transformers are added up through a network of passive power combiners. The outputs of two adjacent transformers are directly averaged on the internal WCT PCBs. Each 5 Ω secondary-side load is followed by a series 95 Ω back-matching resistor to set the source impedance to 100 Ω. Short transmission lines of the same characteristic impedance merge pairs of transformer outputs into four intermediate WCT outputs, thus becoming 50 Ω sources. The intermediate outputs can then be summed with three external resistive power combiners producing a single common WCT output. To mitigate the sensor's sensitivity to the transverse beam position, the internal WCT resistors are matched to within 0.01%.

The passive four-way combiner used at the WCT output generates a thermal noise with a spectral density of $2\,nV/\sqrt{Hz}$. Within the theoretical 1.1 GHz bandwidth of the WCT, this corresponds to the root-mean-square (RMS) noise of 66 µV. Numerical simulations show that the RMS value of an expected WCT output signal shape calculated over a 25 ns window is equal to nearly 15% of the peak voltage. Therefore, the RMS WCT output signal level is 5 mV for a pilot bunch and 94 mV for a nominal bunch. The resulting signal-to-noise ratio (SNR) is 37 dB for a pilot bunch and 63 dB for a nominal bunch which is equivalent to a non-averaged measurement error of 1.3% and 0.07%, respectively. However, as the LHC bunch intensity measurements are averaged over 1 s, i.e., over 11,245 samples, the effective SNR and measurement error are better by some 40 dB, i.e., two orders of magnitude.

Besides the secondary winding generating beam-related signal, each RF transformer of the WCT has an additional one-turn winding which can accept external calibration signals. The WCT is optimised for calibration with long current pulses, the amplitude of which can be precisely measured. The input impedance of its calibration port is 10 Ω which, for a 1 A current, generates a modest voltage of 10 V. To decouple the calibration winding from the high-frequency current flowing through the conductive screw, the calibration signal path includes several ferrite beads with a high impedance at RF frequencies.

Figure 9 shows a schematic of the internal LHC PCBs together with the external combiner network. The schematic does not account for the sensor's conductive screws.

Figure 10 shows the WCT's internal PCBs and some details of RF transformers. All boards were assembled by hand using manually matched resistors. The PCBs only use the top layer for signal routing and the impedance of all traces is controlled. The RF transformers were carefully wound by hand and fixed to the PCB with a small amount of acrylic adhesive which was also used to secure the windings to the core. The four intermediate signal outputs, as well as the calibration inputs, use standard SMA connectors (not visible in the photographs).

A noteworthy feature of the WCT mechanical design is that all its parts are cut in half to allow the assembly and disassembly of the sensor around a closed LHC vacuum chamber. Such a solution was chosen to make it possible to install and remove the first prototype WCT without uninstalling the original FBCT from the accelerator. Instead, the FBCT could be simply slid away from the dielectric insert along the vacuum chamber. Figure 11 show both sensors installed in the LHC side-by-side. However, in a general case, when retrofitting is not required, the WCT mechanical design can be simplified if its parts are not cut in half.

Further technical details about the WCT and a thorough derivation of the sensor's electrical model can be found in [5].

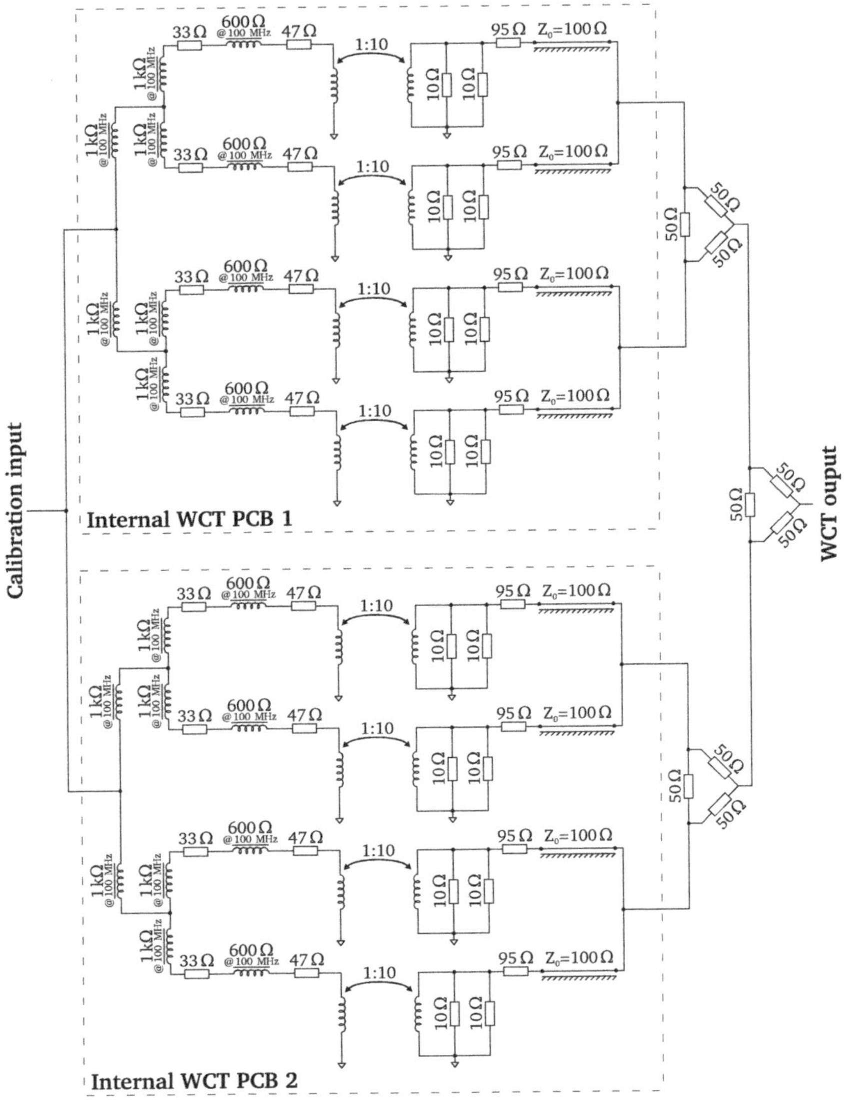

Figure 9. Schematic of the internal WCT PCBs [5].

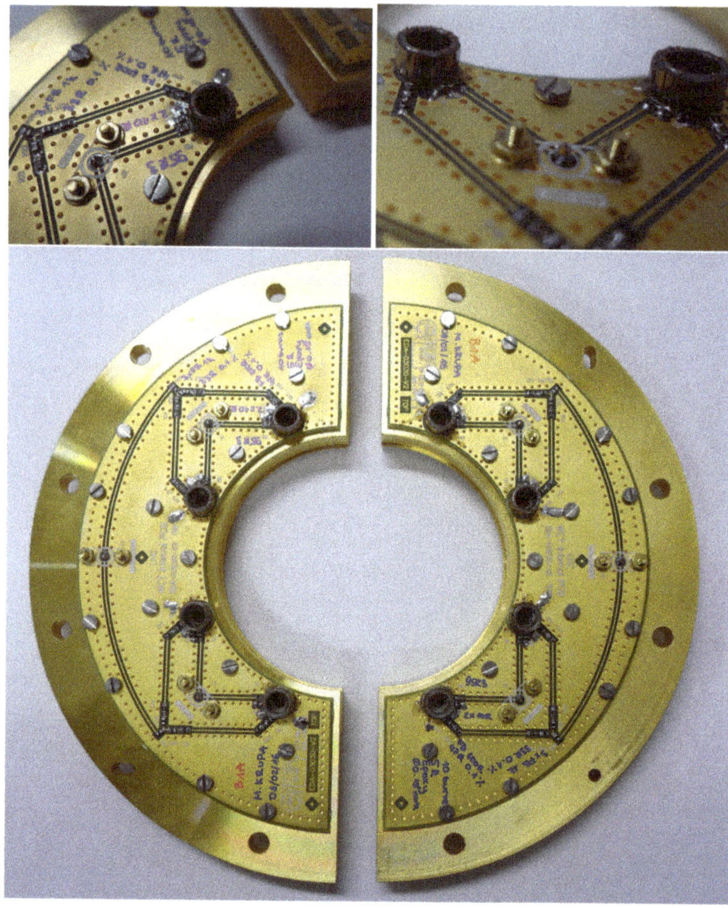

Figure 10. Internal WCT PCBs [5].

Figure 11. The new WCT (left) installed next to a displaced FBCT (right) [5].

Besides the sensor itself, a complete set of custom analogue front–end electronics has been designed and deployed for the LHC WCT with their functional diagram shown in Figure 12. The WCT output signal's bandwidth is first reduced by a non-reflective linear-phase low-pass filter (LPF) directly located after the final signal combiner. The signal then is sent over a short run of low-loss coaxial cable to the head amplifier located closer to the accelerator tunnel floor. The head amplifier provides two copies of the signal with 20 dB amplitude difference, foreseen for low- and high-intensity bunches. The signals are then sent through about 20 m of low-loss coaxial cable to a nearby technical gallery where the remaining electronics are well shielded from ionising radiation present during the LHC operation. A common-mode (CM) choke suppresses any interference picked up on the cable. The signal bandwidth is further reduced by another LPF and the signal is boosted by a distribution amplifier which makes four copies of each signal. Each output of the distribution amplifier is equipped with an LPF and an attenuator to adapt the signal to a given acquisition system. The expected signal levels along the WCT signal path are listed in Table 1. The amplifiers lose linearity with outputs exceeding 2.7 V and saturate at around 3.8 V; therefore, the high gain channel cannot be used for observing high-intensity bunches.

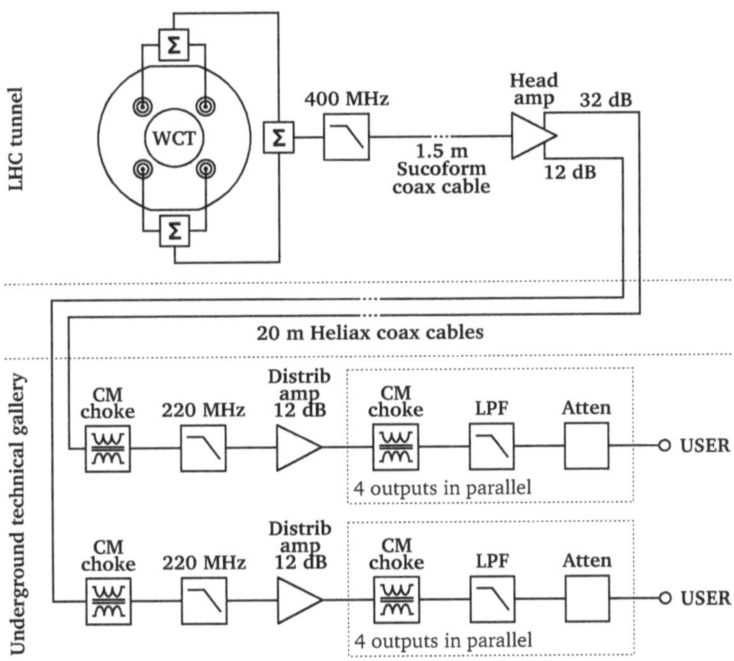

Figure 12. WCT front–end electronics diagram [5].

The WCT's noise performance is dominated by the noise of the head amplifier which has an RMS value of 0.4 mV and 2.5 mV at the output of the "low" and "high gain" channels, respectively. The RMS value of a WCT head amplifier output signal calculated over a 25 ns window equals approximately 25% of the peak voltage. Therefore, from the estimated signal levels listed in Table 1, it is possible to calculate an SNR of 34 dB for a pilot bunch measured with the "high gain" channel and 57 dB for a nominal bunch measured with the "low gain" channel. Similarly to the SNR directly calculated at the sensor output, the effective SNR for measurements averaged over a 1 s window is better by two orders of magnitude. Hence, the noise-related measurement error is well below the required 0.1%.

Further technical details concerning WCT electronics can be found in [5].

Table 1. Signal levels of the WCT and its front–end electronics with typical LHC bunches.

		Pilot	Nominal
Bunch	Bunch intensity (charges)	5×10^9	1.2×10^{11}
	Bunch length (ns)	1.2	1.2
	Peak bunch current (A)	1.1	25.5
WCT output amplitude (mV)	In 1.1 GHz BW	29	694
	In 400 MHz BW	20	460
Head amplifier amplitude (mV)	Low gain (in 220 MHz BW)	51	1184
	High gain (in 220 MHz BW)	517	Saturated
Distribution amplifier amplitude (mV)	Low gain (in 70 MHz BW)	70	1621
	High gain (in 70 MHz BW)	708	Saturated

5. Results

The WCT and FBCT were compared through extensive laboratory tests followed by beam measurements in the LHC. The first prototype WCT was installed in the place originally occupied by the FBCT, which was temporarily slid away from its dielectric insert. As the WCT can be assembled and disassembled around a closed vacuum chamber, such an approach allowed us to thoroughly test the new detector with the real LHC beam, leaving the option of a quick return to the original sensor in case any problems were discovered.

Figure 13 shows the FBCT and WCT amplitude-normalised time response to a real nominal LHC bunch (bunch intensity of 1.1×10^{11} protons, bunch length of 1.2 ns, beam energy of 450 GeV) measured with a Teledyne Lecroy HDO6104 12-bit oscilloscope (procured from Teledyne Lecroy SA, Vernier, Switzerland) with 1 GHz analogue bandwidth sampling at an equivalent rate of 125 GS/s installed close to the sensor. To reduce the noise level, 100 consecutive acquisitions were averaged. The FBCT was equipped with its standard 80 MHz LPF, while the WCT was measured in two configurations: unfiltered full bandwidth and through a 120 MHz non-reflective LPF.

The FBCT produces a pulse of about 12 ns followed by a tail lasting some 40 ns. The FBCT response clearly extends over 25 ns, potentially overlapping with a subsequent bunch.

The unfiltered WCT response is short enough to be easily shaped with external low-pass filters. After passing through a 120 MHz LPF, the signal almost fills the entirety of the available 25 ns bunch window, leaving approximately 2 ns of baseline following the pulse. This proves that the WCT signal allows the individual LHC bunches to be clearly distinguished and that there is no signal leakage to subsequent bunch slots.

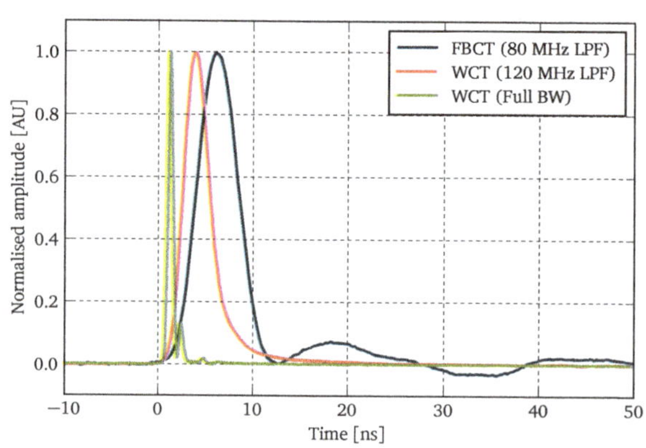

Figure 13. Amplitude-normalised time response of the WCT and the FBCT to a nominal LHC bunch [5].

Figure 14 plots the absolute signal levels measured in the same way at the output of the WCT distribution amplifier for typical LHC bunches. For comparison, the right plot also includes a response of the WCT to the nominal bunch measured directly at the sensor's output. A pilot bunch with an intensity of 5.8×10^9 protons measured through the "high gain" channel generates a pulse with a 702 mV amplitude. This can be translated into a sensitivity of 121 mV/10^9 ppb. The "low gain" channel outputs peaks at 1363 mV for a nominal bunch with an intensity of 1.2×10^{11} protons yielding a sensitivity of 11 mV/10^9 ppb.

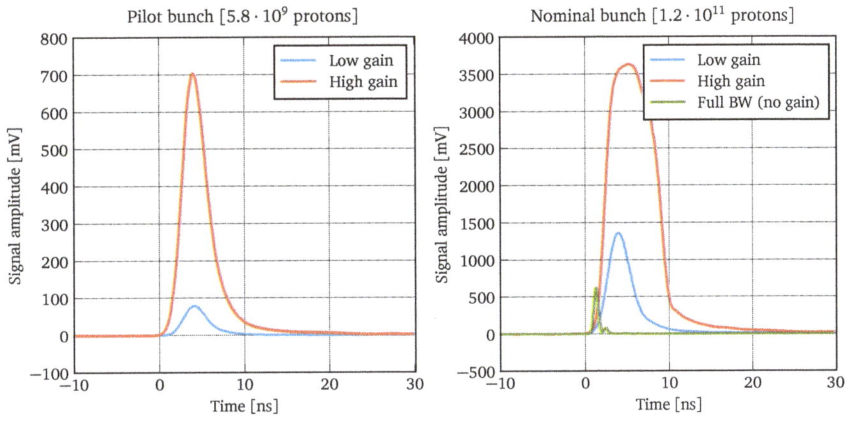

Figure 14. WCT response to pilot and nominal LHC bunches measured at the output of the distribution amplifier [5].

Figure 15 shows the frequency response of the FBCT and the WCT in the range of 1–2000 MHz. The sensors were measured without any additional filtering ("full BW") and with their typical external LPF. The measurements were performed with an Agilent Technologies E5071C Vector Network Analyzer (VNA) (procured from Agilent Technologies Schweiz AG, Basel, Switzerland) on a custom-built laboratory coaxial test setup, as illustrated in Figure 16. The data were normalised to equal 0 dB at 1 MHz.

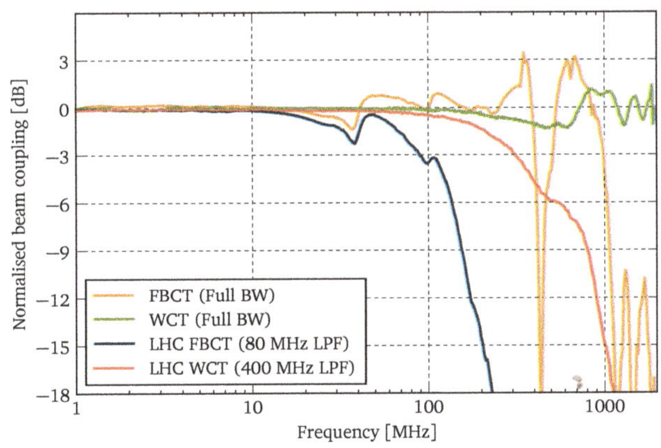

Figure 15. WCT and FBCT frequency response measured on a laboratory coaxial test bench [5].

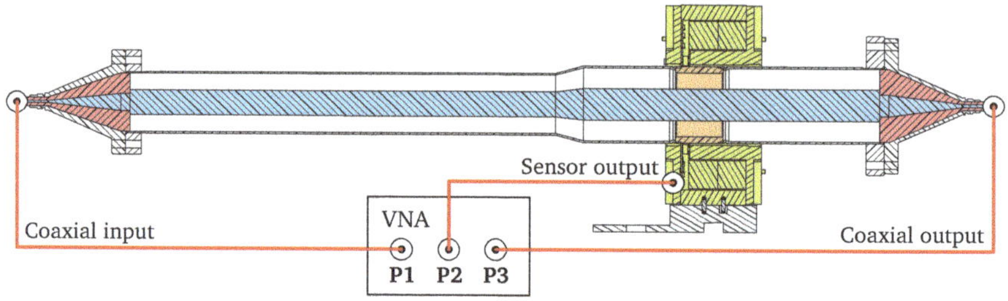

Figure 16. Connection diagram for frequency domain measurements [5].

The non-filtered FBCT has a high cutoff frequency close to 1 GHz, but the sensor's imperfections start being clearly visible already above 30 MHz with a very distinct resonance at 450 MHz. The FBCT's 80 MHz LPF strongly attenuates the high-frequency resonance, but the magnitude fluctuations in the range of 30–100 MHz remain visible.

The frequency–domain measurements of the WCT further substantiate its excellent time–domain performance. The frequency response remains flat within ± 1 dB over the entire measurement range. Even though some mild fluctuation can be seen above 700 MHz, the LHC bunches carry relatively little power at such high frequencies. The standard WCT 400 MHz LPF almost completely mitigates the high-frequency imperfections without compromising the sensor's time–domain response.

The laboratory coaxial test setup was also used to quantify the longitudinal beam-coupling impedance of the FBCT and the WCT. The beam's electromagnetic field unavoidably interacts with any component installed on the vacuum line inside which the beam travels. For components made from certain material and with certain geometries, the amount of electromagnetic energy extracted from the beam might be significant and lead to heat generation and other detrimental consequences. These effects are typically modelled as an additional impedance that the component exerts on the beam, and hence are referred to as the longitudinal beam-coupling impedance [15].

The longitudinal beam-coupling impedance of the FBCT and the WCT, as shown in Figure 17, was measured using the traditional stretched-wire technique [16]. With its well-defined paths for high-frequency currents, the WCT impedance remains below 6 Ω over the entire tested frequency range. On the other hand, the FBCT impedance is significantly higher and features two strong peaks at 380 MHz and 1.1 GHz. Even though both monitors are very minor contributors to the overall LHC impedance budget, the WCT qualitatively demonstrates another improvement over the FBCT design.

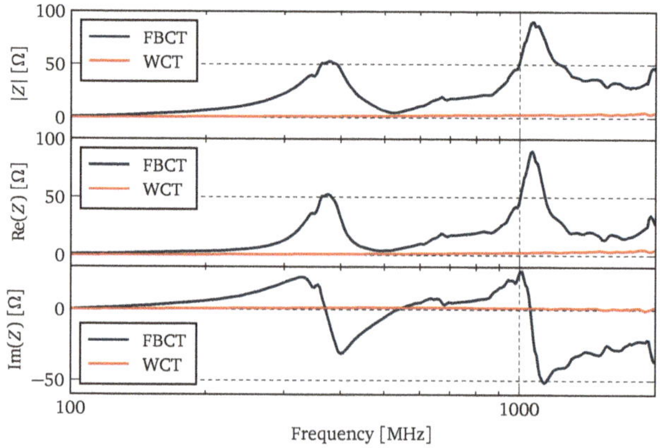

Figure 17. WCT and FBCT beam-coupling impedance measured on a laboratory coaxial test bench [5].

The sensitivity of the WCT and the FBCT to the transverse beam position and the bunch length were quantified during dedicated measurement sessions with the LHC beams. For the former, a beam consisting of five nominal bunches with an intensity of approximately 1.05×10^{11} ppb each was displaced in a series of steps, as plotted in Figure 18. The top plot shows the total beam intensity (i.e., the sum of five individual bunch intensities) measured by the FBCT and the WCT. The bottom plot displays by how much the beam was displaced in the horizontal (H) and vertical (V) plane from its average orbit at the location of the FBCT and the WCT. These values were calculated by interpolating the measurements of the closest upstream and downstream beam position monitors.

The total beam intensity measured by both sensors steadily decreased by about 0.5% over the data collection period. This decay is a natural phenomenon as the bunches lose some of their particles over time. However, the FBCT measurements are also visibly correlated with the transverse beam position and they vary by 0.5–0.8% mm^{-1} depending on the plane in which the beam was moved. Conversely, the WCT displayed no sensitivity to the transverse beam position down to the detection limit of 0.005% mm^{-1}. The measurements prove that the outputs of the WCT's eight internal RF transformers are combined with an excellent symmetry.

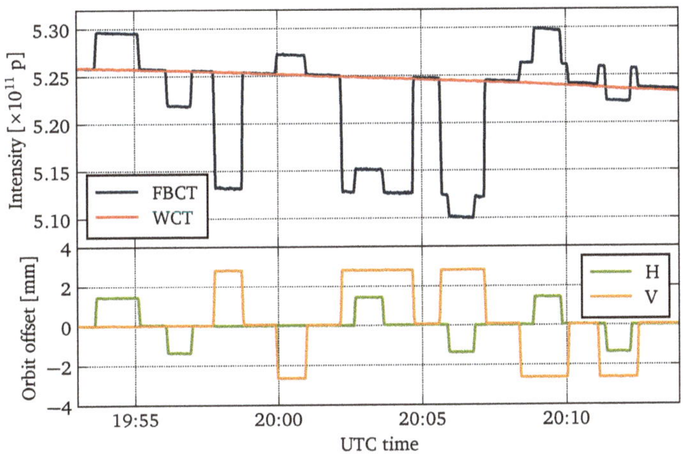

Figure 18. Sensitivity of the WCT and FBCT to the transverse beam position measured with the LHC beam [5].

The sensitivity of both sensors to the bunch length was tested with a beam consisting of 13 nominal bunches with a total intensity of around 13.1×10^{11} protons. The bunches were shortened and lengthened by adjusting the amplitude of the sinusoidal electric field applied to the bunches by the LHC RF cavities. Figure 19 shows the total beam intensity measured by the FBCT and the WCT when the bunch length was intentionally changed. The average bunch length, as shown in the bottom plot, was recorded by the beam quality monitor (BQM). Besides the natural slow beam intensity decay, both monitors also recorded a sharp drop of approximately 0.02% when the bunch length was quickly increased by 200 ps. However, when the bunch length was just as quickly decreased by 200 ps, a minute later, the readings of both sensors remained stable. Therefore, the observed drop is considered to be an observation of a true beam loss rather than a proof of bunch length sensitivity of either monitor. Overall, neither the FBCT nor the WCT displayed any sensitivity to the bunch length within the detection limit of 0.2% ns^{-1}. Such a result obtained for the FBCT is better than what had been previously published [10]. It is assumed that the apparent elimination of the FBCT's bunch length sensitivity might be caused by the 80 MHz low-pass filter installed on the FBCT output, which stretches the signal pulses to the extent that their shapes do not change significantly with the bunch length variations.

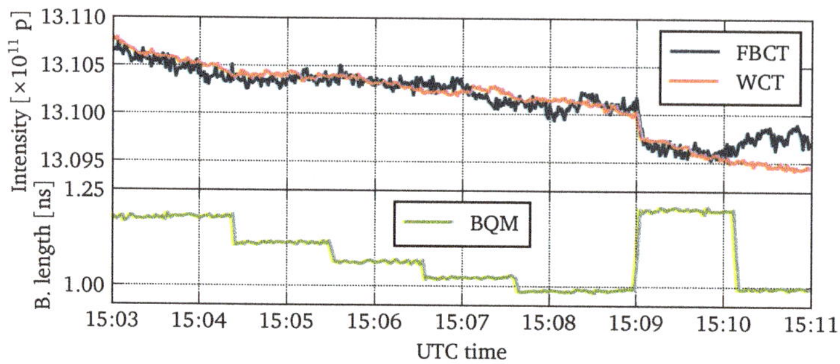

Figure 19. Sensitivity of the WCT and FBCT to the bunch length [5].

A practical example demonstrating the difference in performance of the WCT and the FBCT under real operational conditions are beam chromaticity measurements. Chromaticity links the beam's transverse oscillation frequency and its momentum and is one of the most fundamental parameters to monitor and control in a circular accelerator such as the LHC [17]. Chromaticity is typically measured by modulating the RF frequency used to accelerate the beam while keeping a constant field in the bending magnets. This results in a sinusoidal modulation of the horizontal beam position. Figure 20 shows an example of beam intensity measurements performed by the WCT and the FBCT of a beam consisting of eight pilot bunches at an energy of 450 MeV as the RF frequency is being modulated for beam chromaticity measurements. The FBCT readings are evidently correlated with the RF frequency while the WCT behaves as expected from a good beam intensity sensor.

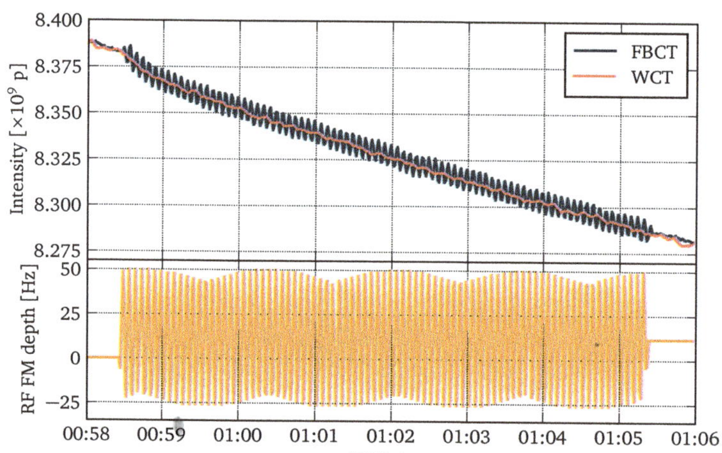

Figure 20. Beam intensity measured by the FBCT and the WCT during RF Frequency Modulation (FM) for beam chromaticity measurements [5].

Another situation in which the beam position can drastically change during operation is the calibration process of one of the beam profile monitors which requires displacing the beam by several millimetres [18]. This procedure is carried out several times each year with Figure 21 showing one such period. The intensity of a beam consisting of a single nominal bunch was measured by the WCT and the FBCT (top plot), while the change in the beam position was recorded by nearby beam position monitors. Once again, WCT's insensitivity to the beam position results in much more accurate intensity measurements than those provided by the FBCT.

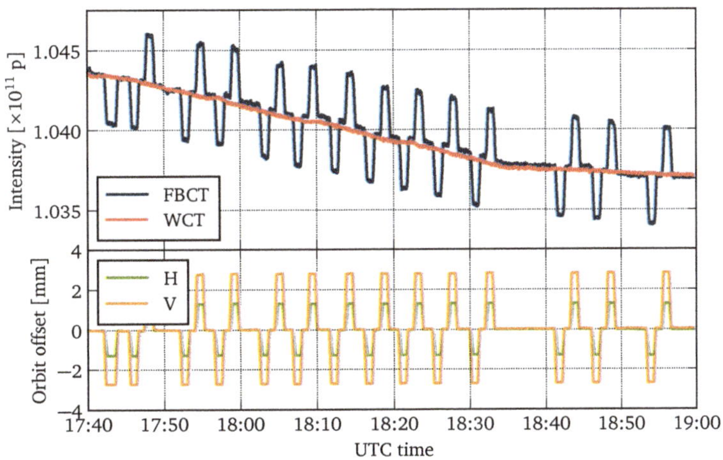

Figure 21. Beam intensity measured by the FBCT and the WCT during beam profile monitor calibration procedure [5].

The prototype WCT either matched or exceeded the FBCT performance in every laboratory and beam test. Based on these results, the WCT became the new reference LHC bunch intensity sensor. The four FBCTs originally installed in the LHC were removed and replaced by four WCTs. Moreover, an additional WCT of the same design was installed in the second largest accelerator at CERN, namely the Super Proton Synchrotron (SPS).

6. Conclusions

The developed WCT has successfully addressed the main performance limitations of the original LHC bunch intensity monitor, the FBCT. Due to its relatively large magnetic toroid, the FBCT's frequency response is insufficient to cleanly distinguish the consecutive LHC bunches. Moreover, the inherent asymmetry of the core secondary winding with respect to the particle beam results in a beam position dependence of the FBCT output signal. The WCT design addresses the former issue by using much smaller RF transformers producing a better frequency response. The latter limitation is overcome by employing eight transformers symmetrically distributed around the accelerator's vacuum chamber. Each of the individual transformers measures a part of the beam image current and their outputs are summed by an external RF power combiner. Consequently, the new sensor's bandwidth, transimpedance, and SNR are adequate for precise measurements of individual LHC bunches spaced by 25 ns. During dedicated measurements sessions with the LHC beam, the WCT has been conclusively demonstrated to be insensitive to the transverse beam position and the bunch length.

Table 2 summarises the most important parameters of the LHC FBCT and WCT. The notable difference of the magnetic core sizes and the secondary winding wire length are the prime factors contributing to the different high-frequency responses of the two monitors.

The WCT's mechanical design allows it to be assembled and removed without the need to vent the accelerator vacuum chamber. Thus, it was possible to install the first prototype WCT without removing the original FBCT from the LHC. Once the prototype was fully validated, the WCT became the new reference bunch intensity monitor and completely replaced the FBCT on both LHC rings. The mechanical design of the WCT can be simplified if its installation around an existing dielectric insert is not required.

Although the WCT described in this manuscript has been optimised for the measurements of the short current pulses generated by high-intensity proton bunches circulating in the LHC, the same technology can find use in other applications requiring precise non-intercepting wideband measurements of current signals. The WCT's low beam-coupling impedance and its relatively small footprint make it a good candidate for facilities dealing with high-power beams, such as spallation sources, or where space is limited, such as medical accelerators.

Table 2. A summary of the most important FBCT and WCT parameters.

	FBCT	WCT
Sensed quantity	Beam field	Image current
Number of magnetic toroids	One large	Eight small
Position component superposition	Toroid	External RF combiner
Beam position sensitivity (% mm^{-1})	0.5–0.8	<0.005
Bunch length sensitivity (% ns^{-1})	<0.2	<0.2
Number of secondary turns	40	10
Secondary-side load (Ω)	50	$\frac{5}{8} = 0.625$
Transimpedance (mΩ)	625	31.25
Low cutoff frequency (Hz)	200	410
Clean response limit (MHz)	≈ 30	≈ 700
Toroid dimensions (mm)	$\approx \varnothing 130 \times \varnothing 100 \times 25$	$\approx \varnothing 10 \times \varnothing 7 \times 5$
Secondary winding wire length (cm)	≈ 240	≈ 8

The final advantage of the WCT is that it is based on inexpensive off-the-shelf small magnetic cores, making it an interesting option for in-house developments.

Author Contributions: Conceptualisation, M.K. and M.G.; methodology, M.K.; validation, M.K.; formal analysis, M.K.; investigation, M.K.; resources, M.G.; data curation, M.K.; writing—original draft preparation, M.K.; writing—review and editing, M.G.; visualisation, M.K.; supervision, M.G. All authors have read and agreed to the published version of the manuscript.

Funding: This research received no external funding.

Data Availability Statement: Data available on request from the corresponding author.

Acknowledgments: The authors would like to thank their colleagues from the CERN Beam Instrumentation Group, especially: S. Bart Pedersen, D. Belohrad, J.J. Gras, and J. Kral (currently at DESY) for their work on the WCT acquisition systems, as well as T. Lefevre and L. Soby for their support during the development of the Wall Current Transformer.

Conflicts of Interest: The authors declare no conflict of interest.

Abbreviations

The following abbreviations are used in this manuscript:

CM choke	Common-mode choke
FBCT	Fast-beam current transformer
LHC	Large Hadron Collider
LPF	Low-pass filter
PCB	Printed circuit board
RF	Radio frequency
RMS	Root mean square
SNR	Signal-to-noise ratio
VNA	Vector network analyser
WCT	Wall current transformer

References

1. Evans, L. *The Large Hadron Collider: A Marvel of Technology*; EPFL Press: Lausanne, Switzerland, 2009.
2. Denard, J.-C. Beam current monitors. In *CAS—CERN Accelerator School: Beam Instrumentation*; Brandt, D., Ed.; CERN: Geneva, Switzerland, 2008; pp. 141–155.

3. Belohrad, D.; Jensen, L.K.; Jones, O.R.; Ludwig, M.; Savioz, J.-J. The LHC Fast BCT System: A Comparison of Design Parameters with Initial Performance. In Proceedings of the 14th Beam Instrumentation Workshop (BIW 10), Santa Fe, NM, USA, 2–6 May 2010; pp. 269–273.
4. Odier, P.; Belohrad, D.; Gras, J.-J.; Ludwig, M. Operational Experience and Improvements of the LHC Beam Current Transformers. In Proceedings of the 10th European Workshop on Beam Diagnostics and Instrumentation for Particle Accelerators (DIPAC 11), Hamburg, Germany, 16–18 May 2011; pp. 467–469.
5. Krupa, M. A Wall Current Transformer for Beam Intensity Measurements in the Large Hadron Collider. Ph.D. Thesis, Lodz University of Technology, Lodz, Poland, 2022.
6. Odier, P.; Ludwig, M.; Thoulet, S. The DCCT for the LHC Beam Intensity measurement. In Proceedings of the 9th European Workshop on Beam Diagnostics and Instrumentation for Particle Accelerators (DIPAC 09), Basel, Switzerland, 25–27 May 2009; pp. 143–145.
7. Belohrad, D.; Jones, O.R.; Ludwig, M.; Savioz, J.-J.; Thoulet, S. Implementation of the Electronics Chain for the Bunch by Bunch Intensity Measurement Devices in the LHC. In Proceedings of the 14th Beam Instrumentation Workshop (BIW 10), Santa Fe, NM, USA, 2–6 May 2010; pp. 137–139.
8. Belohrad, D.; Esperante Pereira, D.; Kral, J.; Pedersen, S. Upgrade of the LHC Bunch by Bunch Intensity Measurement Acquisition System. In Proceedings of the 5th International Beam Instrumentation Conference (IBIC 16), Barcelona, Spain, 11–15 September 2016; pp. 135–138.
9. Fischer, C.; Schmidt, R. *On the Measurements of the Beam Current, Lifetime and Decay Rate in the LHC Rings*; CERN: Geneva, Switzerland, 2005; p. 7.
10. Anders, G.; Gabaldon, C.; Hopchev, P.; Balagura, V.; Kozanecki, W.; Ludwig, M.; Bacchetta, N.; Oyama, K.; Jeff, A.; White, S.; et al. *Study of the Relative LHC Bunch Populations for Luminosity Calibration*; CERN: Geneva, Switzerland, 2005; pp. 5–10.
11. Hofmann, A. Dynamics of beam diagnostics. In *CAS—CERN Accelerator School: Beam Instrumentation*; Brandt, D., Ed.; CERN: Geneva, Switzerland, 2008; pp. 66–67.
12. Satoh, K. New Wall Current Beam Position Monitor. In Proceedings of the 8th Particle Accelerator Conference (PAC 79), San Francisco, CA, USA, 12–15 March 1979; pp. 3364–3366.
13. Cherepanov, V.P. Image Current Monitor for Bunched Beam Parameters Measurements. In Proceedings of the 2nd European Workshop on Beam Diagnostics and Instrumentation for Particle Accelerators (DIPAC 95), Lubeck-Travemunde, Germany, 28–31 May 1995; pp. 142–144.
14. Gasior, M. An Inductive Pick-Up for Beam Position and Current Measurements. In Proceedings of the 6th European Workshop on Beam Diagnostics and Instrumentation for Particle Accelerators (DIPAC 03), Mainz, Germany, 5–7 May 2003; pp. 53–55.
15. Rumolo, G. Beam Instabilities. In *CAS—CERN Accelerator School: Advanced Accelerator Physics Course*; Herr, W., Ed.; CERN: Geneva, Switzerland, 2014; p. 202.
16. Caspers, F. Impedance Determination from Bench Measurements. In *Handbook of Accelerator Physics and Engineering*, 3rd ed.; Chao, A.W., Tigner, M., Eds.; World Scientific: Singapore, 1999; pp. 622–626.
17. Steinhagen, R.J. Feedbacks on Tune and Chromaticity. In Proceedings of the 8th European Workshop on Beam Diagnostics and Instrumentation for Particle Accelerators (DIPAC 07), Venice, Italy, 20–23 May 2007; pp. 43–47.
18. Trad, G. Development and Optimisation of the SPS and LHC Beam Diagnostics Based on Synchrotron Radiation Monitors. Ph.D. Thesis, University of Grenoble, Grenoble, France, 2014; pp. 118–120.

energies

MDPI

Article

New Approach to Evaluate the Transformation Accuracy of Inductive CTs for Distorted Current

Michal Kaczmarek * and Ernest Stano

Institute of Mechatronics and Information Systems, Lodz University of Technology, 90-537 Lodz, Poland
* Correspondence: michal.kaczmarek@p.lodz.pl

Abstract: This paper presents a newly developed method to determine the values of current error and phase displacement for the transformation of distorted current harmonics by the inductive current transformers. This approach eliminates the necessity for the utilization of an expensive, high-current supply system for the measuring setup. In this method, the secondary winding is fed by the distorted voltage with RMS values of harmonics calculated in order to reproduce the operation point of the inductive current transformer on the magnetization characteristic of its magnetic core, as in primary winding excitation conditions. This proposed approach is successfully verified with the typically used primary current excitation method, where the secondary currents of the reference and tested current transformers are compared in the differential measuring setup. It was confirmed that the inductive CT with current error and phase displacement for transformation of distorted current harmonics determined in the rated ampere-turns conditions may be effectively used in the measuring setup as the reference source of the primary current.

Keywords: secondary current excitation; transformation accuracy; distorted current; harmonics; current transformer; phase displacement; current error

check for updates

Citation: Kaczmarek, M.; Stano, E. New Approach to Evaluate the Transformation Accuracy of Inductive CTs for Distorted Current. *Energies* **2023**, *16*, 3026. https://doi.org/10.3390/en16073026

Academic Editor: Wencheng Guo

Received: 30 January 2023
Revised: 9 March 2023
Accepted: 24 March 2023
Published: 26 March 2023

1. Introduction

Inductive current transformers (CTs) are devices that are used to transform high currents into an appropriate level for measuring and as a protection apparatus of the electrical power system [1,2]. CTs are widely used in the monitoring of power generation, transmission and distribution. The transformation accuracy of the CTs is critical to the reliability of the power network and energy billing. Inductive CTs are designed to ensure transformation accuracy over a specific range of the primary current. Its upper value is limited by the knee point on the magnetization characteristic of the magnetic core. This is where its nonlinear part begins. Further increase in the value of magnetic flux density will cause saturation of the magnetic core [3–7]. The secondary current of the inductive CT will be strongly distorted by the self-generation of higher harmonics caused by the nonlinearity of the magnetization characteristic of the magnetic core. This phenomenon is typical for CTs with the magnetic core. The magnetization characteristic, even below the knee point, is nonlinear, and some level of the low order higher harmonics is generated to the secondary current. The pure sinusoidal waveforms of the current never exists in the electrical power system. Its frequency spectrum consists of not only fundamental components but also higher harmonics. This is due to an increased number of loads connected to the power grid with the nonlinear current to the voltage relationship [8,9]. Moreover, the renewable energy sources may introduce additional distortion. Therefore, all devices belonging to the electrical energy system are exposed to the current components other than rated frequencies. The new edition of the standard IEC 61869-1 will be introduced in 2023 and contains the optional requirements for transformation accuracy of higher harmonics by the inductive CTs. Many of the papers [10–16] present behavior and accuracy tests of inductive CTs during transformation of the distorted current. However, most of them are based on the utilization of the high-current test system, which have to be able to generate distorted currents associated with the rated primary current of the tested CT. Other problems concern the devices which may be adopted as a reference source of the primary current. A solution for this problem is presented in the papers [17–19]. In this case, the low primary current transformer (LPCT) was used. Another device, which can be adopted for this purpose, is the inductive CT characterized by a high accuracy of transformation for distorted current higher harmonics. Its accuracy is then tested in its rated ampere-turns conditions [20–22]. The inductive CT must be a bushing or window type. The additional primary winding with the number of turns equal to the rated current ratio has to be made. It is required to be evenly spread on the surface of the magnetic core.

The paper presents a new approach to evaluate the transformation accuracy of inductive CTs for transforming distorted current harmonics. In this method, the secondary winding is fed by the distorted voltage with the RMS values of harmonics calculated in order to reproduce the operation point of the inductive CT on the magnetization characteristic of its magnetic core under primary winding excitation conditions. Therefore, the values of resistance and reactance of the secondary winding and its load must be considered. It has been proven that the value of reactance of the secondary winding may be estimated from the measured value of its DC resistance by multiplying it by the factor equal to 0.1. To determine the value of the current error and phase displacement for a given hk harmonic of the distorted secondary voltage representing the considered distorted primary current, the RMS values of the hk harmonics of the secondary winding excitation current and phase shift between them are measured.

The novelty of this paper is summarized in the following bullet points:

- The method to determine current error and phase displacement for the transformation of distorted current harmonics by the inductive CTs without the high current generation system is proposed;
- It is confirmed that the effect of the nonlinearity of the magnetic core is covered by the secondary current excitation method;
- The proposed approach is successfully verified with the typically used primary current excitation method, where the secondary currents of the reference and tested current transformers are compared in the differential measuring setup;
- Inductive CTs with current error and phase displacement for transformation of distorted current harmonics determined in the rated ampere-turns conditions may be effectively used in the measuring setup as the reference source of the primary current.

The differences between the results obtained by the means of the secondary and primary current excitation methods are the most significant for the main and 3rd harmonics of the considered distorted primary current. The highest difference does not exceed $\pm 0.1\%$ for the current error at the harmonic and $\pm 0.1^\circ$ for phase displacement at the harmonic. These values are the most important for the main component, taking into consideration the limit values of current error and phase displacement defined for a given accuracy class in the standard IEC 61869-2 and IEEE C57.13 for transformation of the sinusoidal current of frequency 50 Hz/60 Hz. Therefore, the equivalent method is found not yet suitable for accuracy tests in accordance with appropriated standards for the transformation of sinusoidal currents. The accuracy requirements for higher harmonic transformation by the inductive current transformer are still not defined by the standards. Nevertheless, the requirements for transformation of distorted current higher harmonics are expected to be several times less restrictive, and the measurement uncertainty of the proposed method would be acceptable.

Presented in this manuscript is an approach and method that can be used for the protection and measuring the types of inductive CTs. It is crucial to determine the correct values of inductance and resistance of the CT's secondary winding. Moreover, this method can be utilized to determine the values of the current error and phase displacement of the distorted current harmonics transformation by the inductive CT characterized by even distribution of the winding turns on the surface of the magnetic core. In the other case, the measurement uncertainty of the method increases. In our laboratory studies, we tested only inductive CTs with even distribution of the primary and secondary winding turns. This is important, because the even distribution of the magnetic field strength in the magnetic core is required in order to reproduce the same conditions from the secondary side as it would be obtained during the primary excitation. To summarize, the advantages of the proposed method are listed below:

- It eliminated the necessity of the utilization of expensive, high-current supply systems of the measuring setup;
- It did not require the utilization of the reference source of the primary current (e.g., reference transducer/transformer);
- It enabled the determination of the values of current error and phase displacement even for very high frequencies;

However, the disadvantages are pointed out below:

- It required the determination of the correct values of inductance and resistance of the CT's secondary winding,
- It was applicable only to the inductive CTs with even distribution of the winding turns on the surface of the magnetic core.

2. Measuring Circuits

To discuss in detail the idea of the developed, secondary current excitation method used to determine the transformation accuracy of distorted current harmonics by the inductive CTs, their equivalent circuit, presented in Figure 1, is analyzed.

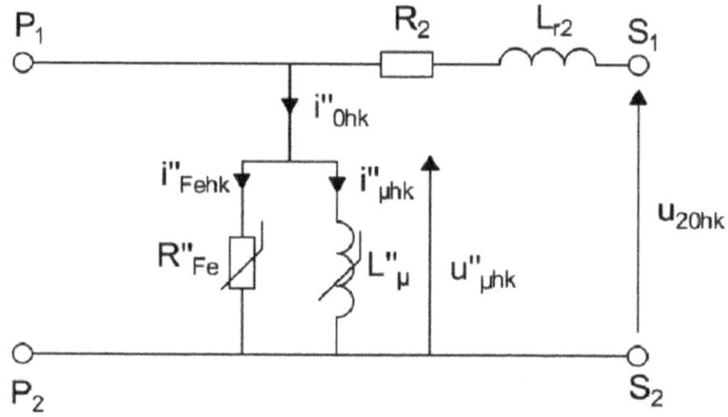

Figure 1. Equivalent circuit of the inductive CT.

where:

i''_{0hk} is the instantaneous value of the secondary excitation current;
$i''_{\mu hk}$ is the instantaneous value of the reactive component of the secondary excitation current;
i''_{Fehk} is the instantaneous value of the active component of the secondary excitation current;
$L''_{\mu hk}$ is the nonlinear mutual inductance between windings of the inductive CT;
R''_{Fehk} is the nonlinear resistance representing the active power losses in the magnetic core;
L_2 is the leakage inductance of the secondary winding;
R_2 is the resistance of the secondary winding;
$u''_{\mu hk}$ is the instantaneous value of the voltage on the mutual inductance between windings;
and u_{20hk} is the instantaneous value of the equivalent secondary winding supply voltage.

In order to determine the values of the current error and phase displacement using the secondary current excitation method, the first stage involves calculating the RMS values of the distorted voltage harmonics required to supply the secondary winding of the TCT (tested current transformer). These values must be determined for the given RMS values of the distorted primary current harmonics converted to the secondary side, taking into account the resistance and reactance of the secondary winding and its load.

These assumptions are necessary to adequately represent the TCT's operation for the transformation of the distorted primary current with a specific secondary winding load. In accordance with the results and method presented in paper [23], the reactance of the secondary winding may be estimated from the measured value of its DC resistance by multiplying it by a factor equal to 0.1. To ensure the same operating point of the tested CT on the magnetization characteristic of its magnetic core, as in the conditions of primary winding excitation, it is necessary to calculate the RMS values of the harmonics of the secondary winding supply voltage. The RMS value of each harmonic is calculated in accordance with the following formula:

$$U_{20hk} = I''_{z1hk} \cdot \sqrt{(R_2 + R_L)^2 + (2 \cdot \pi \cdot f_{hk} \cdot L_2)^2}, \quad (1)$$

where:

I''_{z1hk} is the RMS value of the *hk* higher harmonic of the considered distorted primary current for which the values of the current and phase displacement are determined;
R_L is the load resistance of the secondary winding;
L_L is the load inductance of the secondary winding;
and U_{20hk} is the RMS value of the *hk* higher harmonic of the equivalent secondary winding supply voltage.

In the second stage of the developed method, the TCT is connected to the measuring circuit presented in Figure 2. To determine the value of the current error and phase displacement for a given *hk* harmonic of the considered distorted primary current, the RMS values of the excitation current I''_{0hk} and applied supply voltage U_{20hk} as well as the phase shift between them ω_{hk} must be measured. The value of the *hk* harmonic of the composite error for a given RMS value of the hk higher harmonics of the considered distorted primary current I''_{z1hk} can be calculated with the following formula:

$$\varepsilon_{\%IAhk} = \frac{I''_{0hk}}{I''_{z1hk}} 100\%, \quad (2)$$

where:

I''_{0hk} is the RMS value of the measured *hk* harmonic of the distorted excitation current.

The RMS value of a given voltage harmonic of the mutual inductance $U_{\mu hk}$ is equal to:

$$U_{\mu hk} = \left[U_{20hk}^2 + I''^2_{0hk} \cdot [R_2^2 + (2 \cdot \pi \cdot f_{hk} \cdot L_2)^2] - 2 \cdot U_{20hk} \cdot I''_{0hk} \cdot \sqrt{R_2^2 + (2 \cdot \pi \cdot f_{hk} \cdot L_2)^2} \cdot \cos\left(\omega_{hk} - arctg \frac{2 \cdot \pi \cdot f_{hk} \cdot L_2}{R_2} \right) \right]^{\frac{1}{2}}, \quad (3)$$

The value of the *hk* harmonic of the current error is defined by the following equation:

$$\Delta I_{Ahk} = \varepsilon_{\%IAhk} \cdot \cos\left(arccos\left(\frac{U_{20hk}^2 + U_{\mu hk}^2 - I''^2_{0hk} \cdot (R_2^2 + (2 \cdot \pi \cdot f_{hk} \cdot L_2)^2)}{2 \cdot U_{20hk} \cdot U_{\mu hk}} \right) + \omega_{hk} - arctg\left(\frac{2 \cdot \pi \cdot f_{hk} \cdot (L_2 + L_L)}{R_2 + R_L} \right) \right), \quad (4)$$

The value of the *hk* harmonic of the phase displacement is defined by the following equation:

$$\delta\varphi_{Ahk} = arcsin \left(\varepsilon_{\%IAhk} \cdot \sin\left(arccos\left(\frac{U_{20hk}^2 + U_{\mu hk}^2 - I''^2_{0hk} \cdot (R_2^2 + (2 \cdot \pi \cdot f_{hk} \cdot L_2)^2)}{2 \cdot U_{20hk} \cdot U_{\mu hk}} \right) + \omega_{hk} - arctg\left(\frac{2 \cdot \pi \cdot f_{hk} \cdot (L_2 + L_L)}{R_2 + R_L} \right) \right) \cdot \frac{1}{100\%} \right), \quad (5)$$

The measuring circuit of the proposed secondary current excitation method is presented in Figure 2.

(a)

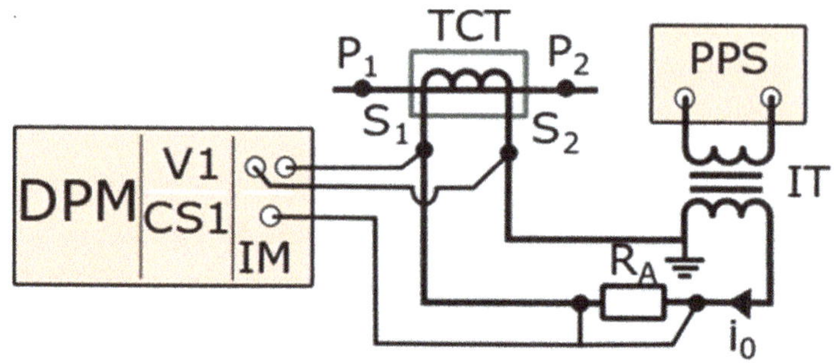

(b)

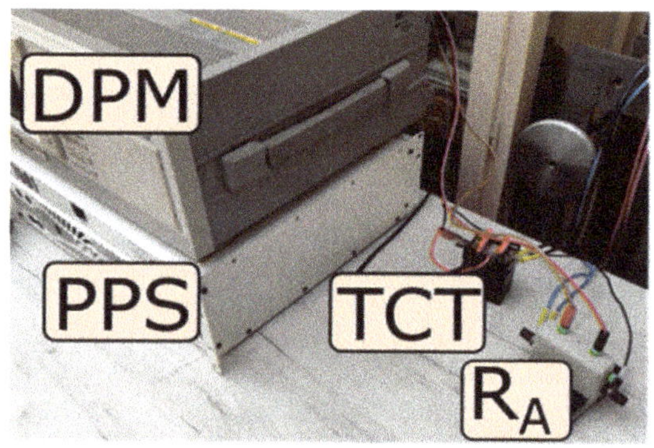

Figure 2. Measuring system (**a**) and photo (**b**) of the secondary current excitation method used to determine the current error and phase displacement of the TCT.

In Figure 2 the following abbreviations are used:

DPM is the digital power meter;
TCT is the tested inductive CT;
PPS is the programmable power supply;
IT is the insulation transformer;
I_0 is the instantaneous value of magnetic core's excitation current;
R_A is the current shunt used to measure value of I_0;
V_1/CS_1 is the input terminal of the DPM;
P_1/P_2 is the primary terminal of the tested CT;
and S_1/S_2 is the secondary terminal of the tested CT.

The verification of the results obtained from the proposed method was performed by comparing the determined values of the current error and phase displacement with the utilization of the primary current excitation method. The measuring setup had to be supplied with a high value of the distorted primary current resulting from the rated current ratio of the tested inductive CT. This may be accomplished using the high current generation system composed of the step-up current transformer (SCT), programmable power source (PPS) and isolation transformer (IT). The reference current transformer (RCT) is also supplied by the same primary current as the tested unit. Therefore, their secondary current may be compared.

The accuracy of the RCT is determined in the rated ampere-turns method described in detail in paper [20]. The differential circuit is used to ensure low measurement uncertainty. The value of the composite error for a given higher harmonics is directly determined from the measured value of voltage on the current shunt R_D. In the measuring circuit presented in Figure 1, the current shunt R_S is used to measure the RCT's secondary current. The load of the secondary winding of the TCT is represented by the current shunt R_L.

In Figure 3, the abbreviations are the same as those in Figure 1, with the addition of the following:

SCT is the step-up current transformer;
R_D is the current shunt used to measure instantaneous value of the differential current;
RCT is the reference CT tested in the ampere-turns conditions;
R_L is the load of the secondary winding of the tested CT;
R_S is the current shunt used to measure the instantaneous value of RCTs secondary current;
i_2 is the instantaneous value of the tested CTs secondary current;
i_{2r} is the instantaneous value of the RCTs secondary current;
and i_D is the instantaneous value of the differential current between the TCT and RCT secondary currents.

The digital power meter (DPM) enables instantaneous measurements of the RMS value of a given higher harmonics of voltages on current shunts R_S and R_D as well as the phase angle between them.

The percentage value of the RCT's secondary current is determined from Equation (6) [22].

$$I_{2hk} = \frac{U_{Shk}}{R_S} \cdot 100\%, \tag{6}$$

where:

U_{Shk} is the RMS value of the hk voltage higher harmonic of the current shunt R_S.

The value of the current error specified for a given hk harmonic of the distorted primary current is equal to [22]:

$$\Delta I_{hk} = \frac{\sqrt{\left(\frac{U_{Shk}}{R_S}\right)^2 + \left(\frac{U_{Dhk}}{R_D}\right)^2 - 2\frac{U_{Shk}}{R_S} \cdot \frac{U_{Dhk}}{R_D} cos\phi_{hk}} - \frac{U_{Shk}}{R_S}}{\frac{U_{Shk}}{R_S}} \cdot 100\%, \tag{7}$$

where:

U_{Dhk} is the RMS value of the hk voltage higher harmonic of the current shunt R_D;
and ϕ_{hk} is the phase angle of the hk higher harmonic measured between voltages of the current shunts R_D and R_S.

The value of the phase displacement is calculated from the following equation [22]:

$$\delta\varphi_{hk} = arcsin\left(\frac{\sqrt{\left(\frac{U_{Dhk} \cdot R_S}{R_D \cdot U_{Shk}} \cdot 100\%\right)^2 - \Delta I_{hk}^2}}{100\%}\right), \tag{8}$$

The tests were performed for the transformation of the distorted primary current with the main component of a 50 Hz frequency and a single higher harmonic frequency from 100 Hz to 5 kHz. The RMS value of the higher harmonic is equal to 10% of the main harmonic. In accordance with the standard IEC 61869-2, both inductive CTs are tested for 5%, 20%, 100% and 120% of the rated primary current RMS value. The CT with an accuracy class of 0.2S was additionally tested for 1%. Due to the different phase angle of higher harmonics in relation to the main harmonic of the distorted primary current, various values of the current error and phase displacement may be determined. Therefore, during the tests in the primary current excitation method, the phase angle of the transformed distorted primary current higher harmonic is changed by 5° in the range from 0° to 355° in relation to the main component. While, in the secondary current excitation method, the phase angle of the secondary winding supply voltage higher harmonic is also changed by 5° in the range from 0° to 355° in relation to the main component. This approach enables the possibility to evaluate the most positive and negative values of the current error and phase displacement that may be obtained for TCT due to the influence of the self-generation phenomenon [14,21,22,24]. In this paper, the highest absolute values of the current error and phase displacement are presented.

(a)
(b)

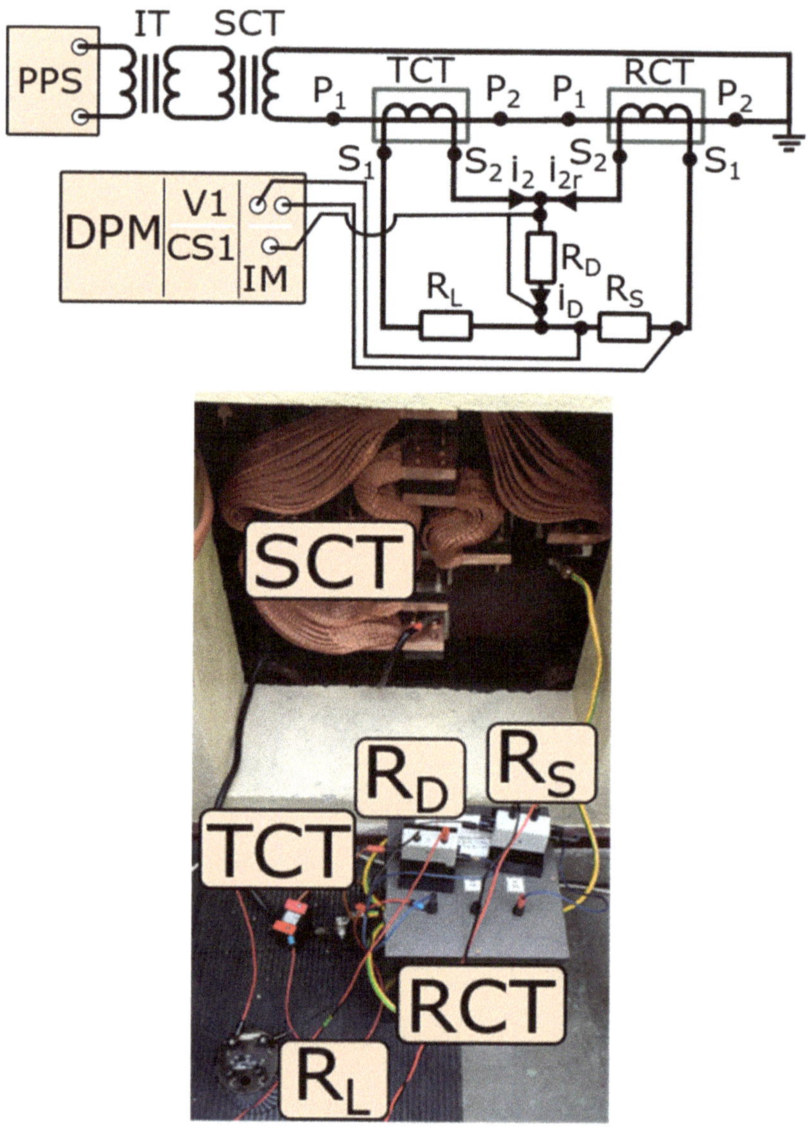

Figure 3. Measuring circuit (**a**) and photo (**b**) of primary current excitation method used to determine the current error and phase displacement of the tested CT.

3. Reference and Tested CTs

The developed, secondary winding excitation current method was applied to two different inductive CTs. The first one had a rated current ratio of 1500 A/5 A, and its accuracy class determined for the transformation of the sinusoidal current, with a frequency of 50 Hz, was 0.2S. The second one had a rated current ratio of 100 A/5 A, and its accuracy class was 0.5. The developed, secondary current excitation method requires specifying the values of inductance and resistance of the secondary winding and parameters of the load for which the RMS values of harmonics of the secondary winding supply voltage in accordance with Equation (1) are calculated. These values for both CTs are given in Table 1.

Table 1. The values of resistance and inductance of the secondary winding and load of tested inductive CTs are shown.

Tested CTs	L_{r2} [mH]	R_2 [Ω]	L_L [mH]	R_L [Ω]
1500 A/1 A, cl. 0.2S	0.074	0.305	-	5
100 A/5 A, cl. 0.5	0.0035	0.0121	-	0.1

The TCT 1500 A/1 A, cl. 0.2S is connected to the measuring circuit presented in Figure 2. To determine the value of the current error and phase displacement, the RMS values of the excitation current I''_{0hk} and applied supply voltage U_{20hk} as well as phase shift between them and ω_{hk} were measured. In Table 2 the results for the equivalent test conditions representing 20% of the rated primary current are reported.

Table 2. The measured values of the *hk* harmonics of the excitation current I''_{0hk} and applied supply voltage U_{20hk} as well as phase shift between them and ω_{hk} for the TCT 1500 A/1 A, cl. 0.2S in the equivalent test conditions, representing 20% of the rated primary current, are shown.

Harm. Order [-]	U_{20hk} [V]	I''_{0hk} [mA]	ω_{hk} [°]
1	1.0610	3.7551	167.79
3	0.1061	0.4353	182.86
5	0.1061	0.3333	192.33
7	0.1061	0.2736	193.44
10	0.1062	0.2605	197.24
15	0.1063	0.2800	202.26
20	0.1065	0.2887	206.37
25	0.1067	0.3078	209.62
30	0.1070	0.3367	212.23
40	0.1077	0.4152	215.61
50	0.1086	0.5254	217.87
60	0.1097	0.6261	219.99
70	0.1110	0.7615	221.65
80	0.1124	0.9671	222.97
90	0.1141	1.1419	225.21
100	0.1158	1.3028	227.41

The values of the voltage U_{20hk} for a given hk harmonic were calculated in order to ensure equivalent conditions for the operation of the TCT magnetic core as obtained for transformation of the distorted primary current. The above presented calculations were conducted to obtain the secondary supply voltage under conditions where 20% of the rated primary current with the RMS value of the higher harmonic equal to 10% of the main harmonic is transformed by the TCT. Similar calculations were performed for 1%, 5%, 100% and 120% of the rated primary current RMS value.

The TCT 100 A/5 A, cl. 0.5 was connected to the same measuring circuit. In Table 3, the results for the equivalent test conditions representing the transformation of the distorted primary current for 100% of its rated value are reported.

Table 3. The measured values of the hk harmonics of the excitation current I''_{0hk} and applied supply voltage U_{20hk} as well as phase shift between them and ω_{hk} for the TCT 100 A/5 A, cl. 0.5 in the equivalent test conditions, representing 100% of the rated primary current, are shown.

Harm. Order [-]	U_{20hk} [V]	I''_{0hk} [mA]	ω_{hk} [°]
1	0.5605	5.9601	167.79
3	0.0561	0.6593	182.86
5	0.0561	0.5146	192.33
7	0.0562	0.4698	193.44
10	0.0563	0.4661	197.24
15	0.0567	0.4636	187.20
20	0.0571	0.4467	187.77
25	0.0577	0.4446	188.53
30	0.0584	0.4422	189.25
40	0.0602	0.4403	191.01
50	0.0624	0.4447	192.77
60	0.0650	0.4598	194.29
70	0.0680	0.4640	195.96
80	0.0712	0.4897	197.17
90	0.0748	0.4899	198.58
100	0.0785	0.5076	199.68

In the above presented case, the calculations were performed to obtain secondary supply voltage under the conditions where the TCT transforms 100% of the rated primary current with the RMS value of the higher harmonic equal to 10% of the main harmonic. Similar calculations were performed for 5%, 20% and 120% of the rated primary current RMS value.

The results presented in Tables 2 and 3 enable the calculation of the values of the current error and phase displacement for the transformation of a given hk harmonic of the distorted primary current by TCTs using Equations (4) and (5).

The developed, inductive RCT is designed for the rated primary current RMS value equal to 300 A. It has two secondary windings for rated secondary currents equal to 1 A and 5 A. The values of the current error and phase displacement of the RCT for transformation of the distorted current harmonics are determined by the means of the rated ampere-turns method described in detail in paper [2]. The tests were performed for the transformation of the distorted primary current with the main component of frequency being 50 Hz and a single higher harmonic frequency ranging from 100 Hz to 5 kHz. The RMS value of the higher harmonic is equal to 10% of the main harmonic. To test the CT, an additional primary winding must be made, and its number of turns results from the rated current ratio of the tested inductive CT. However, this solution is only applicable to window-type CTs. In the differential circuit, the currents in the additional primary winding and in the secondary winding of the tested CTs are compared. The frequency characteristics of the current error and phase displacement of the RCT for the rated current ratio equal to 300 A/1 A are presented in Figure 4.

The presented results in Figure 4 indicate that the values of the current error and phase displacement were the highest for the main harmonic of the transformed distorted primary current. They did not exceed −0.05% and 0.08°, respectively.

The frequency characteristics of the current error and phase displacement of the RCT for the rated current ratio equal to 300 A/5 A are presented in Figure 5.

The presented results in Figure 5 indicate that the values of the current error and phase displacement were the highest for the main harmonic of the transformed distorted primary current. They did not exceed −0.05% and 0.06°, respectively.

(a)

(b)

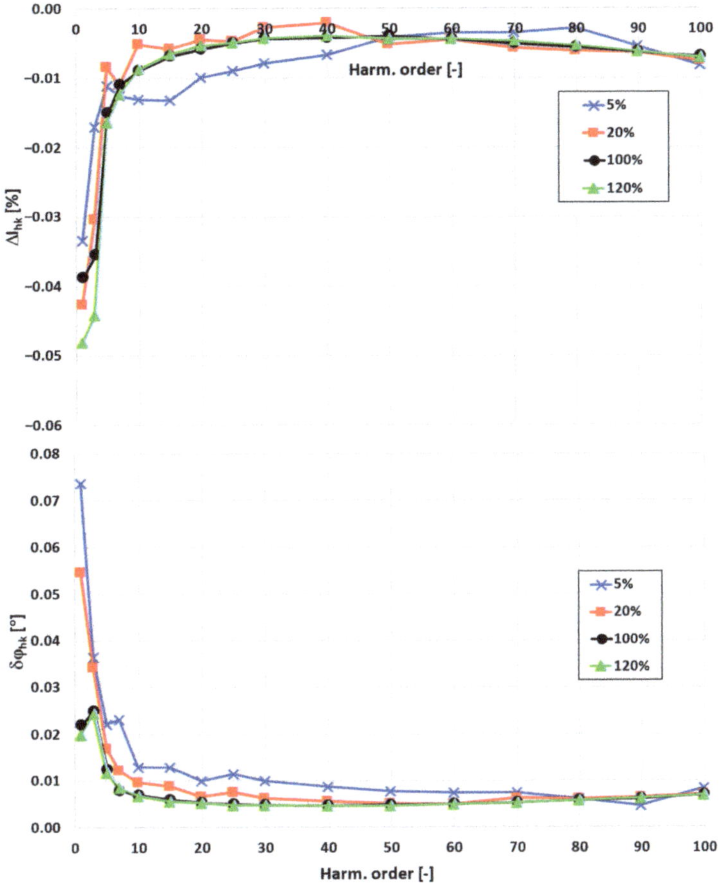

Figure 4. The frequency characteristics of (**a**) current error and (**b**) phase displacement of the RCT for the rated current ratio equal to 300 A/1 A are shown.

(a)

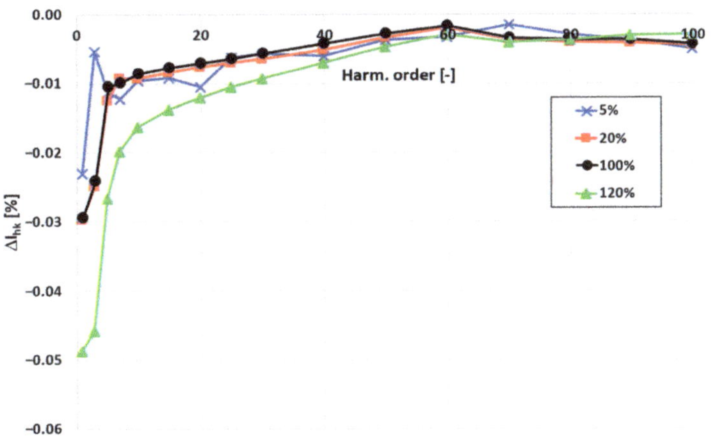

Figure 5. *Cont.*

(b)

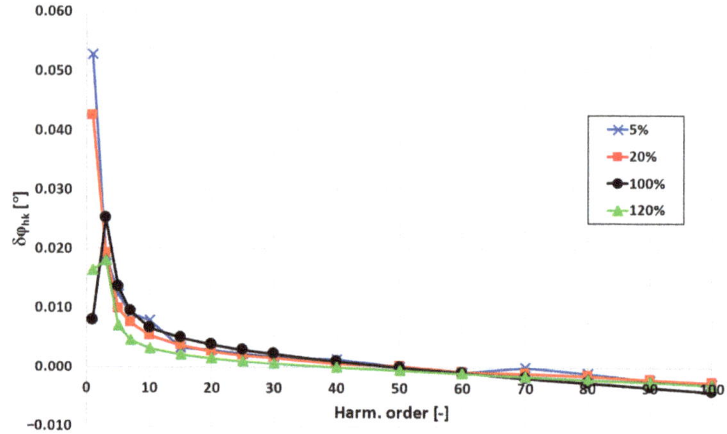

Figure 5. The frequency characteristics of (**a**) current error and (**b**) phase displacement of the RCT for the rated current ratio equal to 300 A/5 A are shown.

4. Results

The results of the values of the current error and phase displacement determined for a given harmonic of the tested CT with a rated current ration equal to 1500 A/1 A in the proposed secondary current excitation method and primary reference excitation method are presented in Figure 6. The dotted lines present the results without the consideration of the correction of the secondary winding turns. The method to determine the turns ratio correction of the tested CT is described in paper [25], where its value of considered TCT is equal to six turns. This value changes the current error by 0.4% towards positive values. The results of the tests are presented for the distorted primary current equal to 120% and 20% of the TCT's rated current value.

The high values of the current errors and phase displacements obtained from the 3rd to 10th order harmonics result from the self-generation of low-order higher harmonics in the secondary current. The proposed method also allows for determining the influence of this phenomenon on the values of the current error and phase displacement. The convergence of the results obtained from the secondary current excitation method and the reference method confirms the effectiveness of the developed approach. However, it is necessary to also consider the turns ratio corrections. The discrepancies between the results are within the extended measurement uncertainty of both methods.

Figure 7 shows a comparison of the obtained values of the current error and phase displacement of the TCT with the rated current ratio of 100 A/5 A in the proposed secondary and primary current excitation methods. The TCT was made with the applied turns ratio correction of the secondary winding by a value of 0.1 turns. This can be achieved when the last turn is divided into 10 wires, 9 of which are made through the window of the magnetic core. The tests were conducted for the distorted primary current equal to 100% and 5% of the TCT's rated current value.

It is important to note that the differences between the results obtained by the secondary current excitation method and the primary current excitation method are significant, especially for the main and 3rd harmonics. The highest difference reaches ±0.1% for current error and does not exceed ±0.1° for the phase displacement. These values are significant for the main component of the distorted primary current, considering the limit values of the current error and phase displacement defined for a given accuracy class in the standards IEC 61869-2 and IEEE C57.13 [26,27]. However, it should be noted that the accuracy requirements for higher harmonic transformation by the inductive current transformer are not yet defined by the standards. The limiting values of the current error and phase displacement at harmonics are proposed in paper [19]. The values defined in the standard IEC 61869-6 for the low-power instrument transformers may be also adopted [28]. Nevertheless, the requirements for the transformation of distorted current higher harmonics are several times less restrictive. The differences between the values of the current error and phase displacement determined by both methods are acceptable and do not have a significant influence on the possible designation of the wide frequency accuracy class, as the requirements for the transformation of the distorted current higher harmonics are several times less restrictive.

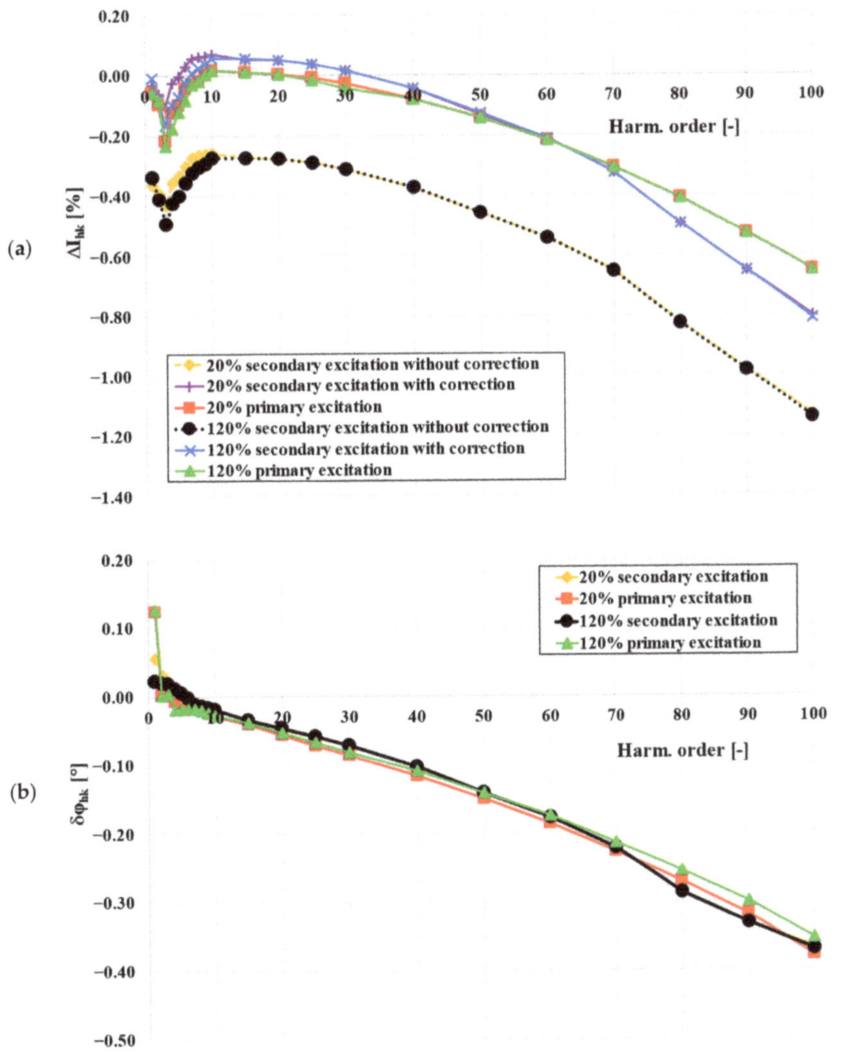

Figure 6. The comparison of obtained values of current error (**a**) and phase displacement (**b**) of the TCT with the rated current ratio equal to 1500 A/1 A in proposed secondary and primary current excitation method are shown.

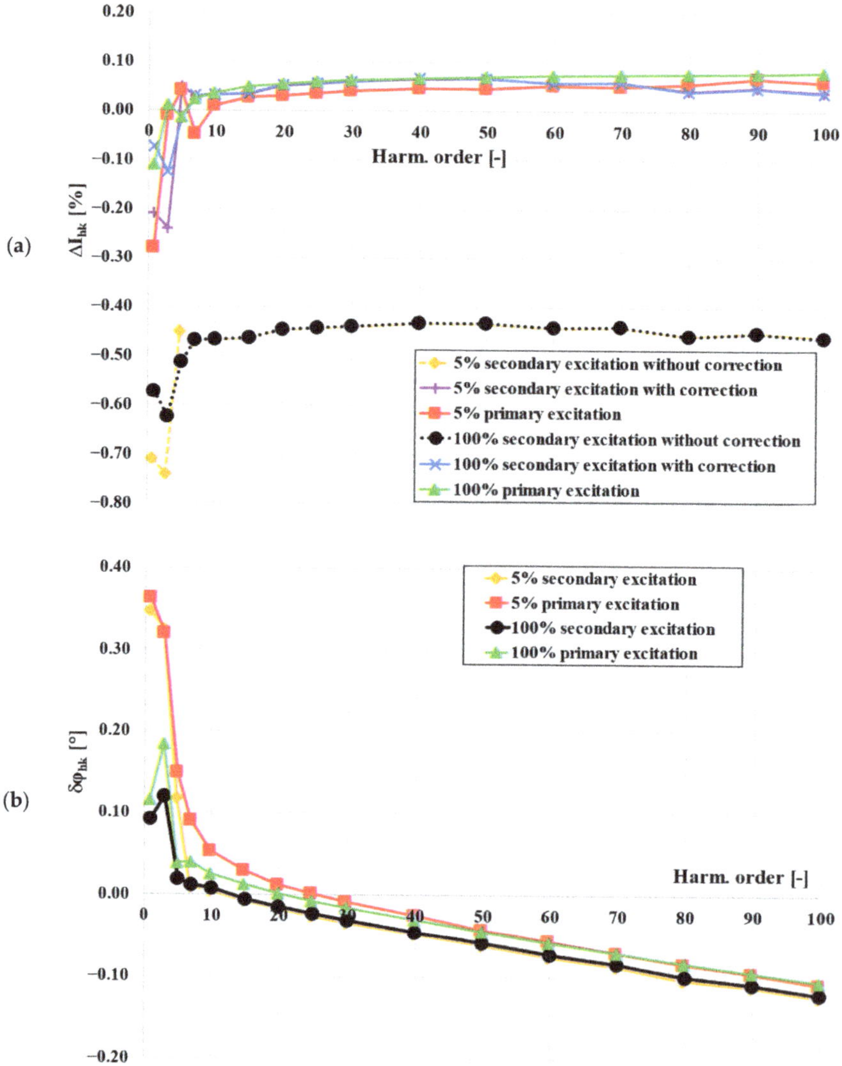

Figure 7. The comparison of obtained values of current error (**a**) and phase displacement (**b**) of the TCT with the rated current ratio equal to 100 A/5 A in proposed secondary and primary current excitation method are shown.

5. Conclusions

This paper presents a newly developed method to determine the values of the current error and phase displacement for the transformation of distorted current harmonics by the inductive current transformers. The proposed method eliminates the need for expensive, high-current supply systems in the measuring setup and provides an efficient way to determine the accuracy of inductive current transformers for the transformation of distorted current harmonics. In this method, the secondary winding is fed by the distorted voltage with RMS values of the harmonics calculated in order reproduce the operation point of the inductive current transformer on the magnetization characteristic of its magnetic core, as in the primary winding excitation conditions. The accuracy of the inductive current transformer depends on the turns ratio correction and the value of the magnetic flux density of the magnetic core resulting from the instantaneous value of the secondary voltage. All of the inductive current transformers suffer from the self-generation problem caused by the nonlinearity of the magnetization characteristic of the magnetic core. This is the main factor that determines their wide frequency transformation accuracy. The efficiency of the secondary current excitation approach was verified with the typically used primary current excitation method, where the secondary currents of the reference and tested current transformers are compared in the differential measuring setup. The convergence of the results from both methods have also confirmed the applicability of the inductive current transformer as the source of the distorted reference current. The values of the current error and phase displacement for transformation of the distorted current harmonics by the reference inductive current transformer were determined in the rated ampere-turns conditions.

Author Contributions: Conceptualization, M.K. and E.S.; methodology, M.K. and E.S.; validation, M.K. and E.S. formal analysis, M.K. and E.S.; investigation, M.K. and E.S.; resources, M.K. and E.S.; data curation, M.K. and E.S.; writing—original draft preparation, M.K. and E.S.; writing—review and editing, M.K. and E.S.; visualization, E.S.; supervision, M.K. All authors have read and agreed to the published version of the manuscript.

Funding: This research received no external funding.

Institutional Review Board Statement: Not applicable.

Informed Consent Statement: Not applicable.

Data Availability Statement: Not applicable.

Conflicts of Interest: The authors declare no conflict of interest.

References

1. Vieira, D.; Shayani, R.A.; De Oliveira, M.A.G. Reactive Power Billing under Nonsinusoidal Conditions for Low-Voltage Systems. *IEEE Trans. Instrum. Meas.* **2017**, *66*, 2004–2011. [CrossRef]
2. Dirik, H.; Duran, I.U.; Gezegin, C. A Computation and Metering Method for Harmonic Emissions of Individual Consumers. *IEEE Trans. Instrum. Meas.* **2019**, *68*, 412–420. [CrossRef]
3. Hajipour, E.; Vakilian, M.; Sanaye-Pasand, M. Current-Transformer Saturation Compensation for Transformer Differential Relays. *IEEE Trans. Power Deliv.* **2015**, *30*, 2293–2302. [CrossRef]
4. Hong, Y.Y.; Wei, D.W. Compensation of distorted secondary current caused by saturation and remanence in a current transformer. *IEEE Trans. Power Deliv.* **2010**, *25*, 47–54. [CrossRef]
5. Bauer, J.; Ripka, P.; Draxler, K.; Styblikova, R. Demagnetization of current transformers using PWM burden. *IEEE Trans. Magn.* **2015**, *51*, 1–4. [CrossRef]
6. Esmail, E.M.; Elkalashy, N.I.; Kawady, T.A.; Taalab, A.M.I.; Lehtonen, M. Detection of Partial Saturation and Waveform Compensation of Current Transformers. *IEEE Trans. Power Deliv.* **2015**, *30*, 1620–1622. [CrossRef]
7. Sanati, S.; Alinejad-Beromi, Y. Avoid current transformer saturation using adjustable switched resistor demagnetization method. *IEEE Trans. Power Deliv.* **2021**, *36*, 92–101. [CrossRef]
8. Mazin, H.E.; Xu, W.; Huang, B. Determining the harmonic impacts of multiple harmonic-producing loads. *IEEE Trans. Power Deliv.* **2011**, *26*, 1187–1195. [CrossRef]
9. Zobaa, A.F.; Abdel Aleem, S.H.E. A new approach for harmonic distortion minimization in power systems supplying nonlinear loads. *IEEE Trans. Ind. Inform.* **2014**, *10*, 1401–1412. [CrossRef]
10. Draxler, K.; Styblikova, R. Using instrument transformers in a wider frequency range. In Proceedings of the Conference Record—IEEE Instrumentation and Measurement Technology Conference, Hangzhou, China, 10–12 May 2011.
11. Li, Z.; Zhang, S.; Wu, Z.; Abu-Siada, A.; Tao, Y. Study of current measurement method based on circular magnetic field sensing array. *Sensors* **2018**, *18*, 1439. [CrossRef]
12. Siegenthaler, S.; Mester, C. A computer-controlled calibrator for instrument transformer test sets. In Proceedings of the IEEE Transactions on Instrumentation and Measurement; 2017; pp. 1184–1190.
13. Frigo, G.; Agustoni, M. Calibration of a Digital Current Transformer Measuring Bridge: Metrological Challenges and Uncertainty Contributions. *Metrology* **2021**, *1*, 93–106. [CrossRef]
14. Kaczmarek, M.; Stano, E. Nonlinearity of Magnetic Core in Evaluation of Current and Phase Errors of Transformation of Higher Harmonics of Distorted Current by Inductive Current Transformers. *IEEE Access* **2020**, *8*, 118885–118898. [CrossRef]

15. Mingotti, A.; Peretto, L.; Bartolomei, L.; Cavaliere, D.; Tinarelli, R. Are inductive current transformers performance really affected by actual distorted network conditions? An experimental case study. *Sensors* **2020**, *20*, 927. [CrossRef] [PubMed]
16. Mingotti, A.; Bartolomei, L.; Peretto, L.; Tinarelli, R. On the long-period accuracy behavior of inductive and low-power instrument transformers. *Sensors* **2020**, *20*, 5810. [CrossRef] [PubMed]
17. Brandolini, A.; Faifer, M.; Ottoboni, R. A simple method for the calibration of traditional and electronic measurement current and voltage transformers. *IEEE Trans. Instrum. Meas.* **2009**, *58*, 1345–1353. [CrossRef]
18. Kaczmarek, M.; Szczęsny, A.; Stano, E. Operation of the Electronic Current Transformer for Transformation of Distorted Current Higher Harmonics. *Energies* **2022**, *15*, 4368. [CrossRef]
19. Kaczmarek, M.; Stano, E. Proposal for extension of routine tests of the inductive current transformers to evaluation of transformation accuracy of higher harmonics. *Int. J. Electr. Power Energy Syst.* **2019**, *113*, 842–849. [CrossRef]
20. Stano, E.; Kaczmarek, M. Wideband self-calibration method of inductive cts and verification of determined values of current and phase errors at harmonics for transformation of distorted current. *Sensors* **2020**, *20*, 2167. [CrossRef]
21. Kaczmarek, M.; Stano, E. The Influence of the 3rd Harmonic of the Distorted Primary Current on the Self-Generation of the Inductive Current Transformers. *IEEE Access* **2022**, *10*, 55876–55887. [CrossRef]
22. Stano, E.; Kaczmarek, P.; Kaczmarek, M. Why Should We Test the Wideband Transformation Accuracy of Inductive Current Transformers? *Energies* **2022**, *15*, 5737. [CrossRef]
23. Stano, E.; Kaczmarek, M. Analytical method to determine the values of current error and phase displacement of inductive current transformers during transformation of distorted currents higher harmonics. *Measurement* **2022**, *200*, 111664. [CrossRef]
24. Stano, E.; Kaczmarek, P.; Kaczmarek, M. Understanding the Frequency Characteristics of Current Error and Phase Displacement of the Corrected Inductive Current Transformer. *Energies* **2022**, *15*, 5436. [CrossRef]
25. Stano, E. The Method to Determine the Turns Ratio Correction of the Inductive Current Transformer. *Energies* **2021**, *14*, 8602. [CrossRef]
26. *IEC 61869-2*; Instrument Transformers—Additional Requirements for Current Transformers. IEC: Geneva, Switzerland, 2012.
27. *IEEE C57.13-2016*; IEEE Standard Requirements for Instrument Transformers. IEEE: New York City, USA, 2016.
28. *IEC 61869-6*; Instrument Transformers—Additional General Requirements for Low-Power Instrument Transformers. IEC: Geneva, Switzerland, 2016.

energies

MDPI

Review

Challenges of Accurate Measurement of Distorted Current and Voltage in the Power Grid by Conventional Instrument Transformers

Michal Kaczmarek * and Ernest Stano

Institute of Mechatronics and Information Systems, Lodz University of Technology, 90-537 Lodz, Poland; ernest.stano@p.lodz.pl
* Correspondence: michal.kaczmarek@p.lodz.pl

Abstract: Power grids are a combined system where the electrical energy produced by the power plants is transmitted to consumers. This forms a specific interdependence where the recipients have a significant impact on the power quality. Therefore, the nonlinear loads connected by households and industrial customers cause current and voltage distortion in the power networks. This creates the need for accurate measurement of nonsinusoidal voltage and current composed not only from the fundamental component but also containing higher harmonics, interharmonics, and subharmonics. In order to ensure high transformation accuracy of distorted current and voltage, the inductive instrument transformers have to be tested in these conditions. Many papers describe their behavior during the transformation of sinusoidal current or voltage. Nowadays, the scientific field in this scope is focused on the evaluation of their exploitation properties for distorted signals. The common problem of inductive instrument transformers is the self-generation of low-order higher harmonics to the secondary current or voltage. In the case of the inductive VTs, an additional problem results from the resonance caused by the parasitic capacitance of the primary winding. The proposed solutions to compensate for the values of current or voltage errors and phase displacement of inductive instrument transformers are also analyzed.

Keywords: transformation accuracy; current transformer; voltage transformer; distorted current; distorted voltage; higher harmonics; current error; voltage error; phase displacement; self-generation; instrument transformer; resonance

Citation: Kaczmarek, M.; Stano, E. Challenges of Accurate Measurement of Distorted Current and Voltage in the Power Grid by Conventional Instrument Transformers. *Energies* **2023**, *16*, 2648. https://doi.org/10.3390/en16062648

Academic Editor: Abu-Siada Ahmed

Received: 21 February 2023
Revised: 8 March 2023
Accepted: 8 March 2023
Published: 11 March 2023

1. Introduction

Power grids are a combined system where the electrical energy produced by the power plants is transmitted to consumers. This forms a specific kind of system with interconnected vessels, where the recipients have a significant impact on the power quality. Therefore, an increasing number of nonlinear loads connected by households and industrial customers cause current and voltage distortion in power networks. This creates the need for the transformation of nonsinusoidal voltage and current composed not only from the fundamental component but also containing higher harmonics, interharmonics, and subharmonics [1–5]. Many papers describe their behavior during the transformation of only sinusoidal current or voltage [6–9]. A new concept is to evaluate the long-period accuracy and influence of the ambient temperature [10,11]. Nowadays, the scientific field in this scope is focused on determining exploitation properties for distorted signals [3,12–15]. The transformation accuracy of the inductive CT and VT is strongly dependent on the value and the power factor of the load of the secondary winding. Another aspect that has a significant influence on the wideband accuracy of inductive CTs is the self-generation of low-order higher harmonics to its secondary current [4,16–18]. This phenomenon is caused by the nonlinearity of the magnetic core's magnetization characteristic. It causes additional distortion in the inductive CT's secondary current, even for transformations of the sinusoidal current. Considering inductive VTs, the main factor influencing their wideband accuracy is resonance [3,19–22]. However, self-generation may also be present during the operation of inductive VTs [1,3,18,19]. Previously, over 15 years ago, the wide frequency operation of inductive ITs was limited by the insufficient magnetic permeability of higher frequency components of the current or voltage [14,23–26]. This was caused by the poor quality SiFe magnetic material. In modern constructions of inductive CTs, cold-rolled SiFe or permalloy tape NiFe is used, as well as composed magnetic cores from electrotechnical steel SiFe and nanocrystalline tape FeCuNbSiB [25,27–29]. To improve the wide frequency transformation characteristic of inductive ITs, compensation techniques are used [30–34]. This requires active components for the control of the secondary current or voltage and generates low-order higher harmonics in counterphase to the self-produced ones as a result of the presence of the magnetic core. The required compensation RMS values and phase angles of the self-generated higher harmonics by the inductive CT/VT are measured for the transformation of the sinusoidal current/voltage with various loads of the secondary winding. This approach has one significant disadvantage, the matrix of the measured results is large and requires a high number of test points. This problem increases if the influence of the low-order higher harmonics needs to be evaluated [17].

2. Nonlinear Behavior of Inductive CTs

The metrological properties of inductive CTs remain unaffected by the leakage inductance and resistance of the primary winding during steady-state operation. This is because the primary winding can be considered a current source. The dissipative field of the primary winding only plays a role during transient conditions or in transformers with a large air gap in the core. By ignoring the primary winding's inductance and resistance, the equivalent circuit comprises solely the elements of the secondary circuit and the magnetizing branch. Figure 1 illustrates this circuit while also considering a particular hk harmonic of the distorted primary current.

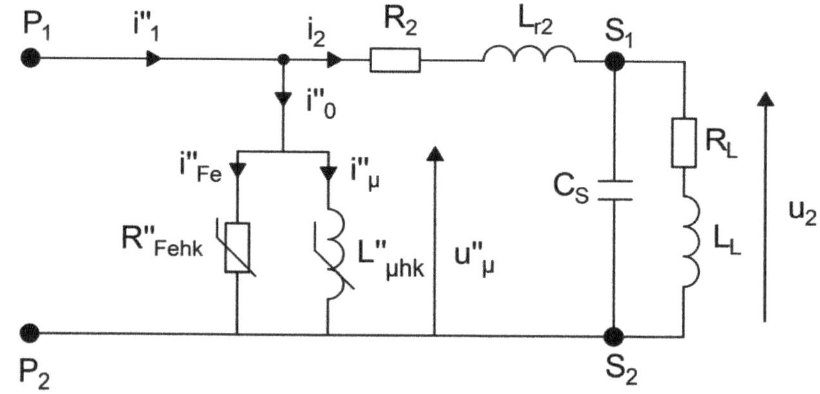

Figure 1. Equivalent circuit of inductive CT for transformation of distorted current. Where: i''_1—instantaneous value of distorted primary current; i_2—instantaneous value of distorted secondary current; i''_0—instantaneous value of distorted excitation current; i''_{Fe}—instantaneous value of distorted current associated with active power losses in the magnetic core; i''_μ—instantaneous value of distorted magnetization current; u''_μ—instantaneous value of distorted voltage on the mutual inductance of the windings; u_2—instantaneous value of distorted secondary voltage; R''_{Fehk}—resistance representing the active power losses in the magnetic core; $L''_{\mu hk}$—mutual inductance of the windings; R_2—resistance of the secondary winding; L_{r2}—leakage inductance of the secondary winding; R_L—resistance of the load connected to the secondary winding; L_L—inductance of the load connected to the secondary winding; C_S—capacitance between terminals to the secondary winding; P1/P2—primary winding terminals; S1/S2—secondary winding terminals.

Transformation errors in inductive CTs are predominantly caused by the excitation current of the magnetic core. This is because the non-linear magnetization characteristics of the core result in transformation accuracy being influenced by the magnetic flux density, as well as the RMS values of primary current harmonics and the load of the secondary winding. Inductive CTs without error compensation, in particular, secondary winding turns correction, are characterized by negative current error values. In order to comply with the requirements of the specific accuracy class, a common practice is to reduce the number of turns of the secondary winding. The turns ratio of the inductive CT is then not equal to the rated current ratio. As a result, the RMS value of the secondary current is increased. Moreover, the magnetization characteristic of the magnetic core obtained in the non-load state of the inductive CT is different from the equivalent magnetization characteristic of the inductive CT determined for various loads of the secondary winding. An example of three magnetization curves is presented in Figure 2.

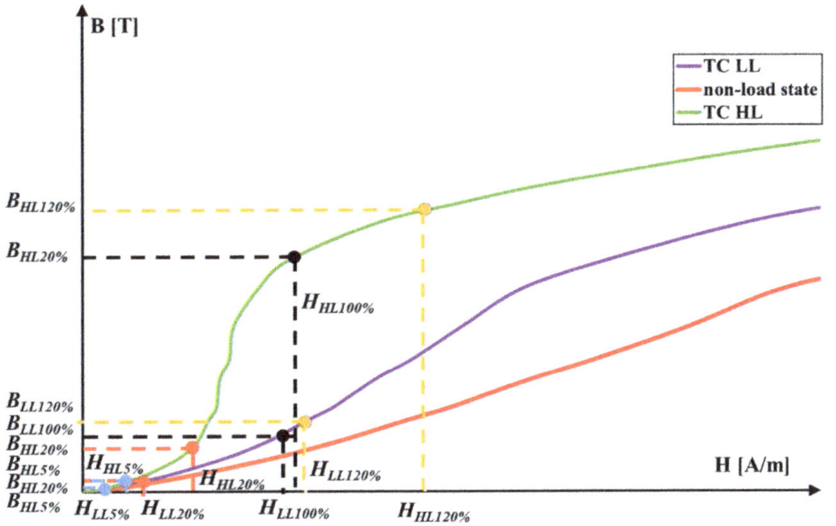

Figure 2. Operation points of the corrected inductive CT on the equivalent magnetization characteristic of the magnetic core [35]. Where: TC LL—equivalent magnetization characteristic of CT with applied turn correction determined for the low value of the load of the secondary winding; TC HL—equivalent magnetization characteristic of CT with applied turn correction determined for the high value of the load of the secondary winding; Non-load state—magnetization characteristic of the magnetic core determined in the non-load state of CT; $B_{HL120\%}$—the value of the magnetic flux density of the inductive CT obtained for 120% of the rated primary current and rated load of the secondary winding; $B_{HL100\%}$—the value of the magnetic flux density of the inductive CT obtained for 100% of the rated primary current and rated load of the secondary winding; $B_{HL20\%}$—the value of the magnetic flux density of the inductive CT obtained for 20% of the rated primary current and rated load of the secondary winding; $B_{HL5\%}$—the value of the magnetic flux density of the inductive CT obtained for 5% of the rated primary current and rated load of the secondary winding; $B_{LL120\%}$—the value of the magnetic flux density of the inductive CT obtained for 120% of the rated primary current and 25% of the rated load of the secondary winding; $B_{LL100\%}$—the value of the magnetic flux density of the inductive CT obtained for 100% of the rated primary current and 25% of the rated load of the secondary winding; $B_{LL20\%}$—the value of the magnetic flux density of the inductive CT obtained for 20% of the rated primary current and 25% of the rated load of the secondary winding; $B_{LL5\%}$—the value of the magnetic flux density of the inductive CT obtained for 5% of the rated primary current

and 25% of the rated load of the secondary winding; $H_{HL120\%}$—the value of the magnetic field strength of the inductive CT obtained for 120% of the rated primary current and rated load of the secondary winding; $H_{HL100\%}$—the value of the magnetic field strength of the inductive CT obtained for 100% of the rated primary current and rated load of the secondary winding; $H_{HL20\%}$—the value of the magnetic field strength of the inductive CT obtained for 20% of the rated primary current and rated load of the secondary winding; $H_{HL5\%}$—the value of the magnetic field strength of the inductive CT obtained for 5% of the rated primary current and rated load of the secondary winding; $H_{LL120\%}$—the value of the magnetic field strength of the inductive CT obtained for 120% of the rated primary current and 25% of the rated load of the secondary winding; $H_{LL100\%}$—the value of the magnetic field strength of the inductive CT obtained for 100% of the rated primary current and 25% of the rated load of the secondary winding; $H_{LL20\%}$—the value of the magnetic field strength of the inductive CT obtained for 20% of the rated primary current and 25% of the rated load of the secondary winding; $H_{LL5\%}$—the value of the magnetic field strength of the inductive CT obtained for 5% of the rated primary current and 25% of the rated load of the secondary winding.

Figure 2 shows the operation points of a corrected inductive CT on the equivalent magnetization characteristic of the magnetic core for four rated primary current values. As the load on the secondary winding rises, the inductive CT's equivalent magnetization characteristic approaches the shape of the magnetic core's magnetization curve obtained in the non-load state. The operating point for a primary current of 120% of the rated value and rated secondary winding load ($B_{HL120\%}$, $H_{HL120\%}$) is positioned within the upper knee region of the equivalent magnetization characteristic. As a result, the RMS values of self-generated low-order higher harmonics increase. The same trend is noticed for low percentage values of the rated primary current, where the operating point is located near the lower knee of the equivalent magnetization characteristic. This occurrence alters the current error and phase displacement values measured for the low-order higher harmonic based on its phase angle relative to the transformed distorted current's main component. Figure 3 shows the range of current error and phase displacement values for the low-order higher harmonics from the 2nd to the 15th, computed for 100% of the rated current and rated load of the secondary winding.

(a)

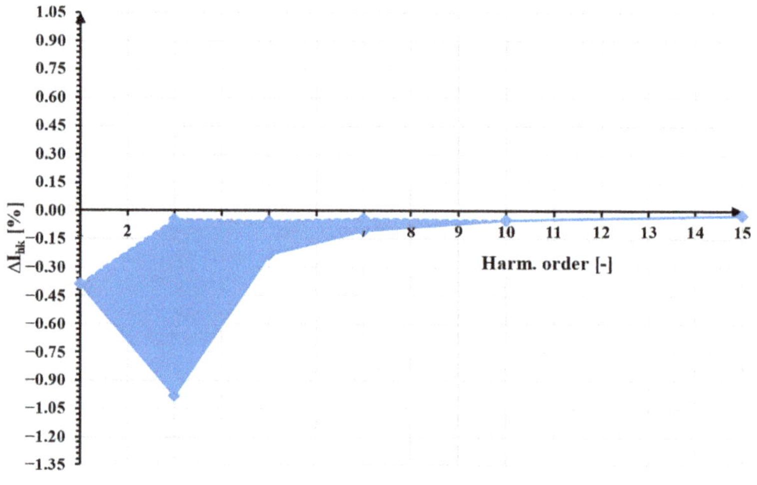

Figure 3. *Cont.*

(b)

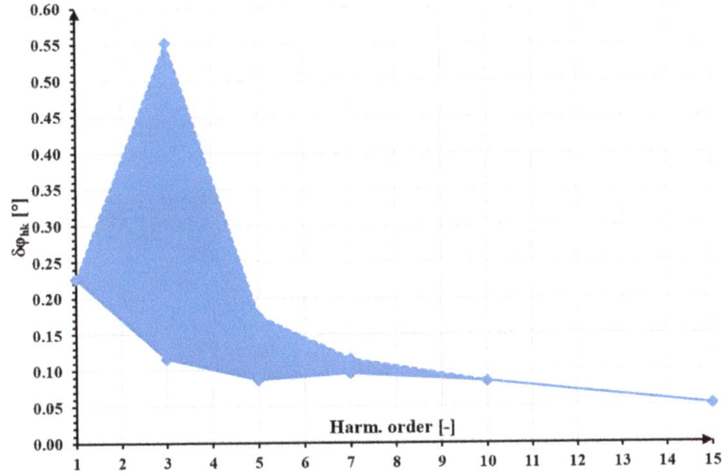

Figure 3. The area of the possible values of (**a**) current error and (**b**) phase displacement of the low-order higher harmonics from the 2nd to 15th determined for 100% of the rated primary current and the rated load of the secondary winding.

Each of the low-order higher harmonics self-generated to the secondary current of the inductive CT are added with the same frequency higher harmonic present in the transformed distorted primary current. Therefore, its RMS value in the secondary current depends on their mutual phase angle, which depends on the phase angle of the transformed higher harmonic in relation to the main component of the distorted primary current. In order to determine all possible values of the current errors and phase displacement for a given higher harmonic, it is necessary to smoothly adjust this phase angle. Obtained values will be in between the most positive (+) and the most negative (−) values of the current error and phase displacement determined for the phase angle when the self-generated and the transformed harmonics are in phase or antiphase, respectively.

In Figure 4, the concept of the most positive (+) and the most negative (−) values of the current error and phase displacement is further explained. In the presented case, the results of the evaluation of the transformation accuracy of the distorted primary current containing a main component of frequency at 50 Hz and one higher harmonic of frequency from 100 Hz to 750 Hz by the inductive CT with a current ratio equal to 300 A/5 A are shown. The conditions of its operation are essential for accuracy. Therefore, it is required to define that, to the secondary winding, its rated resistance is connected. The RMS value of the higher harmonics is equal to 10% of the main component.

The presented results show that for each RMS value of the distorted primary current different most positive (+) and most negative (−) values of current error and phase displacement are determined. In the tested frequency range of higher harmonics of order ranging from the 2nd to 15th, the self-generation of the low-order higher harmonics is the most significant factor that determined the wideband accuracy of the inductive CT. In this case, depending on the phase angle of the transformed higher harmonic in relation to the main component of the value of current error for the third-order harmonic, it changes from −1.2% to 0.9% for 120% of the rated RMS value of the primary current. In these conditions, the value of phase displacement changes from 0.05° to 0.73°. This is a typical situation for inductive CTs.

(a)

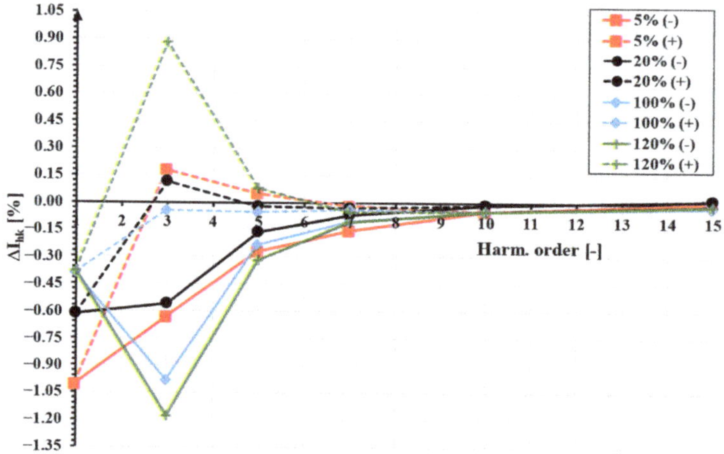

(b)

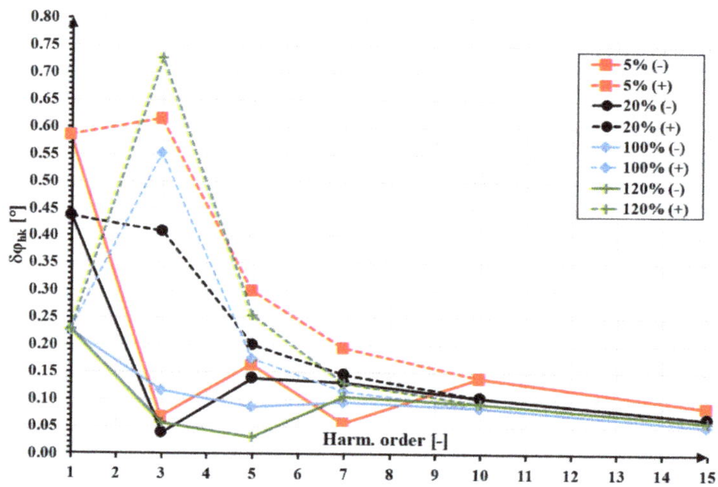

Figure 4. The most positive (+) and the most negative (−) values of (**a**) current error and (**b**) phase displacement determined for inductive CT 300 A/5 A with the rated resistive load of the secondary winding [4].

To show how the value of the secondary winding load resistance influences the results in Figure 5, the most positive (+) and most negative (−) values of (a) current error and (b) phase displacement determined for the same inductive CT with the 25% of the rated resistive load of the secondary winding are presented.

(a)

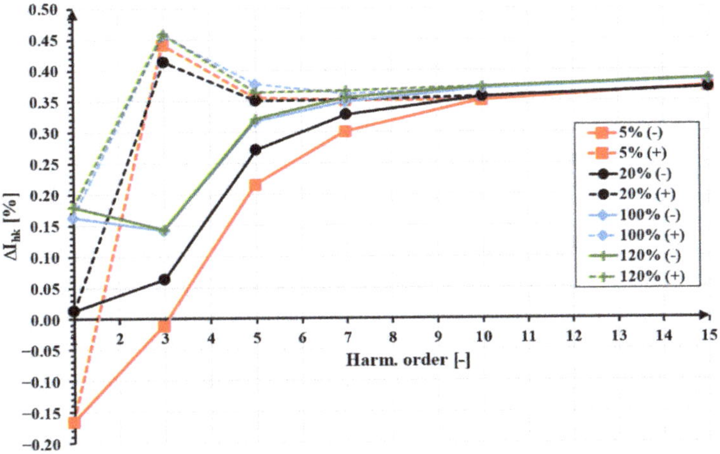

(b)

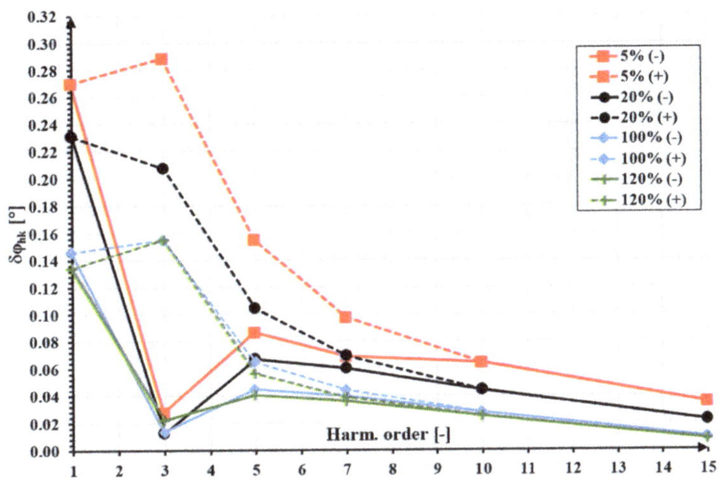

Figure 5. The most positive (+) and most negative (−) values of (**a**) current error and (**b**) phase displacement determined for inductive CT 300 A/5 A with 25% of the rated resistive load of the secondary winding [4].

A decrease in the load resistance of the secondary winding of the inductive CT typically causes a decrease in the RMS values of the self-generated low-order higher harmonics to the secondary current. Therefore, the change between the most positive (+) and the most negative (−) values of current error and phase displacement is also significantly smaller.

To define the worst possible transformation accuracy of the distorted current harmonics by the inductive CT, the maximum absolute values of the current error and phase displacement should be determined, as presented in Figure 6. These are the highest absolute values chosen from the most positive (+) and the most negative (−) values of current error and phase displacement determined for a given RMS value of the primary current and load of the secondary winding.

(a)

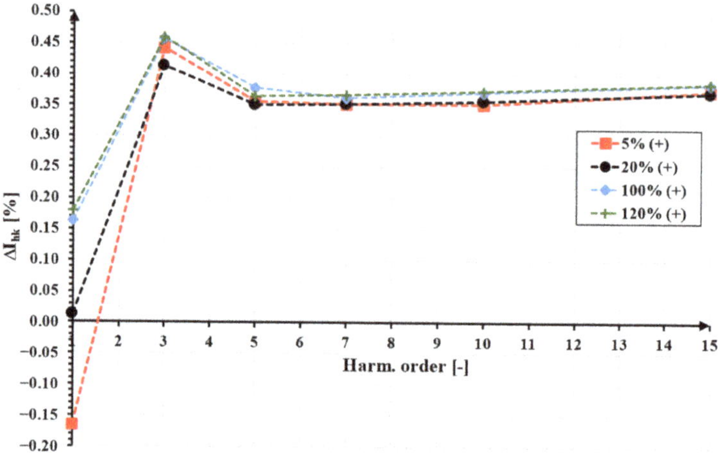

(b)

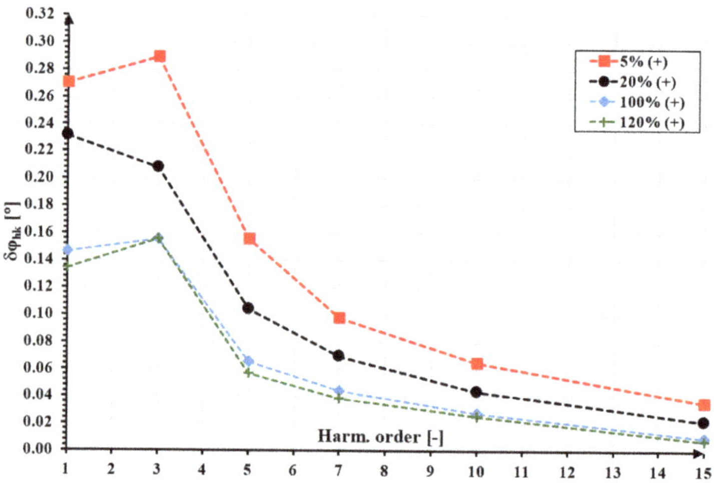

Figure 6. The highest absolute values of (**a**) current error and (**b**) phase displacement determined for inductive CT 300 A/5 A with 25% of the rated resistive load of the secondary winding [4].

The defined maximum absolute values of the current error and phase displacement presented in Figure 6 are the highest absolute values chosen from the most positive (+) and the most negative (−) values of the current error and phase displacement presented in Figure 5. These results are analyzed for 5%, 20%, 100%, and 120% of the rated RMS value of the primary current and 25% of the rated resistive load of the secondary winding. The procedure should be repeated for the rated load of tested inductive CT 300 A/5 A. During the evaluation of the distorted current transformation accuracy, these maximum absolute values of the current error and phase displacement should be considered as ± values because some positive and negative values from 0 up to this limit may be obtained.

Another aspect that may affect the high transformation accuracy of the distorted currents by the inductive CTs is the power factor of the load connected to its secondary winding. The frequency characteristics of the current error (a) and phase displacement (b) for inductive CT with a rated current ratio of 300 A/5 A are shown in Figure 7. The results are presented for two load power factors equal to 0.8 ind. (dashed lines) and 1 (solid lines), The analysis involves four different primary current values, which correspond to 5%, 20%,

100%, and 120% of the rated value. The results of the tests and analyses are focused on the transformation of distorted current containing a main component of 50 Hz frequency and a single higher harmonic ranging from 100 Hz to 5 kHz.

(a)

(b)

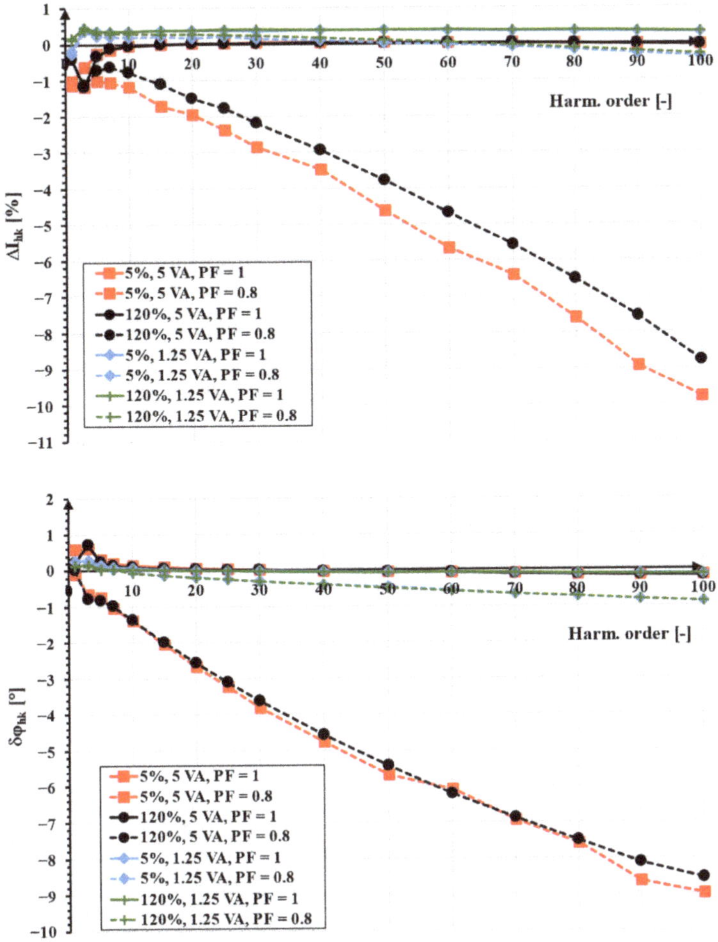

Figure 7. Comparison of the highest absolute values of (**a**) current error and (**b**) phase displacement determined for inductive CT 300 A/5 A with two values and power factors of the load [12].

The analysis of the results from Figure 7 shows that the values of current error and phase displacement determined for higher harmonics increase much more rapidly with frequency when the load power factor is 0.8 inductive. In contrast, when the load has the same impedance but a power factor of 1, the increase in these values is not as significant with increasing frequency. The shift in the inductive CT's operating point on the magnetization characteristic of the magnetic core towards saturation is the main cause of this phenomenon, especially for resistive–inductive load of the secondary winding. However, for the resistive load, the increase in the transformed higher harmonic's leakage reactance of the secondary winding is the only cause of this behavior, as it increases with frequency. If the load power factor is 0.8 inductive, the reactance of the load also increases with the frequency of the transformed higher harmonic. The self-generation of low-order harmonics in the secondary current of the inductive CT can have a major impact on the accuracy of harmonic transformation up to the 13th order. For higher-order harmonics up to 100th, the accuracy depends on the mutual inductance between the CT windings $L''_{\mu hk}$ and the resistance representing the active power losses R''_{Fehk} in the magnetic core. Decreasing these values with increasing harmonic frequency leads to an increase in the RMS values of the reactive $I''_{\mu hk}$ and active I''_{Fehk} components of the core excitation current, resulting in higher current errors and phase displacement.

The presented analysis indicates that using a resistive load for the secondary winding of the inductive CT results in significantly lower values of current error and phase displacement for the transformation of the higher-order harmonics in the investigated frequency range. The resistive load ensures a wider frequency range of operation for the inductive CT with the accuracy class defined for the transformation of sinusoidal currents of 50 Hz (60 Hz) frequency.

Figure 8 presents a comparison of the percentage values of the self-generated low-order higher harmonics for an inductive CT 300 A/5 A with two different power factors and two secondary winding load values.

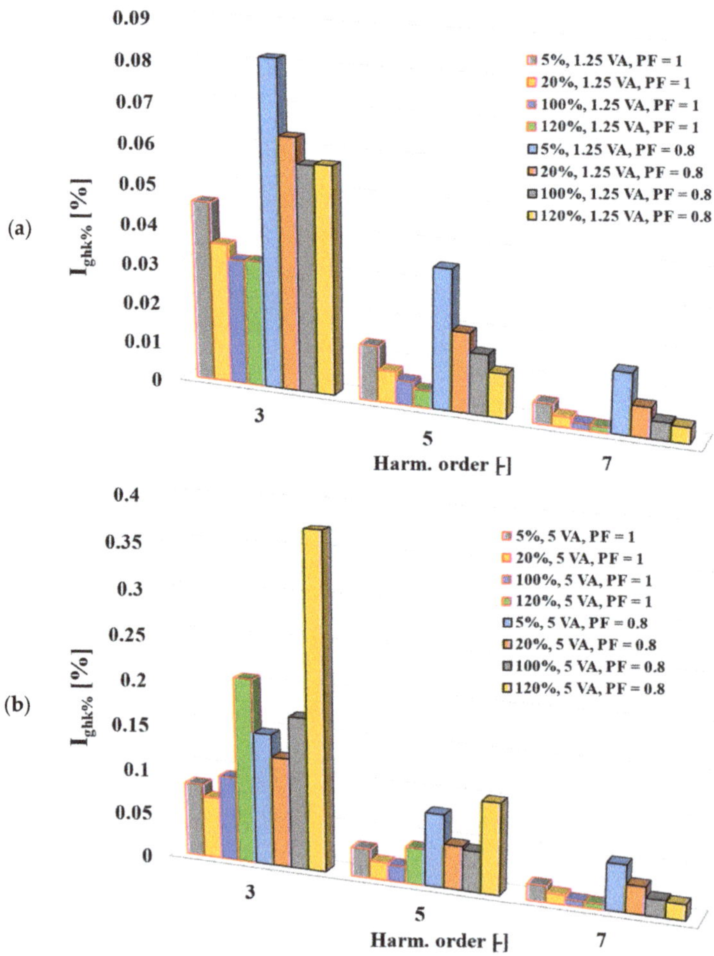

Figure 8. Comparison of the percentage values of the self-generated low-order higher harmonics determined for inductive CT 300 A/5 A with two power factors and (**a**) 25% of the rated load and (**b**) the rated load [12].

The investigation of the percentage values of the self-generated low-order higher harmonics by the inductive CT is important as it corresponds to its non-linear behavior. Therefore, it may not be appropriate for wideband operation due to the low transformation accuracy of the third, fifth, and seventh higher harmonics. This possibility is detected up to the order 13th with the main frequency of the distorted primary current equal to 50 Hz. Typically, the self-generated higher harmonics are significantly lower for the case with the load of the secondary winding decreased to 25% of the rated value. Moreover, the resistive–inductive loads of the secondary winding of the inductive CT cause an increase in the percentage values of the self-generated low-order higher harmonics.

3. Inductive VT's Resonance for Higher Harmonics of the Distorted Voltage

The main factor that determines the wideband accuracy of the inductive VTs is resonance [3,19–22]. Of course, an increase in the leakage reactances with the frequency of the transformed higher harmonics causes an additional increase in the voltage drop on its windings. Therefore, the values of voltage error and phase displacement for the transformation of the distorted primary voltage higher harmonics also increase up to a few or tens of percent values or degrees, respectively. However, the resonance may occur for the low-order higher harmonic of distorted primary voltage, e.g., the fifth, causing the values of voltage error and phase displacement to increase to, e.g., 500% or ±100°. The frequency mainly depends on the number of turns of the primary winding affecting the size of the primary winding cross section and resulting from the RMS value of the rated primary voltage of the inductive VT [3]. It is very important to notice that inductive VT may have more than one resonance frequency. Therefore, its equivalent circuit for transformation of distorted voltage must reflect the partial resistance, leakage inductance, and capacitance of each layer of the inductive VT's primary winding (Figure 9).

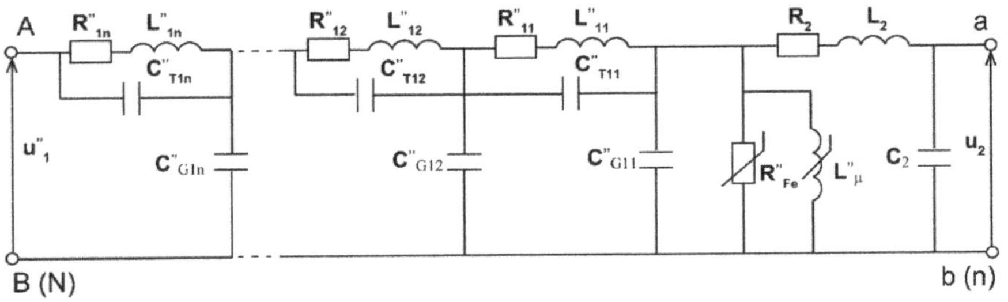

Figure 9. The equivalent circuit of the inductive VT for transformation of distorted voltage [3]. The following abbreviations are used: u''_1—instantaneous value of distorted primary voltage; u_2—instantaneous value of distorted secondary voltage; $R''_{1n}/R''_{12}/R''_{11}$—partial resistance of the primary winding; $L''_{1n}/L''_{12}/L''_{11}$—partial leakage inductance of the primary winding; $C''_{T1n}/C''_{T12}/C''_{T11}$—partial equivalent capacitance between the primary and secondary windings; $C''_{G1n}/C''_{G12}/C''_{G11}$—partial equivalent capacitance of the primary winding to the ground; A/B(N)—terminals of the primary winding; a/b(n)—terminals of the secondary winding; R''_{Fe}—resistance representing the active power losses in the magnetic core; L''_{μ}—mutual inductance of the windings; R_2—resistance of the secondary winding; L_2—leakage inductance of the secondary winding.

According to the analysis presented in papers [3,36], the parasitic capacitance and leakage inductance of the primary winding of the inductive VT are divided into parts to properly reproduce its multi-resonance properties. The conductive path to ground of each layer of the primary winding is formed by the partial equivalent capacitance of the primary winding to ground, which results from the parasitic capacitance between each layer of the winding. The equivalent capacitance between the primary and secondary windings is caused by the parasitic capacitances between turns of the windings connected in series, and its value depends on the thickness of the insulation layer between them. The equivalent circuit indicates that the resistive–inductive load of the secondary winding of the inductive

VT will have a positive influence on its limiting frequency of operation. This is because it will cause a decrease in the secondary current higher harmonic component, limiting the increase in the voltage drop on the reactance of the windings with the frequency of the transformed higher harmonic.

In Figure 10, the comparison of the highest absolute values of the voltage error and phase displacement determined for five medium-voltage (MV) inductive VTs is presented.

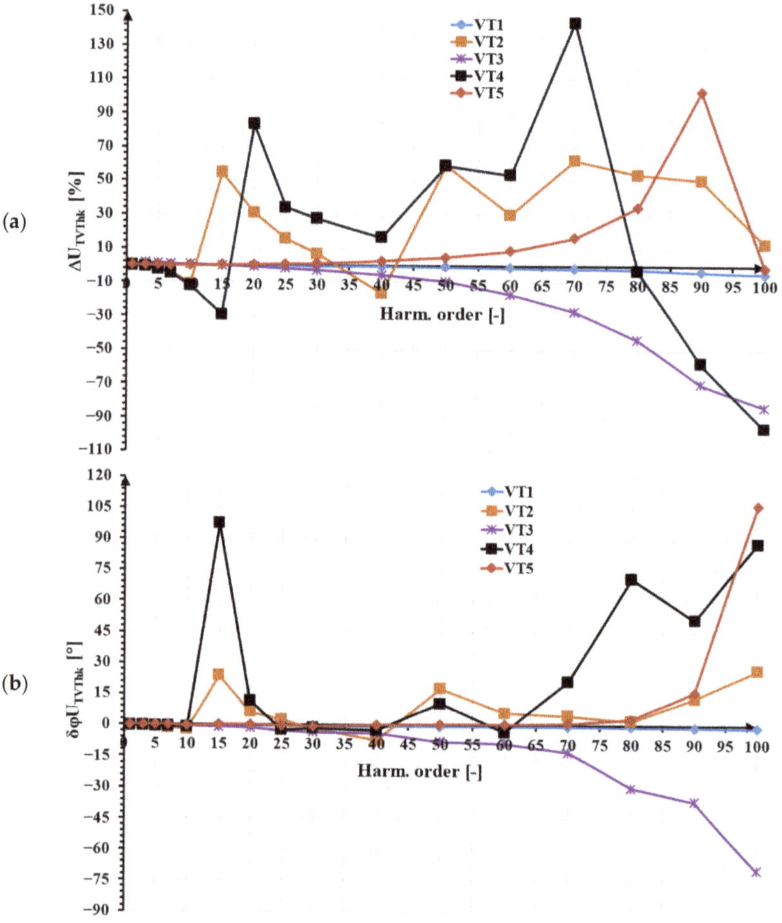

Figure 10. Comparison of the highest absolute values of the (**a**) voltage error and (**b**) phase displacement determined for inductive VTs [3].

In Figure 10, the results labeled VT1 and VT2 concern the transformation accuracy tests of distorted voltage transformation performed on the indictive VTs with rated primary voltage equal to 15 kV/$\sqrt{3}$, while VT3, VT4, and VT5 refer to results obtained for 20 kV/$\sqrt{3}$ units. In the presented case, the distorted primary voltage contains a main component of frequency at 50 Hz and one higher harmonic frequency from 100 Hz to 5 kHz. The conditions of inductive VT operation are essential for the determined accuracy. Therefore, it is required to define that, to the secondary winding, no additional load is connected, except the measuring apparatus with above 1 MΩ impedance. The RMS value of the higher harmonics is equivalent to 10% of the main component. The presented results indicate different levels of accuracy in the transformation of distorted voltage higher harmonics among the MV inductive VTs. For three units of inductive VTs with a rated primary voltage of either 15 kV/$\sqrt{3}$ or 20 kV/$\sqrt{3}$, resonance is the primary factor that affects their wideband accuracy. In the other cases, the increase in the leakage reactances with the frequency of the transformed higher harmonics causes an additional increase in the voltage drop on their windings. Therefore, the values of voltage error and phase displacement for the transformation of the distorted primary voltage higher harmonics also increase up to a few or tens of percent or degrees, respectively.

4. Summary of the Results of the Inductive IT's Transformation Accuracy Evaluation for Distorted Signals

The increase in the frequency of the transformed higher harmonic causes a decrease in the magnetic permeability of a used magnetic core, leading to an increase in current/voltage error and phase displacement [34,37–39]. However, wide frequency operation of inductive voltage transformers (VTs) or current transformers (CTs) up to several or dozens of kHz can be ensured by using silicone steel (VT) or permalloy as well as composed magnetic cores from electrotechnical steel and nanocrystalline material (CT) [25,27–29].

The transformation accuracy of higher harmonics by the inductive VTs is limited by the resonance phenomenon. Its frequency mainly depends on the size (number of turns and layers) of the primary winding resulting from the required RMS value of their rated primary voltage [3,36]. Thus, while 15 kV inductive VTs may ensure wide frequency operation up to 5 kHz, an increase in the RMS value of the primary voltage may result in a decrease in the resonance frequency to only 250 Hz for HV units such as 420 [3,19,20,37,40]. In such cases, the voltage error and phase displacement may increase to large values (e.g., 500%, ±100°). Without resonance, the values of voltage error and phase displacement may increase with frequency up to a few percent and degrees, respectively [3,41]. In inductive VTs, the self-generation phenomenon is less significant than in inductive CTs because the measuring units typically operate in the linear region of the magnetization characteristic of the magnetic core. The values of the voltage error and phase displacement of the transformation of the low-order higher harmonic may increase by about ±0.2% and ±0.2°, respectively [3,18].

In the case of inductive CTs, the transformation accuracy for distorted current may be significantly deteriorated by the self-generation phenomenon of the low-order higher harmonics. The values of current error and phase displacement may reach about ±1.5% and ±1.5°, respectively [4,17,18,32]. However, the high transformation accuracy of distorted currents was confirmed in the case of the inductive CT class 0.2S, where the values of current error and phase displacement in the frequency range of harmonics from 50 Hz up to 5 kHz did not exceed the limits defined for the transformation of sinusoidal current of frequency 50 Hz [4]. Additionally, it is possible to make inductive CTs with values of current error and phase displacement that do not exceed ±0.1% and ±0.1°, respectively, in the frequency range of harmonics from 50 Hz up to 5 kHz [42]. When an inductive CT is operating with a 0.8 inductive power factor of its secondary winding's load, the RMS values of the self-generated low-order higher harmonics may significantly increase. Additionally, as the frequency of transformed higher harmonics increases, the secondary voltage of the inductive CT will also increase. This will cause an increase in the values of current error and phase displacement as the magnetic flux density in the magnetic core increases [12,43]. Therefore, the resistive load ensures a wider frequency range of operation.

5. Error Correction Methods

Many of the papers present solutions to correct the values of current or voltage errors and phase displacements of inductive CTs and VTs during their operation in distorted conditions [33,34,44–46]. The majority of them concentrate on inductive voltage transformers (VTs), where the impact of the magnetic core's non-linear characteristics is less significant, resulting in a relatively low level of self-generated low-order higher harmonics in the secondary voltage. The presented method in paper [33] is based on the linear approximation of the inductive VT's secondary voltage by the defined frequency response function.

In the first step, such function is estimated for the power amplifier and step-up voltage transformer that supply the primary winding of tested inductive VT by injecting a random phase multisine signal [47]. In the next step, the quasi-sinusoidal multisine signals with 50 Hz main frequency are sampled from the defined probability density functions. These signals are used to supply the tested VT while the primary and secondary steady-state voltages are measured. The discrete Fourier transform is used to compute the input and output frequency spectrum. The defined numerical models are used to compensate for the response of the inductive VT. Another approach presented in paper [37] is the application of the digital filter to enable compensation of the frequency response of inductive VTs. This requires an additional device in which the frequency characteristic of voltage error and phase displacement of tested VT are implemented. Therefore, the secondary voltage is modified by the voltage generated by this device with counter-phase harmonics. Their RMS values and phases are measured for various RMS values of transformed sinusoidal voltage and chosen loads of the secondary winding. The paper [45] presents a frequency-domain nonlinear model of the inductive VT based on a simplified Volterra model. The number of coefficients in comparison with the standard model is reduced. This ensures faster identification procedures and computing. The proposed approach evaluates bandwidth limitations and nonlinearity of the magnetic characteristic of the magnetic core. Thus, it can be applied to determine the behavior of any kind of VT. Another approach shown in [34] is based on the implementation of the mapping function between the secondary and primary sides of the tested inductive VT and compensation for its accuracy even in nonlinear operating conditions. In case of the inductive CTs, the compensation methods are more difficult to implement due to the strong influence of the nonlinear behavior of the magnetic core. In paper [46], the compensation method is proposed that is based on the polynomial modeling of the harmonic distortion. Utilization of the least-squares approach results in easier implementation and low computational complexity of this method. However, for the best performance, each inductive CT has to be individually characterized in specified conditions. Therefore, the RMS values of the individual higher harmonics in the secondary current must be determined for all possible conditions, including the change in the RMS values of the primary current, its harmonic content, as well as the value and power factor of the load. The disadvantage of this method is that it requires aggregation of 2500 current waveforms for the compensation lookup table. In paper [32], this compensation method is called SINDICOMP. In the first stage, the tested inductive CT is characterized for the transformation of sinusoidal currents of the rated frequency where the harmonic content of the distorted secondary current is measured. In the second stage, determined phasors of higher harmonics are used to compensate for the frequency response. A similar approach is presented in paper [31], where the compensation factors of the inductive CT are determined using a frequency domain model based on tensor linearization. To obtain a real-value compensation matrix, the tested inductive CT is characterized for the transformation of sinusoidal currents of the rated frequency where the higher harmonics phasors are determined. A similar solution is presented in [48]. However, it uses a developed algorithm with determined coefficients under various operating conditions of the tested inductive CT. The input values are the RMS values and phase angles of individual harmonics recorded for various RMS values of the primary current of the rated frequency. In papers [49,50], the compensation technique of the DC component is studied. The presented solution is based on the measurement of the frequency spectrum of the inductive CT's secondary current, and then required to eliminate the 2nd harmonic DC current is injected to the compensated CT's secondary winding. The proposed idea has two disadvantages: an expensive lock-in amplifier is required, and the presence of a second harmonic in the primary current will be also compensated in the secondary current. An effective solution for this problem is fluxgate sensors [51,52].

6. Conclusions

This review article discusses advanced works devoted to the characterization of the inductive voltage and current transformers' transformation accuracy of distorted currents and voltages. The presented results and cited references show that conventional instrument transformers may ensure the required transformation accuracy not only for the sinusoidal currents of the rated frequency 50 Hz/60 Hz but also for the higher harmonic components of distorted currents/voltages. Proposed compensation techniques are necessary only for poor-quality inductive instrument transformers. Additionally, these methods increase complexity and cost of the measuring circuit. A better solution is to properly design the inductive instrument transformer, including the application of modern magnetic materials, oversizing of the magnetic core (inductive CT), and reduction in the number of turns of the primary winding (inductive VT). Typical inductive instrument transformers designed for the transformation of the sinusoidal current/voltage may ensure, in the frequency range of higher harmonics from 50 Hz to 5 kHz, $\pm0.5\%/\pm0.5^\circ$ transformation accuracy in the case of the inductive CTs and $\pm5\%/\pm5^\circ$ in the case of the MV inductive voltage transformers. The challenges of accurate measurement of distorted current and voltage in the power grid by conventional instrument transformers are self-generation and resonance phenomena. Additional distortion of the secondary current by the inductive CTs is much more significant and important for their transformation accuracy of the low-order higher harmonics than self-distortion of the secondary voltage by the inductive VTs. Recent studies show that the resonance was only detected in the case of the inductive VTs, and it concerns the transformation of higher harmonics of distorted voltage by the medium and the high voltage units. Moreover, the power factor of the secondary winding's load of the inductive IT may significantly affect its limiting frequency of the wideband operation with the required maximum values of ratio error and phase displacement.

Author Contributions: Conceptualization, M.K. and E.S.; methodology, M.K. and E.S.; validation, M.K. and E.S.; formal analysis, M.K. and E.S.; investigation, M.K. and E.S.; resources, M.K. and E.S.; data curation, M.K. and E.S.; writing—original draft preparation, M.K. and E.S.; writing—review and editing, M.K. and E.S.; visualization, E.S.; supervision, M.K. All authors have read and agreed to the published version of the manuscript.

Funding: This research received no external funding.

Data Availability Statement: Not applicable.

Conflicts of Interest: The authors declare no conflict of interest.

References

1. Crotti, G.; D'Avanzo, G.; Letizia, P.S.; Luiso, M. Measuring Harmonics with Inductive Voltage Transformers in Presence of Subharmonics. *IEEE Trans. Instrum. Meas.* **2021**, *70*, 1–13. [CrossRef]
2. Crotti, G.; D'Avanzo, G.; Landi, C.; Letizia, P.S.; Luiso, M. Evaluation of Voltage Transformers' Accuracy in Harmonic and Interharmonic Measurement. *IEEE Open J. Instrum. Meas.* **2022**, *1*, 1–10. [CrossRef]
3. Kaczmarek, M.; Stano, E. Why should we test the wideband transformation accuracy of medium voltage inductive voltage transformers? *Energies* **2021**, *14*, 4432. [CrossRef]
4. Stano, E.; Kaczmarek, P.; Kaczmarek, M. Why Should We Test the Wideband Transformation Accuracy of Inductive Current Transformers? *Energies* **2022**, *15*, 5737. [CrossRef]
5. Kaczmarek, M. The effect of distorted input voltage harmonics rms values on the frequency characteristics of ratio error and phase displacement of a wideband voltage divider. *Electr. Power Syst. Res.* **2019**, *167*, 1–8. [CrossRef]
6. Draxler, K.; Hlavacek, J.; Styblikova, R. Calibration of Instrument Current Transformer Test Sets. In Proceedings of the 2019 International Conference on Applied Electronics (AE), Pilsen, Czech Republic, 10–11 September 2019. [CrossRef]
7. Draxler, K.; Styblikova, R. Calibration of instrument current transformers at low currents using lock-in amplifier. In Proceedings of the 2017 International Conference on Applied Electronics (AE), Pilsen, Czech Republic, 5–6 September 2017; pp. 18–21. [CrossRef]
8. Van Den Brom, H.E.; Jol, L.; Rietveld, G.; So, E. High-current AC current transformer calibration using an automated sampling system. In Proceedings of the 2012 Conference on Precision Electromagnetic Measurements, Washington, DC, USA, 1–6 July 2012; pp. 134–135. [CrossRef]
9. Siegenthaler, S.; Mester, C. A computer-controlled calibrator for instrument transformer test sets. *IEEE Trans. Instrum. Meas.* **2017**, *66*, 1184–1190.

10. Mingotti, A.; Peretto, L.; Tinarelli, R.; Ghaderi, A. Uncertainty analysis of a test bed for calibrating voltage transformers vs. Temperature. *Sensors* **2019**, *19*, 4472. [CrossRef] [PubMed]
11. Mingotti, A.; Bartolomei, L.; Peretto, L.; Tinarelli, R. On the long-period accuracy behavior of inductive and low-power instrument transformers. *Sensors* **2020**, *20*, 5810. [CrossRef]
12. Kaczmarek, M.; Kaczmarek, P.; Stano, E. The Effect of the Load Power Factor of the Inductive CT's Secondary Winding on Its Distorted Current's Harmonics Transformation Accuracy. *Energies* **2022**, *15*, 6258. [CrossRef]
13. Mingotti, A.; Peretto, L.; Bartolomei, L.; Cavaliere, D.; Tinarelli, R. Are inductive current transformers performance really affected by actual distorted network conditions? An experimental case study. *Sensors* **2020**, *20*, 927. [CrossRef]
14. Kondrath, N.; Kazimierczuk, M.K. Bandwidth of current transformers. *Proc. IEEE Trans. Instrum. Meas.* **2009**, *58*, 2008–2016. [CrossRef]
15. So, E.; Bennett, D. Compact wideband high-current (≤ 1000 A) multistage current transformers for precise measurements of current harmonics. *Proc. IEEE Trans. Instrum. Meas.* **2007**, *56*, 584–587. [CrossRef]
16. Stano, E.; Kaczmarek, P.; Kaczmarek, M. Understanding the Frequency Characteristics of Current Error and Phase Displacement of the Corrected Inductive Current Transformer. *Energies* **2022**, *15*, 5436. [CrossRef]
17. Kaczmarek, M.; Stano, E. The Influence of the 3rd Harmonic of the Distorted Primary Current on the Self-Generation of the Inductive Current Transformers. *IEEE Access* **2022**, *10*, 55876–55887. [CrossRef]
18. Cataliotti, A.; Cosentino, V.; Crotti, G.; Giordano, D.; Modarres, M.; Di Cara, D.; Tinè, G.; Gallo, D.; Landi, C.; Luiso, M. Metrological performances of voltage and current instrument transformers in harmonics measurements. In Proceedings of the I2MTC 2018—2018 IEEE International Instrumentation and Measurement Technology Conference: Discovering New Horizons in Instrumentation and Measurement, Houston, Texas, USA, 14–17 May 2018; pp. 1–6.
19. Letizia, P.S.; Signorino, D.; Crotti, G. Impact of DC Transient Disturbances on Harmonic Performance of Voltage Transformers for AC Railway Applications. *Sensors* **2022**, *22*, 2270. [CrossRef]
20. Filipović-Grčić, D.; Filipović-Grčić, B.; Krajtner, D. Frequency response and harmonic distortion testing of inductive voltage transformer used for power quality measurements. *Procedia Eng.* **2017**, *202*, 159–167. [CrossRef]
21. Kaczmarek, M.; Stano, E. Application of the Sinusoidal Voltage for Detection of the Resonance in Inductive Voltage Transformers. *Energies* **2021**, *14*, 7047. [CrossRef]
22. Kaczmarek, M.; Stano, E. Measuring system for testing the transformation accuracy of harmonics of distorted voltage by medium voltage instrument transformers. *Measurement* **2021**, *181*, 109628. [CrossRef]
23. Chen, W.; Ma, J.; Huang, X.; Fang, Y. Predicting iron losses in laminated steel with given non-sinusoidalwaveforms of flux density. *Energies* **2015**, *8*, 13726–13740. [CrossRef]
24. Lesniewska, E. Influence of the selection of the core shape and winding arrangement on the accuracy of current transformers with through-going primary cable. *Energies* **2021**, *14*, 1932. [CrossRef]
25. Fritsch, M.; Wolter, M. High-Frequency Current Transformer Design and Construction Guide. *IEEE Trans. Instrum. Meas.* **2022**, *71*, 1–9. [CrossRef]
26. Yu, X.; Li, Y.; Yang, Q.; Yue, S.; Zhang, C. Loss Characteristics and Model Verification of Soft Magnetic Composites under Non-Sinusoidal Excitation. *IEEE Trans. Magn.* **2019**, *55*, 18–21. [CrossRef]
27. Pluta, W.A.; Swieboda, C.; Leszczynski, J.S.; Soinski, M. Some remarks on metrological properties and production technology of current transformers made of nanocrystalline cores. *Meas. J. Int. Meas. Confed.* **2017**, *97*, 38–44. [CrossRef]
28. Swieboda, C.; Walak, J.; Soinski, M.; Rygal, J.; Leszczynski, J.; Grybos, D. Nanocrystalline oval cut cores for current instrument transformer prototypes. *Meas. J. Int. Meas. Confed.* **2019**, *136*, 50–58. [CrossRef]
29. Lesniewska, E. Modern Methods of Construction Problem Solving in Designing Various Types of Instrument Transformers. *Energies* **2022**, *15*, 8199. [CrossRef]
30. Faifer, M.; Laurano, C.; Ottoboni, R.; Toscani, S.; Zanoni, M. Characterization of voltage instrument transformers under nonsinusoidal conditions based on the best linear approximation. *IEEE Trans. Instrum. Meas.* **2018**, *67*, 2392–2400. [CrossRef]
31. Collin, A.J.; Femine, A.D.; Gallo, D.; Langella, R.; Luiso, M. Compensation of current transformers' nonlinearities by tensor linearization. *IEEE Trans. Instrum. Meas.* **2019**, *68*, 3841–3849. [CrossRef]
32. Cataliotti, A.; Cosentino, V.; Crotti, G.; Femine, A.D.; Di Cara, D.; Gallo, D.; Giordano, D.; Landi, C.; Luiso, M.; Modarres, M.; et al. Compensation of Nonlinearity of Voltage and Current Instrument Transformers. *IEEE Trans. Instrum. Meas.* **2019**, *68*, 1322–1332. [CrossRef]
33. Faifer, M.; Laurano, C.; Ottoboni, R.; Toscani, S.; Zanoni, M. Harmonic Distortion Compensation in Voltage Transformers for Improved Power Quality Measurements. *IEEE Trans. Instrum. Meas.* **2019**, *68*, 3823–3830. [CrossRef]
34. Castello, P.; Laurano, C.; Muscas, C.; Pegoraro, P.A.; Toscani, S.; Zanoni, M. Harmonic Synchrophasors Measurement Algorithms with Embedded Compensation of Voltage Transformer Frequency Response. *IEEE Trans. Instrum. Meas.* **2021**, *70*, 9001310. [CrossRef]
35. Stano, E. The Method to Determine the Turns Ratio Correction of the Inductive Current Transformer. *Energies* **2021**, *14*, 8602. [CrossRef]
36. Kaczmarek, M.; Brodecki, D. Transformation of Transient Overvoltages by Inductive Voltage Transformers. *Sensors* **2021**, *21*, 4167. [CrossRef] [PubMed]

37. Crotti, G.; Gallo, D.; Giordano, D.; Landi, C.; Luiso, M. Compensation of Complex Frequency Errors of Voltage Instrument Transformers. In Proceedings of the 2018 Conference on Precision Electromagnetic Measurements (CPEM 2018), Paris, France, 8–13 July 2018; pp. 3–4. [CrossRef]
38. Kaczmarek, M.; Stano, E. Proposal for extension of routine tests of the inductive current transformers to evaluation of transformation accuracy of higher harmonics. *Int. J. Electr. Power Energy Syst.* **2019**, *113*, 842–849. [CrossRef]
39. Crotti, G.; D'Avanzo, G.; Giordano, D.; Letizia, P.S.; Luiso, M. Extended SINDICOMP: Characterizing MV Voltage Transformers with Sine Waves. *Energies* **2021**, *14*, 1715. [CrossRef]
40. *IEC 61869-103*; Instrument Transformers—The Use of Instrument Transformers for Power Quality Measurement. IEC: Geneva, Switzerland, 2010.
41. Lesniewska, E.; Kaczmarek, M.; Stano, E. 3D Electromagnetic Field Analysis Applied to Evaluate the Accuracy of a Voltage Transformer under Distorted Voltage. *Energies* **2021**, *14*, 136. [CrossRef]
42. Stano, E.; Kaczmarek, M. Wideband self-calibration method of inductive cts and verification of determined values of current and phase errors at harmonics for transformation of distorted current. *Sensors* **2020**, *20*, 2167. [CrossRef]
43. Kaczmarek, M.; Stano, E. Nonlinearity of Magnetic Core in Evaluation of Current and Phase Errors of Transformation of Higher Harmonics of Distorted Current by Inductive Current Transformers. *IEEE Access* **2020**, *8*, 118885–118898. [CrossRef]
44. Wath, M.G.; Raut, P.; Ballal, M.S. Error compensation method for current transformer. In Proceedings of the 2016 IEEE 1st International Conference on Power Electronics, Intelligent Control and Energy Systems (ICPEICES), Delhi, India, 4–6 July 2016; Volume 11, pp. 21–24. [CrossRef]
45. Faifer, M.; Laurano, C.; Ottoboni, R.; Toscani, S.; Zanoni, M.; Crotti, G.; Giordano, D.; Barbieri, L.; Gondola, M.; Mazza, P. Overcoming Frequency Response Measurements of Voltage Transformers: An Approach Based on Quasi-Sinusoidal Volterra Models. *IEEE Trans. Instrum. Meas.* **2019**, *68*, 2800–2807. [CrossRef]
46. Laurano, C.; Toscani, S.; Zanoni, M. A simple method for compensating harmonic distortion in current transformers: Experimental validation. *Sensors* **2021**, *21*, 2907. [CrossRef]
47. Pintelon, R.; Schoukens, J. *System Identification: A Frequency Domain Approach*, 2nd ed.; John Wiley & Sons: Hoboken, NJ, USA, 2012; ISBN 9780470640371.
48. Ballal, M.S.; Wath, M.G.; Suryawanshi, H.M. A novel approach for the error correction of ct in the presence of harmonic distortion. *IEEE Trans. Instrum. Meas.* **2019**, *68*, 4015–4027. [CrossRef]
49. Ripka, P.; Draxler, K.; Styblíková, R. DC-compensated current transformer. *Sensors* **2016**, *16*, 114. [CrossRef]
50. Ripka, P.; Draxler, K.; Styblikova, R. DC-compensated current transformer. In Proceedings of the Conference Record—IEEE Instrumentation and Measurement Technology Conference, Montevideo, Uruguay, 12–15 May 2014; pp. 212–215.
51. Velasco-Quesada, G.; Roman-Lumbreras, M.; Conesa-Roca, A.; Jerez, F. Design of a low-consumption fluxgate transducer for high-current measurement applications. *IEEE Sens. J.* **2011**, *11*, 280–287. [CrossRef]
52. Yang, X.; Zhang, B.; Wang, Y.; Zhao, Z.; Yan, W. The optimization of dual-core closed-loop fluxgate technology in precision current sensor. *J. Appl. Phys.* **2012**, *111*, 07E722. [CrossRef]

energies MDPI

Article

Effect of Magnetic Shunts on Shell-Type Transformers Characteristics

Bronisław Tomczuk [1,*] and Dawid Weber [2]

[1] Department of Electrical Engineering & Mechatronics, Opole University of Technology, Prószkowska 76, 45-758 Opole, Poland
[2] Technology Center, Hitachi Energy, Aleksandrowska 67/93, 91-205 Lodz, Poland; dawid.weber@hitachienergy.com
* Correspondence: b.tomczuk@po.edu.pl

Abstract: A three-dimensional analysis of the leakage magnetic field in a transformer with beveled edges in a magnetic shunt is described. This paper contains a mathematical description of the finite element method used in performing numerical calculations. Formulas allowing to determine the differential and integral parameters of the magnetic field were presented. Magnetic flux density distributions and the points with high values were determined in the shell-type core and movable shunt sub-areas. For different positions of the magnetic shunt, the short-circuit transformer reactance was calculated as the integral parameter of the above-mentioned magnetic field analysis. Their amounts for the extreme positions of the shunt were given, as well. The movable magnetic shunt geometry impact on the short-circuit current value is given. In order to validate the calculation results, the research included some point measurements of magnetic flux density vectors and the winding reactances as well, and good compliance with the simulation data has been obtained. Thus, this analysis can also be applied to other objects with movable shunts in their magnetic systems.

Keywords: shell-type core transformer; high leakage reactance; magnetic shunt with beveled edges; 3D magnetic field; finite element method; short-circuit current regulation; verification by measurements; physical model testing

check for updates

Citation: Tomczuk, B.; Weber, D. Effect of Magnetic Shunts on Shell-Type Transformers Characteristics. *Energies* **2023**, *16*, 6814. https://doi.org/10.3390/en16196814

Academic Editor: Abu-Siada Ahmed

Received: 19 June 2023
Revised: 11 September 2023
Accepted: 21 September 2023
Published: 26 September 2023

1. Introduction

1.1. Introduction

In the era of electronic static converters used in both low and high-voltage electric networks, many harmonics related to current and voltage distortions have appeared in the power net, which should be eliminated as much as possible. Therefore, one should use methods of changing voltage and current parameters without digitizing their waveforms. First, the systems operating with sinusoidal waves of the main electrical quantities should be introduced in designing. Among them are regulating transformers (RT) that adjust the external characteristics to specific loading requirements. Such structures can be called "natural" and apply to rectifier systems for charging energy sources (batteries) and supplying an electric arc. Due to the above, such special transformers (RT) operation enables obtaining the demanded operating characteristics [1–3]. For example, we need high leakage reactance (HLR) transformers that supply the electric arc [4]. Their characteristics are as well as those for the devices that operate close to short-circuit state [5]. Field analysis is needed to obtain the demanded characteristics or optimize their constructions [6,7].

Knowledge of the magnetic field distribution allows for determining the transformer key parameters without the need to build expensive prototypes or perform time-consuming and highly approximate analytical calculations [8,9]. In addition, knowledge of the magnetic flux density vector values at individual points in the area surrounding the core and windings makes it possible to determine the windings leakage reactance for different configurations of the core and windings arrangement [10]. In addition, it is possible to determine the impact of changes in these configurations on the transformer parameters. The results obtained through analytical calculations are often malformed with a large error, primarily due to the need for significant simplifications of the object's physical geometry. The use of many correction factors requires knowledge of the approximate flux distribution, which is not always possible due to the complexity of the tested object structure [11].

The finite element method used in the calculations performed in this paper can be successfully adapted not only to electromagnetic field distribution but also to thermal [12], mechanical simulations [13–15], or transformer loss determination [16,17]. In addition, good compatibility is obtained not only for transformers but also for other objects, e.g., inductive sensors [18].

1.2. Historical Outline of Calculations of Hlr Transformers with Movable Magnetic Shunts

The variety of regulating transformers' (RT) construction [5] makes it difficult to classify them. This also applies to high leakage reactance transformers (HLR) with shunt regulation of their current intensities. Therefore, the most general division seems to be in terms of the main leakage flux direction in relation to the windings axes. As a result of many analyses of the magnetic field in these transformers, B. Tomczuk proposed such a division in his Ph.D. [19]. They were divided into transformers with axial (yoke) leakage and those with radial ones.

Calculations for high leakage reactance transformers (HLR) with extreme magnetic shunt positions were performed repeatedly. For the transformer with yoke leakage, the superposition method of 2D magnetic field solutions was given in [19], which was published later [20]. In addition, for these transformers, but without a shunt, the method of integral equations developed by B. Tomczuk was also used [21,22].

Knowledge of leakage inductance is a particularly important parameter for transformer designers since it has a principal effect on the transformer's short-circuit voltage. For objects with adjustable magnetic shunt positions, the use of 2D numerical computational methods is insufficient and can only be used at extreme magnetic shunt positions, due to the object asymmetry. The novelty of the presented work is to carry out a 3D analysis of the magnetic field distribution considering intermediate states of the magnetic shunt position and its different geometries. The results of 3D magnetic field analyses for many fixes of the transformer shunt were presented by the authors of this paper but only for a transformer with yoke dispersion.

2. Three-Dimensional Mathematical Modeling of Magnetic Flux Distribution in Radial Dispersion HLR Transformer

2.1. The Object under Consideration

In the present work, the authors describe three-dimensional magnetic field modeling in an HLR transformer with radial leakage field dissipation (Figure 1) and beveled edges in its magnetic shunt. The movable shunt packages are fastened with an armature made of stainless steel structural elements (Figure 1). Due to this, within our field analysis, we considered the magnetic system of the shell-type core and only shunt packages without non-magnetic structural components as well as the two windings made of different materials. The primary winding is made from a copper profile wire of rectangular cross-section (1.6 mm × 5 mm) and the secondary one is a two-section made of aluminum strip of cross-section (2.5 mm × 17 mm).

The transformer's primary and secondary windings turn numbers equal 180 and 34, respectively. The transformer magnetic circuit with dimensions of 213 mm × 240 mm × 100 mm is a shell-type core structure [23]. Table 1 presents the rated parameters of the studied object.

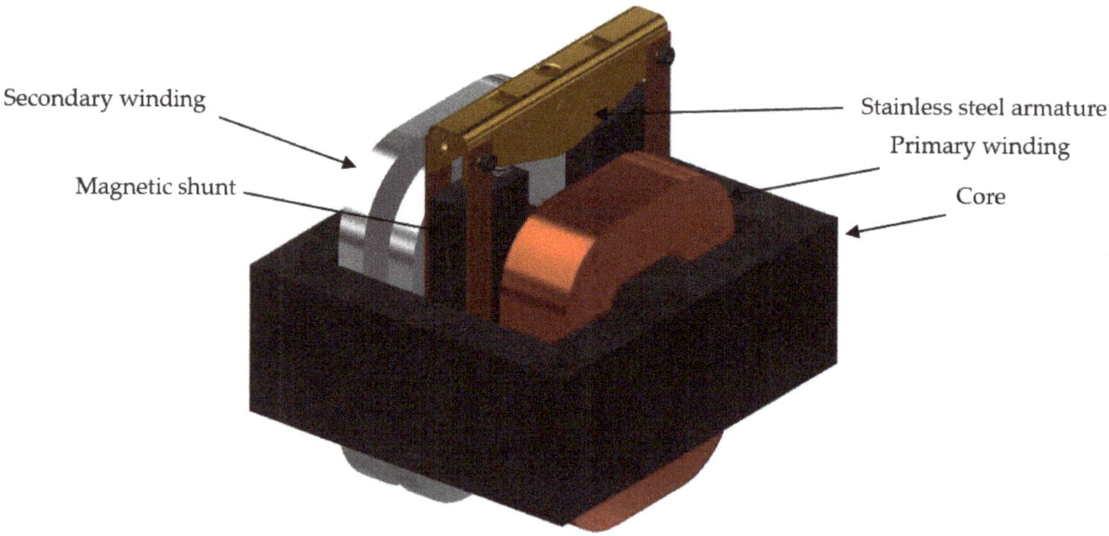

Figure 1. The HLR transformer with radial magnetic leakage.

Table 1. Parameters of the analyzed transformer.

Parameter	Value
Rated voltage	400 [V]
Maximum primary winding current	41 [A]
Maximum secondary winding current	225 [A]
Power	16 [kVA]

2.2. The Analysed Magnetic Field Area and Its Boundary Conditions

Due to the variable nature of the magnetic field distribution depending on the extraction of the shunt in relation to the transformer core, the mathematical model considers the appropriate adjustment of the analyzed area (Figure 2). The movable shunt providing a variable magnetic coupling between the primary and secondary windings in the object above allows for step-less regulation of the secondary voltage to maintain a secondary current.

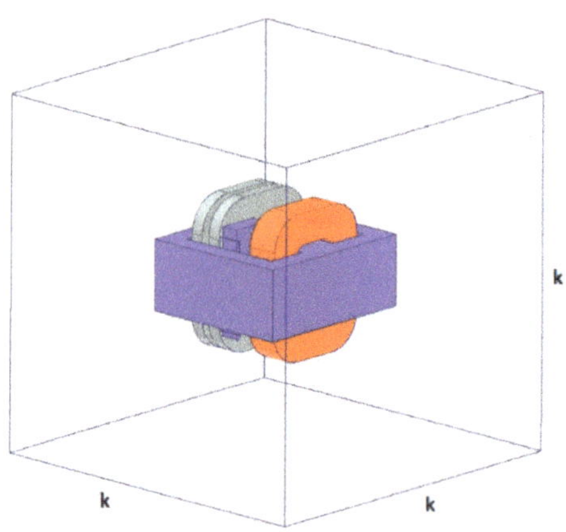

Figure 2. Transformer in the computing area.

For systems with a highly spatial distribution of the leakage fluxes, it is necessary to conduct a three-dimensional (3D) analysis of the unbounded magnetic field, which is the principal goal of this work. One of the most frequently used and well-programmed numerical methods is finite element one (FEM). Thus, this method has been used in this work. The magnetic potentials and flux distribution calculations were performed using FEM, applied in the commercial ANSYS Maxwell 2022 R1 software [24].

Our numerical calculations of the magnetic field distribution take place in a finite and defined by the boundary conditions space. The limitation of the computational region Ω is implemented by the Dirichlet and Neuman conditions. In the Dirichlet condition, the solution function takes known values on the edges of its domain, as in Equation (1) [25,26].

$$u(P) = g(P) \text{ for } P \in S \tag{1}$$

where:

- u(P)—the function sought at points inside the region Ω;
- g(P)—specified function for points P belonging to the S edge of the Ω area.

In the case of the Neumann boundary condition, the normal derivative on the S edge of the region takes known values, Equation (2) [25,26].

$$\frac{\partial u}{\partial n}(P) = g(P) \text{ for } P \in S \tag{2}$$

The HLR transformers with increased leakage reactance, depending on the magnetic system structure and position of the movable shunt, can be characterized by an extensive magnetic field. The analysis of the computational area can be limited to the two extreme magnetic shunt positions, fully retracted and extended. In the case of a fully retracted shunt, the magnetic flux closes through the core-shunt gaps, so the magnetic flux density values outside the transformer window will be relatively small compared to those contained in the transformer window.

For the field analysis, we assumed a cubic computing space in which the transformer is located. Figure 2 shows the object under consideration, placed in the center of the cube area space of the "k" edge length. Therefore, before performing the proper field calculations, the influence of the cube's edge length on the total energy accumulated in the leakage zone was analyzed (Figure 3). Although the transformer shell-type structure analyzed in this article characterizes relatively small leakage flux density values outside the transformer window, their influence on the leakage reactance must be considered.

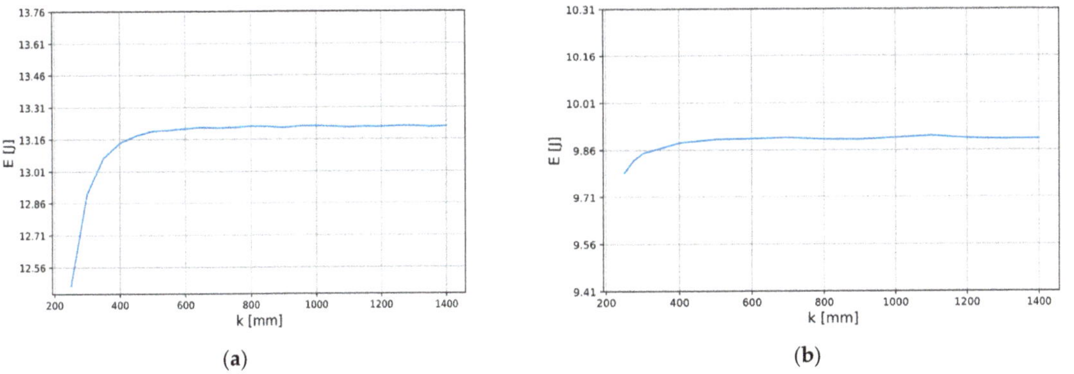

Figure 3. Dependence of the calculated energy stored in the outer leakage field area on the calculation cube edge length "*k*", for the extracted shunt (**a**), and the fully inserted shunt (**b**).

Figure 3a shows that the increase in the calculation cube edge length above 0.6 m does not translate into a significant increase in the energy values of the outer air zone while causing a rise in the magnetic field analysis time. Based on Figure 3b, one can see that for the fully inserted magnetic shunt, the leakage energy outside the core has stabilized for the computational cube of a smaller edge value of 0.45 m. In conclusion, for the transformer without the magnetic shunt, a part of the magnetic field lines goes outside the transformer window, which requires consideration of a higher expanse. To increase the certainty of the correctness of the simulations, in further calculations described in this paper, the length of the edge of the computational area cube was assumed to be 1 m, regardless of the location of the magnetic shunt.

Symmetry and antisymmetry planes make it possible to reduce the computational area and thus reduce the number of discretization grid elements, which significantly reduces the size of the equations necessary to solve and the calculation time while maintaining the solution's accuracy. Due to the movable magnetic shunt position and the object's unsymmetrical winding execution, it is possible to include only one symmetry plane in the analysis, as in Figure 4. The conditions of symmetry and antisymmetry can be described by vector potential identities on both sides of these planes using Equations (3) and (4), respectively [25,26].

$$A_{1(x2,\ y2,\ z2)} = A'_{1(x2,\ y2,\ -z2)} \quad (3)$$

$$A_{2(x2,\ y2,\ z2)} = A'_{2(x2,\ y2,\ -z2)} \quad (4)$$

In transformers with adjustable magnetic shunt position, the area requiring special attention during discretization is the core-shunt gap, which is characterized by a high magnetic field variation; therefore, at least two mesh layers were forced in this area, and the maximum tetrahedron edge length did not exceed 2 mm. During the analysis, current excitation was used, and the non-linear magnetization characteristics of the transformer core sheets were considered, as in Figure 5.

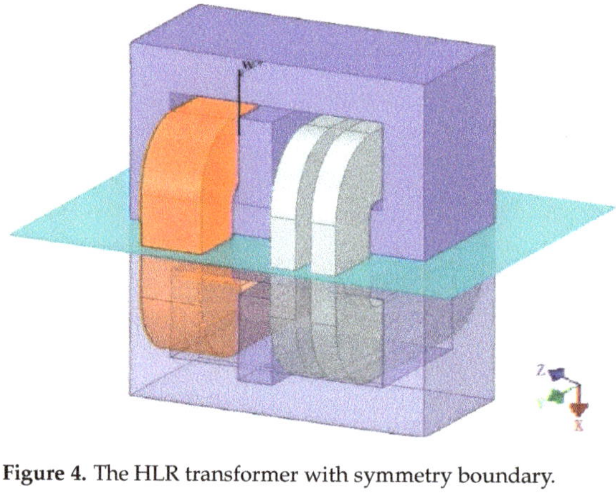

Figure 4. The HLR transformer with symmetry boundary.

The magnetic shunts used to regulate the current can be in cuboidal or tetrahedral shapes. For the shell-type transformer structure, the chamfering (beveling) magnetic shunt (Figure 6a) allows smooth current regulation in the secondary winding. The advantage of this shape is to increase the current regulation linearity by pulling out the magnetic shunt relative to the transformer window at the expense of hotspots in the corners of the core and shunt, which causes construction heating. Significantly increasing the beveling affects the maximum adjustment of the leakage reactance range. The paper examines the influence of the shunt chamfering on the transformer leakage reactance for different positions, with a constant of both primary and secondary ampere-turns windings, equal to Θ = 1800 A.

The simulation was performed by considering the transformer symmetry. Angle α was changed in the range of 0–50°, and the shunt extraction was in the range of 5 to 180 mm.

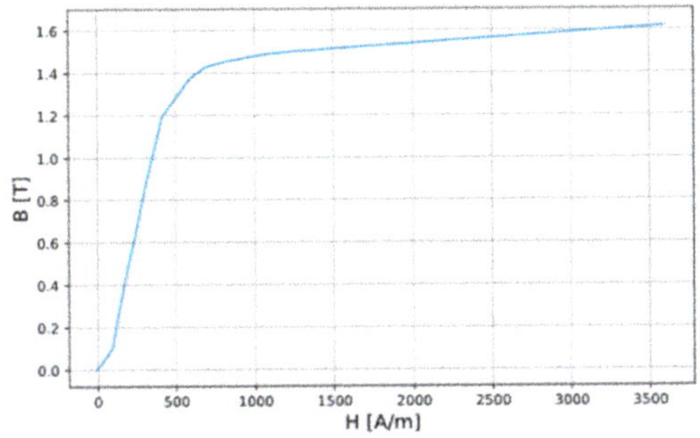

Figure 5. B-H Curve of the M530—50A steel.

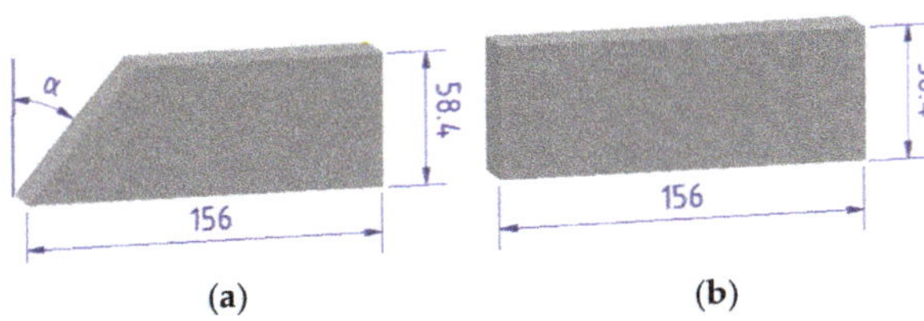

Figure 6. Magnetic shunt with chamfering angle (**a**), and without chamfering angle (**b**).

2.3. *Partial Differential Equations of the Field and Its Integral Parameters*

Performed magnetic field analysis assumed its quasi-stationarity, where time-varying phenomena and the eddy currents effects were omitted. This allows for the calculation time reduction, which is of great importance due to the multi-variant calculation's nature, and simultaneously allows for compliance satisfactory for technical purposes. This limits the equations given in the second half of the 19th century by J.C. Maxwell and describes the electromagnetic field with Equations (5) and (6) given below [27].

$$\nabla \times \vec{H} = \vec{j} \tag{5}$$

$$\nabla \times \vec{E} = 0 \tag{6}$$

The electromagnetic field equations can be solved using the $\vec{T} - \Omega$ potential method, where $\vec{T}$ describes the electric potential vector and Ω the scalar magnetic potential. In the magnetostatic analysis, the potential $\vec{T}$ is omitted, which simplifies the equation describing the intensity of the electromagnetic field to the form of Equation (7) [28].

$$\vec{H} = \vec{H}_S - \nabla\Omega \tag{7}$$

$$\oint_l \vec{H}_S \cdot d\vec{l} = \theta \tag{8}$$

The vector $\vec{H}_S$ (Equation (8)) describes the field intensity coming from the current density $\vec{J}$ of the external source's excitation and can be determined from the circuital law. The $-\nabla\Omega$ component is related to the ferromagnetic core material's magnetization. Given the zero rotation value of this term, it can be expressed by the reduced scalar potential in the form $-\nabla\phi$. Considering the above dependencies and substituting them to Equation

(7), Expression (9) is obtained, which is solved in the sub-areas of air and the transformer windings [29,30].

$$-\nabla\cdot(\mu\nabla\phi)+\nabla\cdot\left(\mu\vec{H}_S\right)=0 \tag{9}$$

In the transformer ferromagnetic area, due to the zero value of the current density external sources, Equation (10) is solved.

$$\nabla\cdot(\mu\nabla\Omega)=0 \tag{10}$$

Knowledge of the distribution of magnetic potentials allows the calculation of their spatial derivatives in order to determine the components of the cartesian magnetic flux density vectors. However, the most relevant parameters are those calculated from the flux density vectors' field integration. These are fluxes in the magnetic core and the leakage ones, stored energy, as well as windings inductances linked with the fluxes and magnetic forces. Assuming the flux density vector $\vec{B}$ distribution in a surface S, flux linkage with this surface can be determined via Equation (11).

$$\Psi=\int_S \vec{B}\cdot d\vec{S} \tag{11}$$

Using the leakage energy stored in non-magnetic areas (Equation (12)), the transformer static leakage inductance can be determined from Expression (13). The dynamic leakage inductance in a linear environment can be determined using Equation (14) [31].

$$W=\frac{1}{2\mu}\left(\int_V B^2 dV\right) \tag{12}$$

$$L_{rs}=\omega\frac{2W}{I^2} \tag{13}$$

$$L_{rd}=L_{rs}+I\frac{\partial L_{rs}}{\partial I} \tag{14}$$

The magnetic shunt inserted into the transformer window causes the division of the flux generated by the primary winding into two parts, in both the short-circuit and the no-load state. One of them closes through the magnetic shunt and is unrelated to the secondary winding, and the other part passes through this winding. Considering the above, the static (Equation (15)) and dynamic (Equation (16)) mutual inductance were calculated by forcing the current excitation only within the primary winding, simultaneously determining the value of the magnetic flux $\Psi(I)$ linked with the secondary winding.

$$L_{\mu s}=\frac{\Psi(I)}{I} \tag{15}$$

$$L_{\mu d}=\frac{\partial\Psi}{\partial i} \tag{16}$$

3. Results of Field Analysis for Various Constructions and Partial Protrusion of the Magnetic Shunt

3.1. Magnetic Field in the Transformer with the Same Shunt Construction, but Its Different Protrusion from the Window

The performed magnetic field distribution analysis considered the primary winding current intensity in the range of 1–41 A and various positions of the beveled shunt fixes in the range of w = (20–180) mm. Figures 7 and 8 show the magnetic flux density distribution for the two exemplary positions of the same shunt under the same ampere-turns for both the primary and secondary sides of the transformer windings. Due to the object symmetry, the distributions have been presented only for half of the transformer core, i.e., for one part of the shunt system. The current intensity value in the primary winding was equal to 9 A.

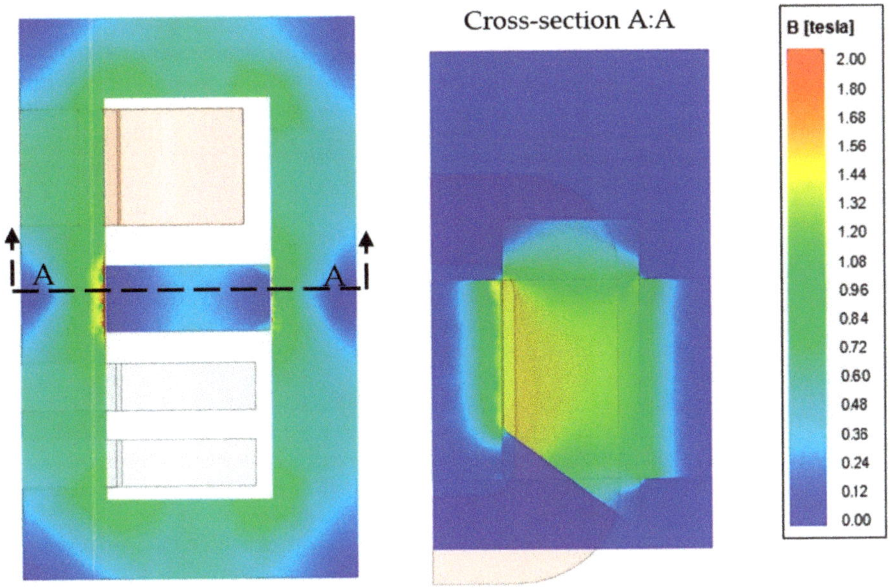

Figure 7. Magnetic flux density distribution in the transformer core and the leakage zone, for the beveled edges of magnetic shunt, pulled out by w = 20 mm and the current intensity in the primary winding of 9 A.

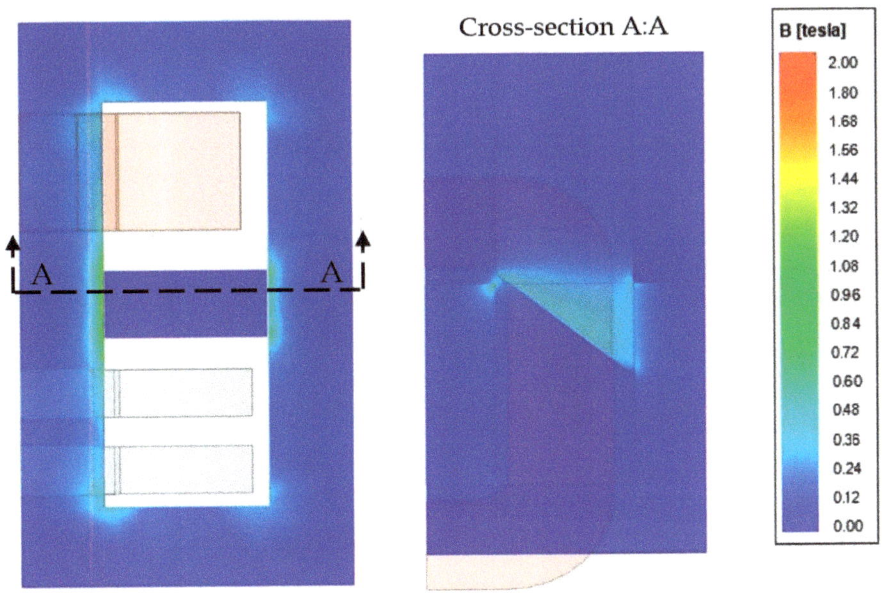

Figure 8. Flux density for pulled the beveled shunt out of the transformer window by 110 mm, and the current intensity in the primary winding of 9 A.

For the transformer short-circuit state, with no magnetic shunt presence or its complete pulling out, the flux generated by the primary winding is almost entirely compensated in the core by the secondary winding flux. Inserting the magnetic shunt into the transformer window allows some of the magnetic flux to flow through it. In Figures 7 and 8, one can see the flux division into two parts: one closed by a magnetic shunt and the other one related to the secondary winding. The amount of flux flowing through the shunt depends on the ratio of the equivalent reluctance for the magnetic shunt with its two air gaps and the reluctance of the core part with the secondary winding. At a small shunt extraction, the flux divides into almost equal parts. With its pulling out the so-called hot spots, i.e., areas of shunt saturation, are visible. This is due to a relatively small cross-section of the shunt's active parts.

3.2. *Field Simulation for Different Chamfering (Beveling) of the Magnetic Shunt*

Figures 9–12 show the magnetic flux density distribution in the half of the transformer core, magnetic shunt, and air gaps for beveled shunt exemplary positions. The first one presents the flux density histogram for the simple cuboidal shunt, which is the shunt geometry special case ($\alpha = 0°$) mentioned above. The shunt extraction from the transformer window was equal to $w = 30$ mm. Figure 10 shows the flux density histogram for the same shunt extraction but beveled with the angle $\alpha = 45°$.

Pulling the magnetic shunt out causes a part inside the transformer window and the core edges slightly saturation. For the cuboidal shunt ($\alpha = 0°$) only, its slight saturation is visible on both sides; see Figure 11. For the beveled shunt, at this value of its protrusion, the magnetic "hot spot" (saturation) occurs only in the place of the shunt's shorter surface inserted into the transformer window (Figure 12). This has a direct impact on the local construction overheating, so the temperature field analysis could be undertaken in further research.

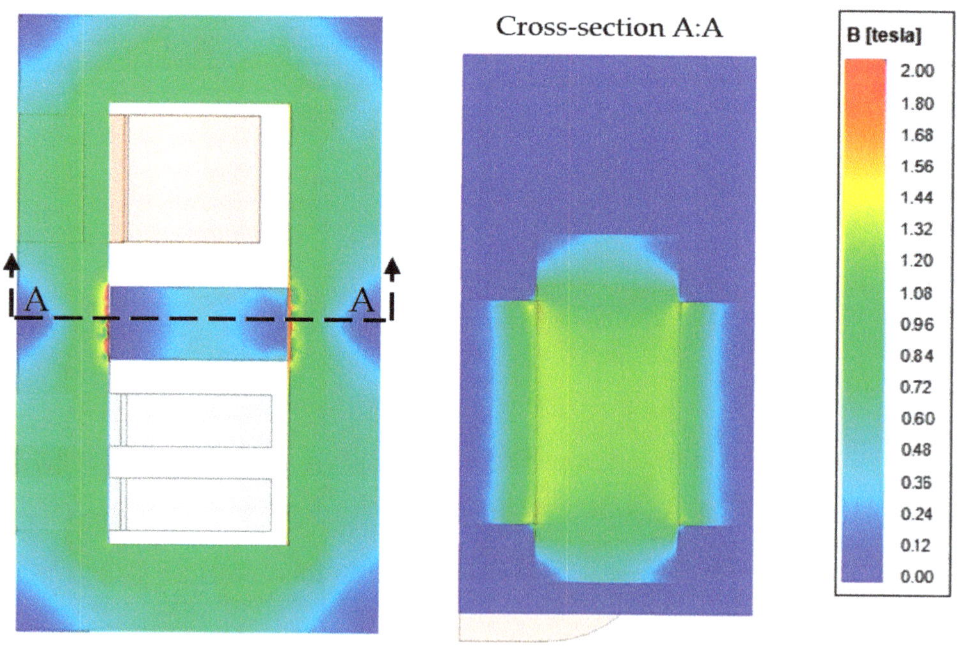

Figure 9. Flux density in the transformer and leakage zone for 30 mm extraction of the cuboidal magnetic shunt ($\alpha = 0°$).

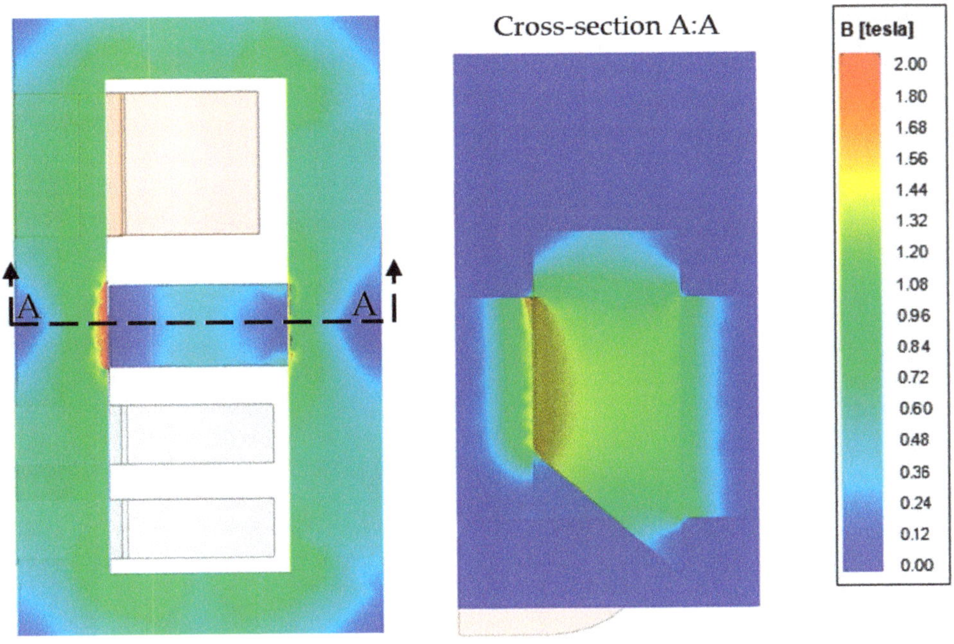

Figure 10. Flux density in the transformer core and leakage zone for 30 mm magnetic shunt extraction and $\alpha = 45°$.

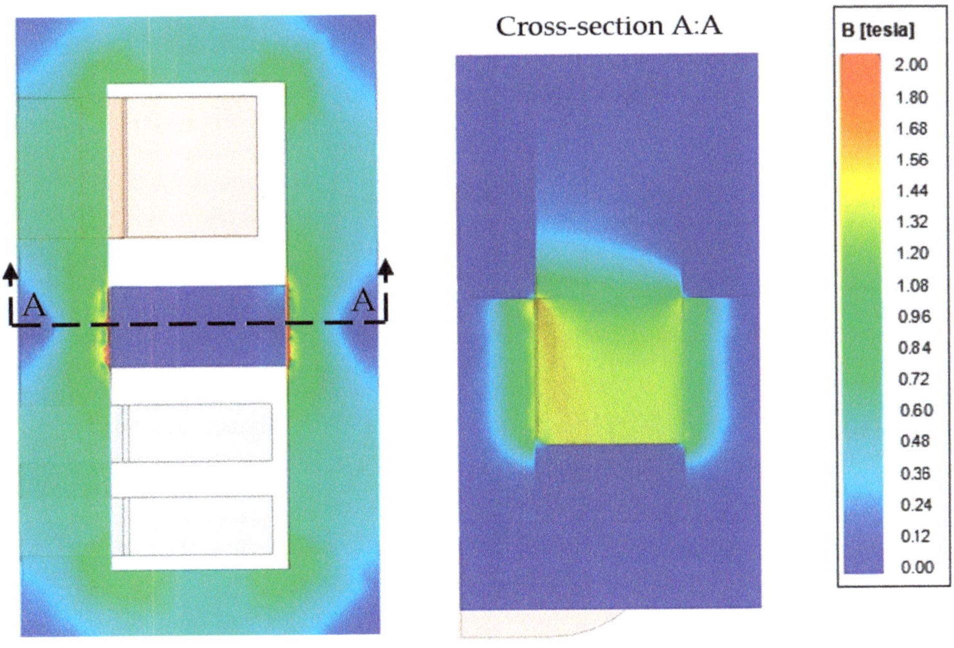

Figure 11. Flux density in transformer and leakage zone for $w = 90$ mm magnetic shunt extraction without beveling ($\alpha = 0°$).

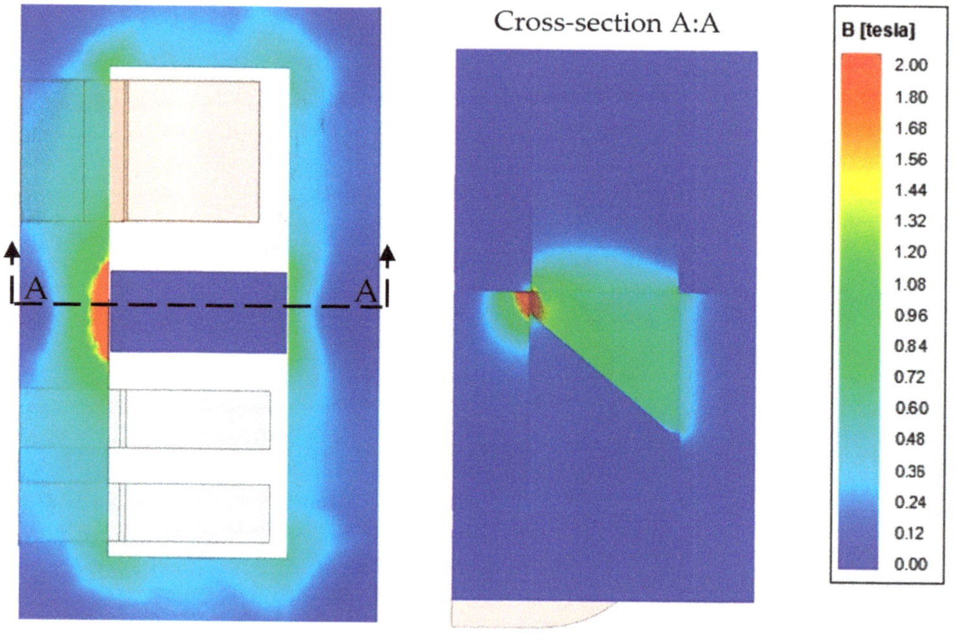

Figure 12. Leakage flux density for w = 90 mm magnetic shunt extraction and the shunt (α = 45°).

3.3. *Short-Circuit Current and Leakage Reactance for Various Positions and Shunt Chamfering (Beveling)*

Based on the magnetic field distribution, the transformer leakage reactance was determined for each beveling and shunt position. Based on this, the current I_2 related to the secondary winding was calculated, with the primary side powered with a voltage equal to 380 V. When the shunt is fully inserted into the transformer window, its beveling in the range of 0–20° does not affect transformer leakage reactance. Further angle α increasing shortens the active length of the shunt width located in the transformer window, which translates into a decrease in maximum leakage reactance value. This situation could be interpreted as an intermediate shunt position in the transformer window. On the opposite, for the extreme case when the magnetic shunt is fully extracted, the shunt beveling will have no significant effect on the leakage reactance, which is almost constant. Considering the above, at small bevel angles, the short-circuit reactance characteristic in the initial stage of shunt extraction is somewhat flattened. However, in the final phase of shunt extension, the reactance values change significantly. Beveling the shunt below 20 degrees does not affect the leakage reactance adjustment range. For bevel values above 20 degrees, the maximum leakage reactance value decreases, reducing the adjustment range while increasing the smoothness of the secondary winding short-circuit current regulation. The reactance changes impact strongly the short-circuit current intensity I_2, which is visible in Figure 13.

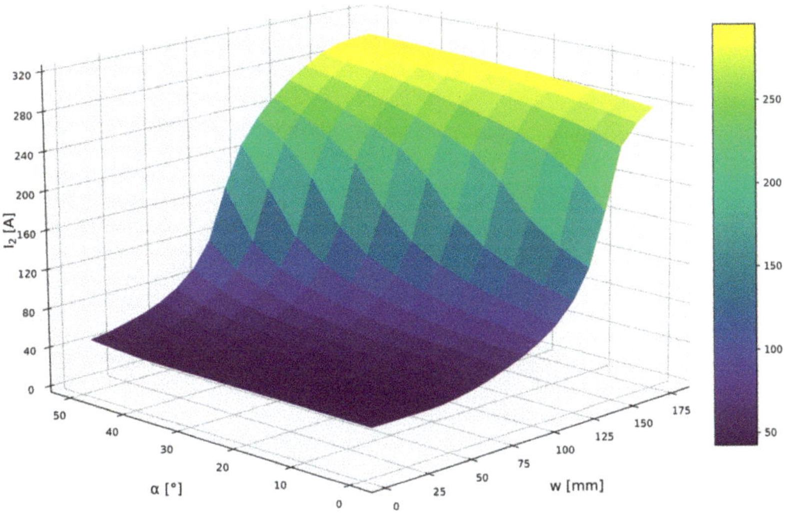

Figure 13. Short second current intensity in function of beveling and shunt extraction.

4. Measurement Verification of the Simulations

The proposed simulation model of the described transformer was verified by comparing the flux density distributions at the chosen leakage zone points and by comparing the most important integral parameter of the electromagnetic field—transformer leakage reactance. The first verification allows for a detailed assessment of the quality of the mathematical model and works very well in finding the so-called magnetic hot spots. However, a comprehensive local errors assessment would require measurements of all magnetic flux density vector components for the entire transformer leakage zone. Therefore, due to the multitude of variants of shunt positions, and current excitation, the comparison of the calculated and measured leakage reactance values was chosen as the main criteria for evaluating the model quality. Figure 14 shows the tested object with instruments that enable performing precise tests related to the magnetic shunt positioning in the transformer window.

4.1. Comparison of Three-Dimensional Magnetic Flux Density Distribution

The magnetic flux density vector components distribution comparison was limited to the magnetic shunt outer plane; see Figures 15 and 16. Their values in this area reflect the field's asymmetric nature concerning the XZ plane. The flux density values comparison was made in the short-circuit state at 4.5 mm from the shunt ferromagnetic surface, under the constant ampere-turns of the windings current flow equal θ = 1020 A. The current value was selected considering the possibility of forcing the same flux density for a transformer with a fully inserted shunt, as well as for a transformer without a magnetic shunt. The comparison was limited to the normal component since it has a major impact on the field's nature. The principal differences between the measured values and those obtained during simulation were at the outer edges of the magnetic shunt and close to the transformer core edges. These differences are due to the higher contribution of the remaining components of the magnetic flux density vector due to the inaccuracy of the probe mounting. This problem is highlighted at a small magnetic shunt extraction, where there is a large variability in the direction of the magnetic field force lines and the flux density values on a small core segment area.

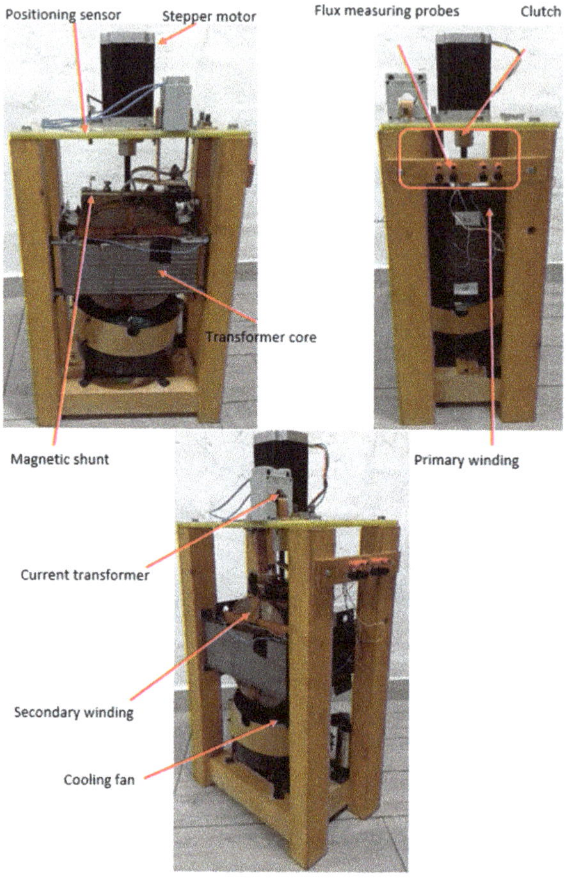

Figure 14. Object test setup with measurement instruments.

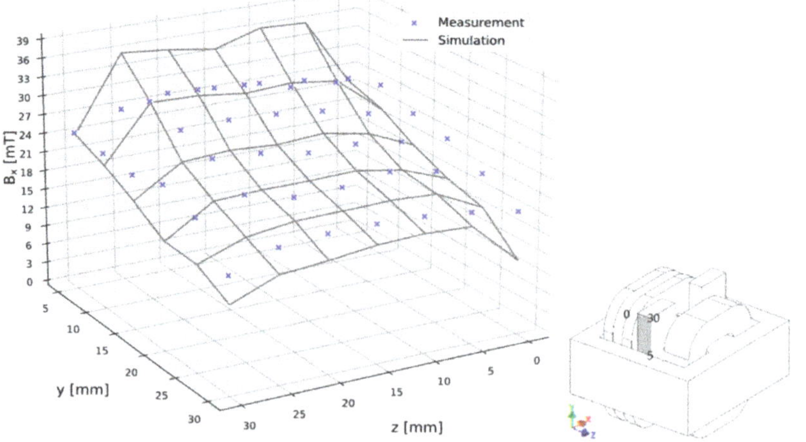

Figure 15. Comparison of the simulated and measured B_x values, of the flux density vector at the points for 4.5 mm from the shunt surface for w = 30 mm of the magnetic shunt extraction.

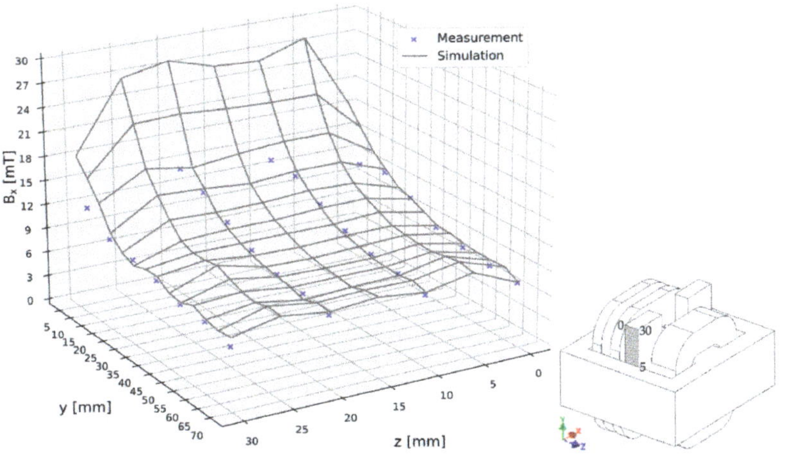

Figure 16. Comparison of the simulated and measured Bx values, of the flux density vector at the points for 4.5 mm from the shunt surface for w = 70 mm of the magnetic shunt extraction.

Figures 15 and 16 show the distribution of the normal to the magnetic shunt, Bx components on the outer shunt surface, obtained on the simulation basis along with the marked values at the measure points, with the pulling shunt out from the transformer window equal to 30 mm and 70 mm, respectively.

4.2. Comparison of the Leakage Reactances

One of the transformer's magnetic field integral parameters is its leakage inductance associated with each winding. By calculating the leakage energy stored in nonmagnetic objects, it is impossible to directly determine the part of the energy associated with a specific winding. So, these inductances could be calculated by assuming that the leakage reactance of the primary and secondary transformer windings are equal. Based on the magnetic field distribution, the static leakage inductance related to the primary winding was determined from the Expression (13), Figure 17. The static inductance associated with the secondary winding was calculated with the square of the turn's ratio $(N_2/N_1)^2$ and is considerably smaller, see Figure 18.

The leakage reactance values comparison allows for a comprehensive evaluation of the field model. Figure 19a,b show the calculated and measured values of the windings' leakage reactance transferred to the primary side.

Table 2 shows the relative error calculated from Expression (17), depending on magnetic shunt extraction w in relation to the transformer window and the primary winding current I_1.

$$\delta_{X_r} = \left| \frac{X_{rsym} - X_{rmes}}{X_{rmes}} \right| \cdot 100\% \tag{17}$$

where:

- X_{rsym}—calculated leakage reactance;
- X_{rmes}—measured leakage reactance.

Table 2. Percentage relative error value δ_{X_r} calculated at each measurement point.

Extraction w [mm]	Primary Winding Current Intensity in Amps									
	1	3	5	7	9	11	13	15	17	19
25	7.4%	2.2%	1.3%	5.7%	11.1%	-	-	-	-	-
35	8.6%	3.5%	0.5%	4.9%	9.7%	-	-	-	-	-

Table 2. *Cont.*

Extraction w [mm]	Primary Winding Current Intensity in Amps									
	1	3	5	7	9	11	13	15	17	19
45	5.6%	2.9%	0.7%	4.7%	9.1%	10.6%	-	-	-	-
55	5.6%	3.1%	0.7%	4.5%	8.3%	9.5%	-	-	-	-
65	3.5%	3.7%	0.2%	3.3%	6.7%	8.5%	9%	8.7%	-	-
75	4.3%	3.5%	0.8%	2.2%	5.3%	7.1%	8.4%	8.6%	8.5%	8.4%
85	5.1%	4.5%	2%	0%	2.8%	4.7%	6.6%	7.8%	8.1%	8.4%

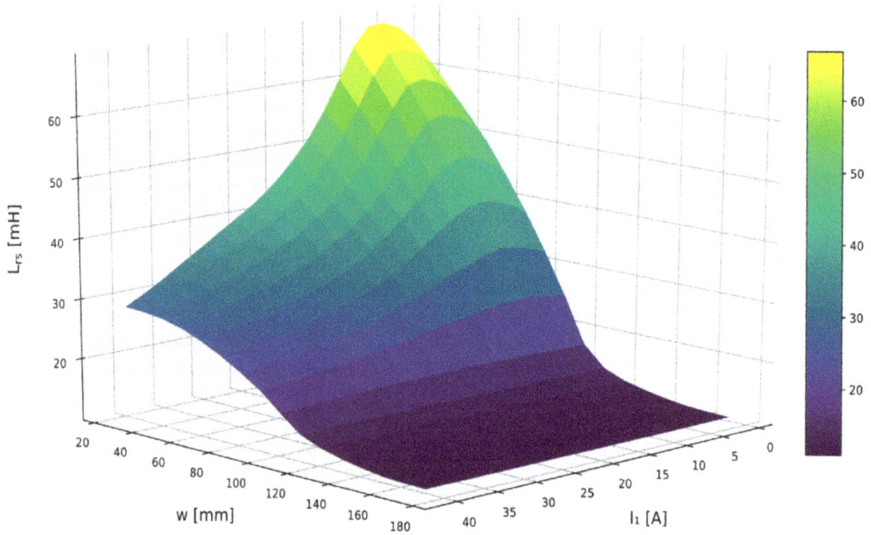

Figure 17. Calculated transformer primary winding static leakage inductance.

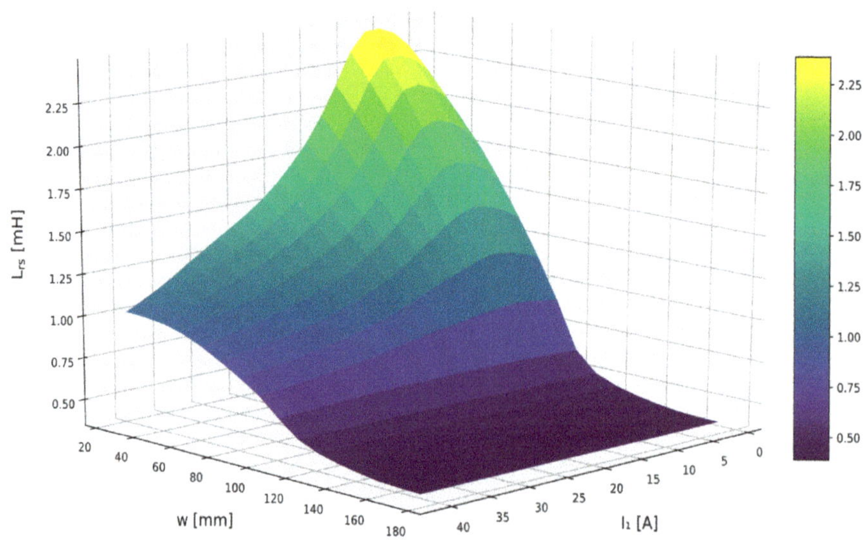

Figure 18. Calculated transformer secondary winding static leakage inductance.

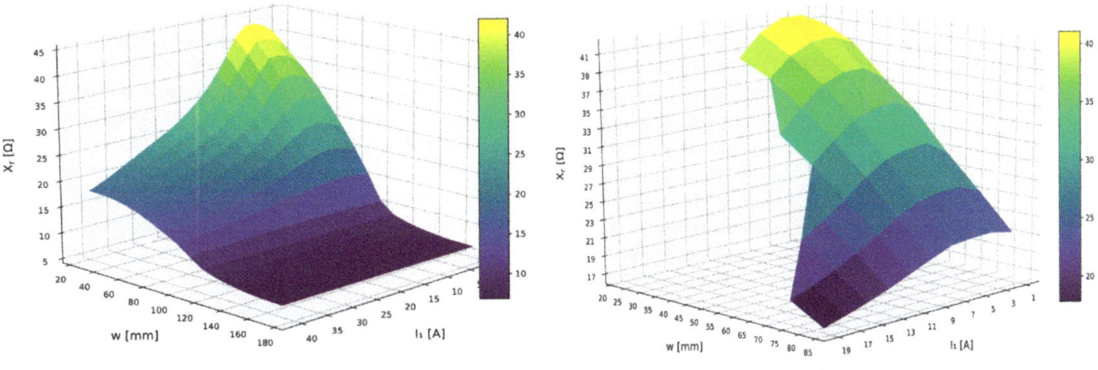

Figure 19. Leakage reactance transferred to the transformer primary winding, calculated values (**a**), measured ones (**b**).

5. Discussion

The field distribution analysis was started by determining (based on the simulations) the smallest size of three-dimensional computational space that one should consider during transformer field distribution calculation, as in Figure 3. Then, to increase the field analysis accuracy and reduce the computational costs, the symmetry plane was applied, which reduced the field analysis to half the area of the leakage region.

A comparison of the magnetic flux density components distribution confirmed the asymmetric nature of the magnetic field for intermediate magnetic shunt positions. Pulling the shunt out of the transformer window pulls the magnetic field behind it, causing its asymmetric distribution, which makes it impossible to use the symmetry with respect to the XZ plane; see Figures 15 and 16. The above is confirmed by both the simulation and measurement results.

In the leakage field sub-areas characterized by relatively lower magnetic flux density values, good agreement between simulation and performed measurements was obtained, mainly due to smaller measurement errors related to the influence of other magnetic flux density vector components, which are presented in Figures 15 and 16. Magnetic flux density components values comparison allows for assessing the object's mathematical model quality. However, a comprehensive assessment would require measurements of all magnetic flux density vector components for the entire transformer leakage area. Therefore, due to variants multitude, shunt positions, and current excitation, the determined and measured leakage reactance value comparison was chosen as the principal criterion for evaluating field model quality, performed as part of measurement verification.

The average percentage relative error between calculated and measured transformer leakage reactance values equals 5.4%. Considering performed simplifications, which included physical model dimensions averaging, assumption of air gap equality in the entire magnetic shunt positions range, as well as computational area limitation, a good agreement with the measured values was obtained, which confirms the correctness of the used method and the assumed symmetry conditions for the leakage zone.

In the HLR transformers with increased leakage reactance and the movable magnetic shunt, the main factor limiting the short-circuit current is its leakage reactance, which mainly depends on the magnetic shunt position in the transformer window, Figure 19. It reaches a maximum value for a fully inserted into the transformer window magnetic shunt and the minimum value when the transformer is without the shunt. The maximum value of the leakage reactance for the considered transformer equals 43.3 Ω and the minimal value 6.8 Ω, which gives a six-fold change in the short-circuit current in the entire adjustment range.

Increasing the shunt bevel angle α improves secondary winding current regulation linearity. Changes in this angle from 0° to 20° have almost no effect on the maximum leakage reactance value, but at the same time, they only slightly improve the above-mentioned regulation linearity. Due to the constant magnetic shunt length, further angle α increases, causing the maximum reactance value to decrease, while transformer short-circuits current control linearity is improved, Figure 13.

6. Conclusions

Three-dimensional field analysis enables the magnetic field distribution simulation of an HLR transformer with a movable magnetic shunt. This makes it possible to determine its static and dynamic characteristics over the entire range of magnetic shunt positions. The numerical calculations results give good (for technical purposes) compliance with the measurements, which confirms the developed model's correctness. This allows for determining the mentioned key parameters without the need to build expensive prototypes or perform time-consuming analytical calculations. Although the magnetic flux flowing through the magnetic shunt, inserted in the transformer window, forces the non-linear characteristic of the leakage inductance, the short circuit current control is effective. The magnetic shunt enables smooth regulation of the secondary winding current in a wide range.

Analysis of the magnetic field distribution showed that pulling the magnetic shunt out from the transformer window results in saturation of its core and shunt corner. This leads to the appearance of so-called "hot spots". A thermal analysis covering the above phenomenon will allow the design of an appropriate cooling system and maybe a conducted research continuation.

Author Contributions: Conceptualization, B.T. and D.W.; methodology, B.T. and D.W.; software, D.W.; validation, B.T. and D.W.; formal analysis and simulation, B.T. and D.W.; resources, B.T. and D.W.; data curation, D.W.; writing—original draft preparation, B.T. and D.W.; writing—review and editing, B.T.; visualization, D.W.; supervision, B.T. All authors have read and agreed to the published version of the manuscript.

Funding: This research received no external funding.

Data Availability Statement: The data presented in this study are available on request from the corresponding author.

Conflicts of Interest: The authors declare no conflict of interest.

References

1. Alves, B.S.; Kuo-Peng, P.; Dular, P. Contribution to power transformers leakage reactance calculation using analytical approach. *Int. J. Electr. Power Energy Syst.* **2019**, *105*, 470–477. [CrossRef]
2. Dawood, K.; Cinar, M.A.; Alboyaci, B.; Sonmez, O. A new method for the calculation of leakage reactance in power transformers. *J. Electr. Eng. Technol.* **2017**, *12*, 1883–1890. [CrossRef]
3. Jamali, S.; Nasir, K.; Abbaszadeh, K.; Arand, S.J. The Study of Magnetic Flux Shunts Effects on the Leakage Reactance of Transformers via FEM. *Majlesi J. Electr. Eng.* **2010**, *4*, 47–52. [CrossRef]
4. Hayes, J.G.; O'Donovan, N.; Egan, M.G.; O'Donnell, T. Inductance characterization of high-leakage transformers. In Proceedings of the Conference Proceedings—IEEE Applied Power Electronics Conference and Exposition—APEC, Miami Beach, FL, USA, 9–13 February 2003; Volume 2, pp. 1150–1156.
5. Grossner, N.R. The Geometry of Regulating Transformers. *IEEE Trans. Magn.* **1978**, *14*, 87–94. [CrossRef]
6. Koteras, D.; Tomczuk, B.; Waindok, A. Implicit iteration calculations using 3D field analyses, to predict the power loss in powder ferromagnets, with their measurement tests. *Measurements* **2023**, *207*, 112311. [CrossRef]
7. Tomczuk, B.; Koteras, D. Magnetic flux distribution in the amorphous modular transformers. *J. Magn. Magn. Mater.* **2011**, *323*, 1611–1615. [CrossRef]
8. O'Connor, P. *Test Engineering: A Concise Guide to Cost-Effective Design, Development and Manufacture*; John Wiley & Sons: Hoboken, NJ, USA, 2001.
9. Güemes-Alonso, J.A. A new method for calculating of leakage reactances and iron losses in transformers. In Proceedings of the ICEMS'2001, Fifth International Conference on Electrical Machines and Systems, Shenyang, China, 18–20 August 2001; Volume 1, pp. 178–181.

10. Kashtiban, A.M.; Hagh, M.T.; Milani, A.R.; Haque, M.T. Finite Element Calculation of Winding Type Effect on Leakage Flux in SinglePhase Shell Type Transformers. In Proceedings of the 5th WSEAS International Conference on Applications of Electrical Engineering, Prague, Czech Republic, 12–14 March 2006; pp. 39–43.
11. Jahromi, A.N.; Faiz, J.; Mohseni, H. A fast method for calculation of transformers leakage reactance using energy technique. *IJE Trans. B Appl.* **2003**, *16*, 41–48.
12. Mechkov, E.; Tzeneva, R.; Mateev, V.; Yatchev, I. Thermal analysis using 3D FEM model of oil-immersed distribution transformer. In Proceedings of the 19th International Symposium on Electrical Apparatus and Technologies (SIELA), Bourgas, Bulgaria, 29 May–1 June 2016.
13. Ahmad, A.; Javed, I.; Nazar, W.; Mukhtar, M.A. Short Circuit Stress Analysis Using FEM in Power Transformer on H-V Winding Displaced Vertically & Horizontally. *Alex. Eng. J.* **2018**, *57*, 147–157.
14. Naranpanawe, L.; Saha, T.; Ekanayake, C. Finite element analysis to understand the mechanical defects in power transformer winding clamping structure. In Proceedings of the IEEE Power & Energy Society General Meeting, Chicago, IL, USA, 16–20 July 2017; pp. 1–5.
15. Gowri, N.V.; Akthar, Z. Finite Element Analysis of Power Transformer. *Int. J. Mech. Eng. Technol. (IJMET)* **2019**, *10*, 1384–1391.
16. Sieradzki, S.; Rygal, R.; Soiński, M. Apparent Core Losses and Core Losses in Five-Limb Amorphous Transformer of 160 kVA. *IEEE Trans. Magn.* **1998**, *3*, 1189–1191. [CrossRef]
17. Mousavi, S.; Shamei, M.; Siadatan, A.; Nabizadeh, F.; Mirimani, S.H. Calculation of Power Transformer Losses by Finite Element Method. In Proceedings of the IEEE Electrical Power and Energy Conference (EPEC), Toronto, ON, Canada, 10–11 October 2018; pp. 1–5. [CrossRef]
18. Zaitsev, I.; Bereznychenko, V.; Bajaj, M.; Taha, I.B.M.; Belkhier, Y.; Titko, V.; Kamel, S. Calculation of Capacitive-Based Sensors of Rotating Shaft Vibration for Fault Diagnostic Systems of Powerful Generators. *Sensors* **2022**, *22*, 1634. [CrossRef] [PubMed]
19. Tomczuk, B. Calculation of the Magnetic Field and Short-Circuit Reactance of Dissipation Transformers. Ph.D. Thesis, Lodz University of Technology, Lodz, Poland, 1985.
20. Zakrzewski, K.; Tomczuk, B. Application of the finite difference method for the windings reactance calculations of the stray transformers. *Arch. Elektrotechniki PWN* **1990**, *37*, 43–50. (In Polish)
21. Tomczuk, B. Three-Dimensional Leakage Reactance Calculation and Magnetic Field Analysis for Unbounded Problems. *IEEE Trans. Magn.* **1992**, *28*, 1935–1940. [CrossRef]
22. Tomczuk, B. Analysis of 3-D Magnetic Fields in High Leakage Reactance Transformers. *IEEE Trans. Magn.* **1994**, *30*, 2734–2738. [CrossRef]
23. Kulkarni, S.V.; Khaparde, S.A. *Transformer Engineering: Design, Technology, and Diagnostics*, 2nd ed.; CRC Press: Boca Raton, FL, USA, 2012.
24. Ansys Maxwell 3D User's Guide. Available online: https.www.ansys.com/ (accessed on 13 June 2023).
25. Binns, K.J.; Lawrenson, P.J.; Trowbridge, C.W. *The Analytical and Numerical Solution of Electric and Magnetic Fields*; John Wiley and Sons: New York, NY, USA, 1993.
26. Sykulski, J.K. *Computational Magnetics*; Chapman & Hall: London, UK, 1995.
27. Bandelier, B.; Rioux-Damidau, F. Mixed finite element method for magnetostatics in ℝ3. *IEEE Trans. Magn.* **1998**, *34*, 2473–2476. [CrossRef]
28. Li, Y.; Berthiau, G.; Feliachi, M.; Cheriet, A. 3D finite volume modeling of ENDE using electromagnetic T-formulation. *J. Sens.* **2012**, *2012*, 785271. [CrossRef]
29. Specogna, R.; Suuriniemi, S.; Trevisan, F. Geometric T-Ω approach to solve eddy currents coupled to electric circuits. *Int. J. Numer. Methods Eng.* **2008**, *74*, 101–115. [CrossRef]
30. Codecasa, L.; Dular, P.; Specogna, R.; Trevisan, F. A perturbation method for the T- ω geometric eddy-current formulation. *IEEE Trans. Magn.* **2010**, *46*, 3045–3048. [CrossRef]
31. Farahani, H.F.; Zare, M.; Mohammad, S.; Razi, P.; Khodakarami, A. Finite Element Analysis of Leakage Inductance of 3-Phase Shell-Type and Core Type Transformers. *Res. J. Appl. Sci. Eng. Technol.* **2012**, *4*, 1721–1728.

energies

Article

Selection of the Winding Type of Solid-State Transformers in Terms of Transmitting the Greatest Possible Power in the Frequency Range from 500 Hz to 6000 Hz

Elzbieta Lesniewska [1,*] and Daniel Roger [2]

1 Institute of Mechatronics and Information Systems, Lodz University of Technology, 90-924 Lodz, Poland
2 LSEE ER4025, University of Artois, 62400 Béthune, France; daniel.roger@univ-artois.fr
* Correspondence: elzbieta.lesniewska-komeza@p.lodz.pl; Tel.: +48-426312694

Abstract: Solid-state transformer (SST) is an emerging technology that integrates with transformer power electronics converters and control. The most noticeable advantages of SST are the size and weight reduction compared with low-frequency transformers. Since this device is used in many devices such as smart grids, traction systems, systems with renewable energy sources (RESs) and electric vehicle charging devices, it is important to build a high-efficiency device at a low cost. The article evaluates a medium frequency transformer (SST) operating at a frequency of 500 Hz to 6000 Hz with coils wound with aluminum foil or Litz windings and of a grain-oriented electrical steel (GOES) core. The calculations were made using the 3D field method using the numerical finite element method, and the results were compared with the tests of the real model. The measurement method based on the Fourier analysis of real signals was used for the research.

Keywords: medium frequency transformer (SST); wound core; design methods; FEM; Litz winding; aluminum foils winding

Citation: Lesniewska, E.; Roger, D. Selection of the Winding Type of Solid-State Transformers in Terms of Transmitting the Greatest Possible Power in the Frequency Range from 500 Hz to 6000 Hz. *Energies* **2023**, *16*, 6528. https://doi.org/10.3390/en16186528

Academic Editors: Abu-Siada Ahmed and Sérgio Cruz

Received: 31 July 2023
Revised: 28 August 2023
Accepted: 7 September 2023
Published: 11 September 2023

1. Introduction

Medium Frequency Solid State Transformers (SST) are part of a new technology that integrates transformers with power electronics converters and controls.

The importance of this topic is demonstrated by articles dealing with SST transformers. Examples are article [1] for 166 kW and 20 kHz transformer with copper windings, very high copper losses and cooling, and [2] for material compatibility design, insulation, leakage inductance, and temperature limits for medium frequency transformer with Litz winding and nanocrystalline core rated at 1 kV and 200 kVA, and frequency 10 kHz. In article [3], however, the relevant SST topologies for various voltage levels and with different degrees, their control action, and multiple trends in applications were comprehensively reviewed; this did not apply to foil windings but only to the Litz type.

These are special high-power transformers that operate practically at short circuits.

The idle state for this type of transformer is an emergency state, not an operating state. Therefore, it is tested at a measurement short-circuit, and the parameters of the transverse branch tested in the idling state are unimportant.

The most important parameter that needs to be limited to transmit the greatest power possible is leakage inductance. Because the leakage flux, which does not pass through the secondary windings, partially closes through the core and causes unnecessary power loss.

With high-power Solid-State Transformers (SST) above 500 kVA, losses in electronic components are very high [4,5]. For a long time, efforts have been made to achieve higher power densities in energy conversion units, especially in applications such as wind farms and electric traction [4,6]. Reducing the weight and volume of the device parts is the simplest solution here. The use of higher frequencies allows for higher power densities. A 500 Hz transformer is more than three times lighter than a 50 Hz equivalent [7].

Higher frequencies increase the density of losses in the core and the windings [4–8], but the advantage is the reduced size and weight of the device. Medium high-frequency transformers are the central components of SSTs; they are placed between two electronic converters. The leakage inductances of these transformers must be as small as possible [8].

The article presents a preliminary windings design of a 500 kVA medium frequency transformer with a core made of grain-oriented electrical steel (GOES). Research using GOES wound cores to construct high-power medium-frequency SST transformers began several years ago [9–11]. The first step was to define the field of application of such cores before ferrite and nanocrystalline materials were explicitly developed for high-frequency transformers. Initial studies showed that it was interesting to raise the core temperature of GOES, as the magnetic properties of GOES are maintained up to 500 °C. Higher temperatures correspond to lower losses [12]. The analytical study also proves that in GOES cores at medium frequencies, core losses are lower for rectangular voltages than sinusoidal voltages [11]. A special thermal model was also developed to enable a hot core inside the coils at much lower temperatures [13]. Experimental parts of these studies show that with 0.1 mm thick GOES strips, it is not possible to operate above 6000 Hz.

On the other hand, below 500 Hz SST technologies are not attractive. Therefore, the test is carried out between these frequencies. This first step focuses on the hot core of the GOES, not the windings.

The article considered two winding technologies: coils made of Litz wire and coils wound with aluminum foil. Contacts with manufacturers show that implementing large-section Litz wires is very expensive; therefore, it has been decided to study interleaved aluminum foil windings adapted to skin and proximity effects and able to provide very low leakage inductance. A significant article is the article [14] concerning the use of foil windings and related problems. However, the article concerns a power transformer, but these are phenomena occurring usually when using foil windings.

Previously used two-dimensional methods FEM requires thousands of finite element simulations on foil and round wires to account for the edge effect to solve a similar problem [15]. Therefore, 3D simulation was used to calculate the leakage resistance and inductance for the proposed structures.

The first part of the paper is a numerical study that provides a first idea of leakage inductance. The small values obtained allow us to build the interleaved foil coils with a transformer manufacturer's standard tools. The objective was not to make an ideal winding but a realistic one at a reasonable cost.

The work aimed to investigate the behavior of transformer windings made of thin aluminum foil instead of Litz windings at different power frequencies. A prototype winding made of aluminum foil was built in the second stage. The short-circuit test was carried out at a current of 100 A and a frequency of 500 Hz to 6000 Hz.

The experimental part gives the actual values of the leakage inductances considering the actual shapes of the prototype. Measurements were made at a relatively low current (100 A peak) to limit the temperature increase during tests. On the other hand, the available Tektronix A621 precision AC probe gives maximum accuracy at a peak current of 10 A, 100 A, or 1000 A.

Previously, transformer tests were carried out based on approximate analytical models [4,16–18], which did not allow for considering the physical phenomena and dependencies resulting from a three-dimensional system. Many authors currently applied the three-dimensional finite element method for the design of electromagnetic devices [19,20], both power transformers [20–22], instrument transformers [23], motors [24], etc. It is used to design magnetic circuits in steady [23] and transient states [25] when supplied with various currents and voltages (also distorted [26]) and insulation of devices [25,27].

The work of a medium-frequency transformer with a GOES wound core and aluminum foil wound coils was assessed in the article. The standard winding technology, based on Litz-type copper wires, is compared with a low-cost solution based on interwoven aluminum foils. The criteria for comparison are the leakage inductances. The calculations were made using the 3D field-circuit method using the numerical finite element method, considering the real three-dimensional phenomena occurring in the core and windings of the foil. This method allows for high modeling accuracy. During the design, it was also checked by the field method whether the use of a nanocrystalline material for the core instead of GOES electrical steel for the core would reduce the leakage induction. Experimental studies on a real model confirmed the results of the method used. The simulation results show much lower leakage inductances for the winding made with interleaved aluminum foils. This result helped the authors build a full-size prototype that applies the principle of interleaved aluminum-foil winding.

The following parts of this article are structured as follows. Section 2 presents a structure of a geometric 3D model of the medium frequency transformer and magnetization characteristics of cores. Section 3 describes the mathematical model of the transformer. Section 4 describes the result of computer simulations of tested transformer construction examples, including various core materials, foil, and Litz windings. Section 5 describes the prototype and results of the measurement method based on the Fourier analysis of real signals. Section 6 compares the simulation and experimental studies results and explains the higher leakage inductances of the prototype by considering the winding building processes. Practical conclusions from the research are presented in Section 7.

2. Construction of Medium Frequency Transformer

The core of this special transformer differs significantly from the design of power transformer cores. It consists of two parts wound from thin strips of ferromagnetic sheets and is not composed of sheets (like cores from power transformers). The GOES wound cores are made with strips wound on a steel mandrel whose shape defines the internal dimensions of the core. Core losses are acceptable thanks to using a GOES sheet with a thickness of 0.1 mm. After a thermal annealing cycle that eliminates the internal stresses, the mandrel is removed, and the wound core is cut. The cores are cut so that the windings can be put on. The cutting planes are polished with great accuracy with the range of 0.01 to 200 μm. Thus, the value possible for the air gap is from 0.02 to 400 μm. The half-cores are delivered in pairs and must be mounted together in the direction defined by the colored dots affixed by the manufacturer. The two half-cores are held together by a clamp band, which imposes a well-defined mechanical tension (70 kg for the experimental GOES transformer). Figure 1 presents the general shape of such wound cores, and Figure 2 shows the magnetization curves of both materials. Litz wire winding has a standard construction of two concentric coils, without interleaving or with interleaving turns of bundles, which needs no further explanation. The coil structure made of aluminum foil spacers is shown in Figures 3 and 4.

Figure 1. The wound core considered for both technologies (nanocrystalline material or grain-oriented electrical steel) of the considered transformer.

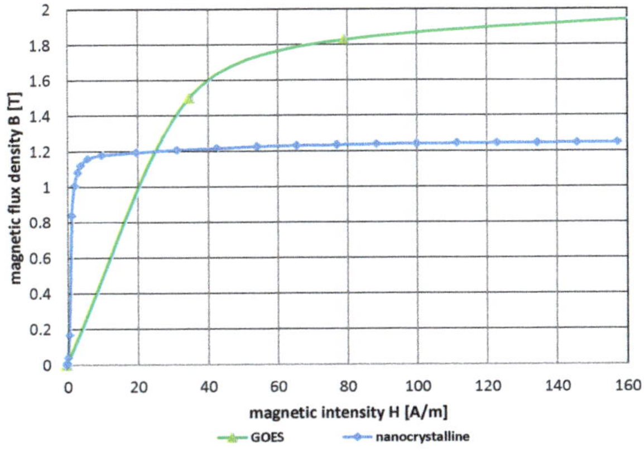

Figure 2. Magnetization curve of the nanocrystalline magnetic material [28] and grain-oriented electrical steel (GOES) [29].

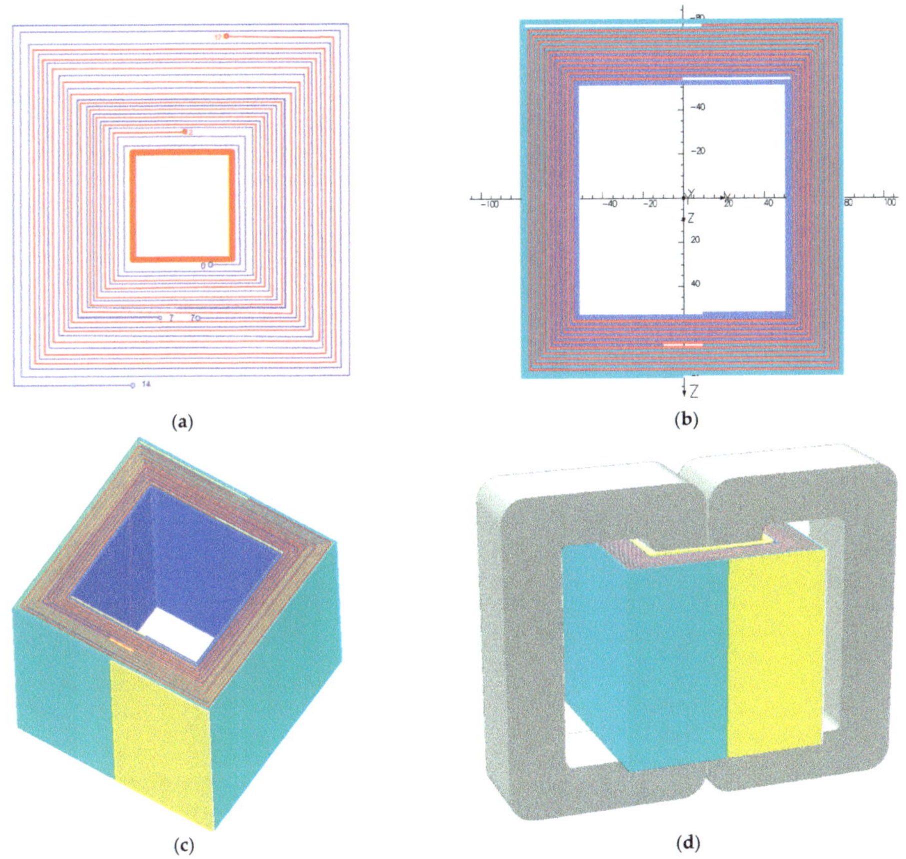

Figure 3. Interleaved foil coils of primary and secondary windings, (**a**) design (**b**) simulation—top view (**c**) 3D—interleaving foil coils (**d**) 3D model of the medium frequency transformer, colors: primary coil-red, secondary coils-blue and aquamarine, insulation-yellow, wound core-grey.

The tested transformer has windings made of aluminum foil, a primary coil of 10 turns, and two secondary coils of 7 turns. Primary and two secondary winding are interleaving together (Figures 3 and 4). The windings are made of 1 mm thick aluminum foil with 0.2 mm insulation.

The work aimed to compare the behavior of transformer windings made of thin aluminum foil or Litz windings. Two variants of the Litz winding were considered without interleaving (sequentially wound primary winding and then two secondary windings) and with interleaving of the primary and two secondary windings (Figure 5).

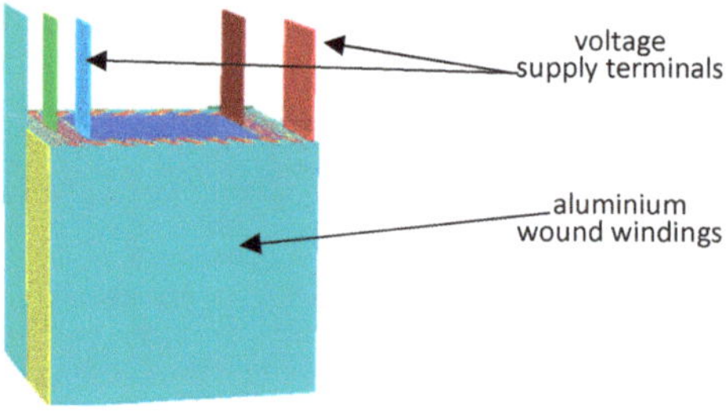

Figure 4. 3D view of the interleaving aluminum foil winding with voltage supply terminals, colors: primary coil-red, secondary coils-blue and aquamarine, insulation-yellow, potential zero-green.

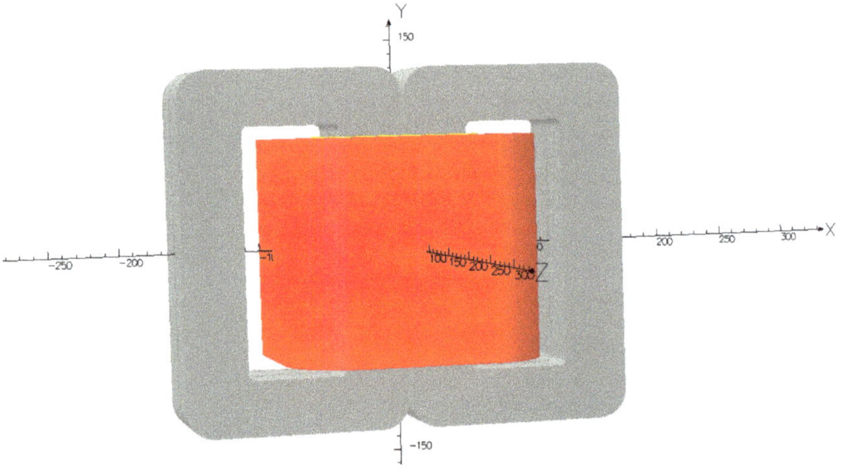

Figure 5. 3D model of the medium frequency transformer with the Litz winding (dimensions in millimeters).

Calculations were also made for a transformer with the same core of grain-oriented electrical steel but with Litz windings.

Windings type Litz with technical data for frequencies up to 10 kHz were used:

- wire diameter—0.2 mm
- number of copper wires in a bundle—100
- cross-sectional area of copper in a bundle—3.14 mm^2
- bundle outer diameter 2.89 mm.

3. Mathematical Model of the Medium Frequency Transformer

The application of the aluminum foil windings forming the medium frequency transformer coils was analyzed using the 3D field method based on Maxwell's equations. Using this method, it was possible to calculate both the currents and voltages induced in the windings of the transformer and the distribution of the magnetic field. The equation of the Helmholtz type for the harmonic electromagnetic field was used. The complex potential of the magnetic vector $\underline{\boldsymbol{A}}$ ($\underline{\boldsymbol{B}} = \nabla \times \underline{\boldsymbol{A}}$) applied in the steady state of the transformer leads to the equation of the Helmholtz-type time-harmonic equation

$$\nabla^2 \underline{\boldsymbol{A}} - \mu \nabla \left(\frac{1}{\mu} \right) \times \nabla \times \underline{\boldsymbol{A}} - j\omega\mu\sigma\underline{\boldsymbol{A}} = -\mu \underline{\boldsymbol{J}_w} \tag{1}$$

where $\underline{\boldsymbol{A}}$ is the complex magnetic vector potential, $\underline{\boldsymbol{B}}$ complex magnetic induction, $\underline{\boldsymbol{J}_w}$ is the complex vector of primary current density, ω is the angular frequency of voltage, μ is the permeability, σ is the conductivity of materials.

At the boundary of the area, i.e., the transformer with the surrounding air, there are the following boundary conditions for the magnetic vector potential $\underline{\boldsymbol{A}} \times \boldsymbol{n} = 0$, which means that only the tangential part of the vector $\underline{\boldsymbol{B}}$ exists on the boundary and $\boldsymbol{n}$ is a normal vector to the surface boundary.

This mathematical model (1) was used, with the nonlinear magnetic characteristics B-H of the ferromagnetic materials (nanocrystalline magnetic material and GOES) considered. As in previous research, a method was considered for power losses in laminated cores [30,31].

The commercial software OPERA 3D was used to compute field distributions. For steady state, the ELEKTRA STEADY-STATE module was used. The numbers of subdivision mesh elements for both models were determined by accuracy analyses performed by refining the mesh until the results did not change. The aluminum foil, 1 mm thick, was divided into four elements of 0.25 mm each and the height (where it is not of special importance) into 6 mm. The thickness of the insulation was 0.2 mm. As a result of the accuracy analysis, a mesh of 2,072,724 elements was adopted for all model variants with foil windings and 757,994 elements for variants with Litz windings. In the second case, it was possible to consider 1/4 of the system due to symmetries. Calculations for all variants of 3D models of transformers were made on the same principles.

Due to the possibility of performing tests, both calculations and tests were performed in the short-circuit state of the transformer (Figure 6). In this state, it is possible to measure the leakage inductance of a transformer; this inductance represents a magnetic field that does not connect both primary and secondary windings and is closed by air.

$$L_z = L_{1r} + L'_{2r} \tag{2}$$

where L_z short-circuits inductance, L_{2r}, L'_{1r} leakage inductance appropriately of the primary and secondary windings.

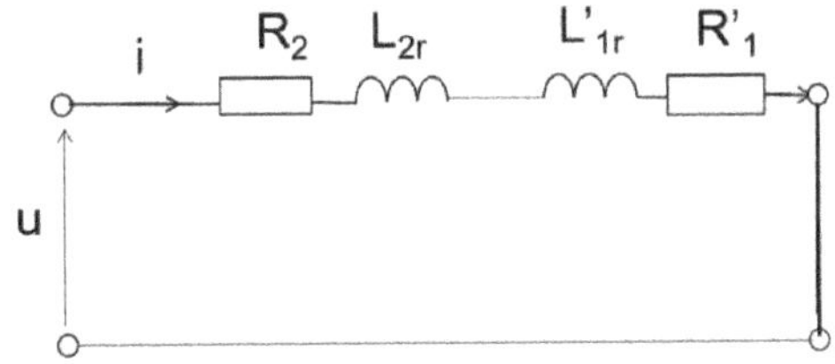

Figure 6. The equivalent circuit of a short-circuited transformer seen from the secondary winding.

The short-circuit inductance is the sum (2) of the leakage inductances of the primary and secondary windings. The value of this inductance should be reduced to a minimum. Therefore, the research concerns this transformer parameter.

4. The Result of the Computation

4.1. Comparison of the Materials Used for the Core

The calculations and tests present an assessment of the operation of a medium-frequency transformer with a wound core of grain-oriented electrical steel (GOES) (Figure 2) and coils wound with aluminum foil at a short circuit and frequency of 500 Hz to 6000 Hz.

The computation and tests were carried out in the same state of the transformer. The 10-turn primary coil of the tested transformer was short-circuited, while the 14-turn coil from the connection of two secondary coils of 7 turns was connected to a sinusoidal voltage of different frequencies.

Verification of previous tests using the field method confirmed that the use of nanocrystalline material cores instead of grain-oriented electrical steel (GOES) in the leakage inductance calculations gave practically the same result (Table 1). Differences in the distribution of magnetic induction result only from slight differences in the magnetic permeability of both materials (Figure 2).

Table 1. The short-circuit inductance of transformers with interleaved foil coils.

Frequency	GOES	Nanocrystalline Material
6000 Hz	0.1987 μH	0.1965 μH
4000 Hz	0.2308 μH	0.2080 μH
3000 Hz	0.2739 μH	0.2468 μH
2000 Hz	0.2732 μH	0.2459 μH
500 Hz	0.3101 μH	0.2804 μH

During the measuring short circuit, the core is unsaturated (Figure 7). However, significant differences will occur when the transformer is loaded with high power (i.e., high current will flow) because the nanocrystalline material, despite its high magnetic permeability, saturates very quickly (Figure 2). Then, the transformer fails to fulfill its task. Therefore, grain-oriented electrical steel is not only cheaper but also a better material for cores.

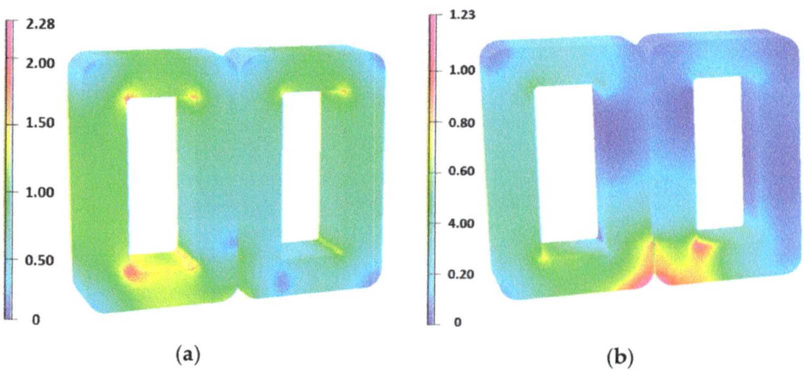

Figure 7. Distributions of magnetic induction (mT) in the cores of transformers with interleaved foil coils in the short-circuit state at 6000 Hz power supply, (**a**) core made of GOES (**b**) core made of nanocrystalline material.

4.2. *Comparison of the Use of Aluminum Foil Windings or Litz Windings*

To obtain the same flow in the Litz winding as in the foil windings, the area occupied by the windings in the core window must be increased (Figures 3d and 5). This is because each interleaving copper wire in a bundle has its own insulation.

During the measurement short-circuit, when using foil windings, the core is very slightly saturated (Figure 7), and when using Litz windings, the core induction decreases even more (Figure 8). In contrast, leakage inductance increases because most of the magnetic flux passes through non-magnetic materials such as air and insulation (Table 2). Also, the use of interlaced Litz windings makes the induction in the core higher than non-interlaced, and the leakage inductance is slightly smaller. However, in relation to the leakage inductance, it is still an order higher (Table 2).

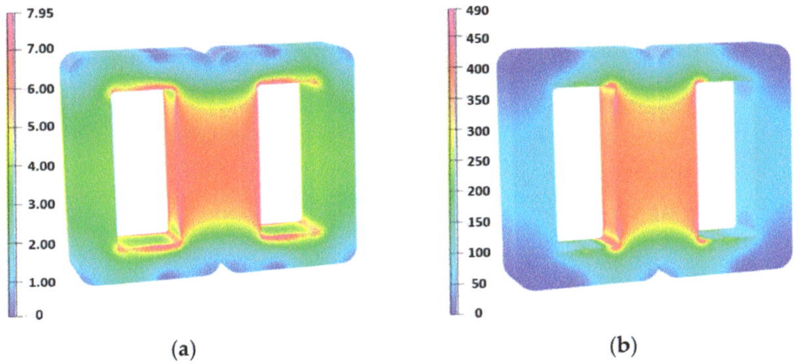

Figure 8. Distributions of magnetic induction (μT) in the cores of transformers with cores made of GOES in the short-circuit state at 6000 Hz power supply, (**a**) Litz windings without interleaving (sequentially wound primary winding and then two secondary windings), (**b**) Litz windings with interleaving of the primary and two secondary windings.

Table 2. The short-circuit inductance of transformers GOES core with Litz windings and foil windings at 6000 Hz.

Litz Windings		Foil Windings
Without Interleaving	**With Interleaving**	**With Interleaving**
9.1056 μH	3.8245 μH	0.1987 μH

The short-circuit inductance, and thus the leakage inductance, are much higher in the case of Litz windings. This allows us to conclude that the interleaving windings made of aluminum foil are much more advantageous in this respect (Table 2).

In addition, it is not without significance that copper, although it has a higher conductivity of approx. 1.6 times 5600 S/mm than aluminum 3600 S/mm, the winding is much more expensive. The price of copper is about 5.4 times higher than aluminum, and the bundle Litz is more labor-intensive, which also affects its price.

4.3. *Influence of Power Frequency on the Short-Circuit Reactance and Inductance of Transformers with GOES Core and Interleaved Foil Coils*

The results of resistance and inductance calculations at different voltage frequencies, based on field calculations for transformer with GOES core and interleaved foil coils, are given in Table 3. The leakage reactance decreases with increasing supply frequency.

Table 3. The short-circuit inductance and reactance of transformers with GOES core and interleaved foil coils.

Frequency	500 Hz	2000 Hz	3000 Hz	4000 Hz	6000 Hz	8000 Hz	10,000 Hz
Inductance	0.31015 μH	0.27323 μH	0.27395 μH	0.23082 μH	0.19873 μH	0.17670 μH	0.16176 μH
Reactance	3.4603 mΩ	4.1988 mΩ	4.8324 mΩ	5.5133 mΩ	5.9340 mΩ	7.4873 mΩ	8.0932 mΩ

The change in resistance with increasing frequency is noticeable only at very high frequencies and results from the skin effect and the phenomenon of the proximity of conductors (Figure 9, Table 3).

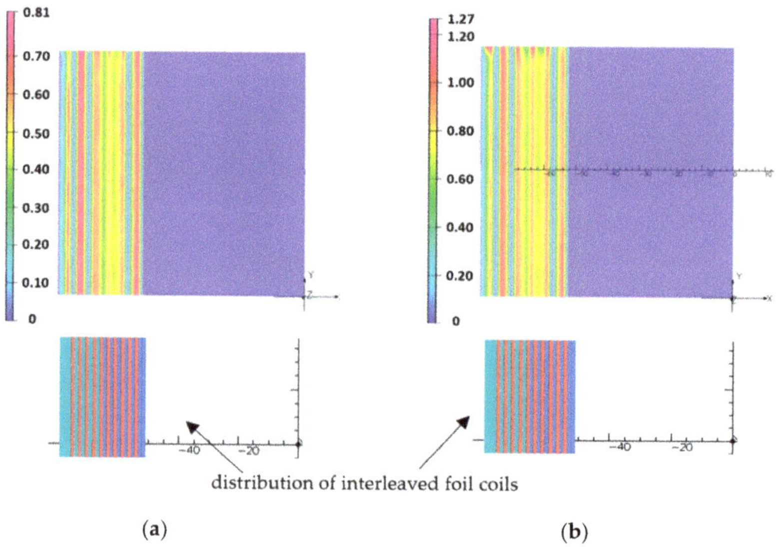

Figure 9. Distributions of current density (A/mm^2) in the cross-section of foil windings in a transformer with a GOES core and interleaved foil coils at a supply frequency of (**a**) 6000 Hz and (**b**) 3000 Hz.

5. Prototype Description and Measurements

The 3D finite element simulation shows that the interleave aluminum foil winding structure yields a lower leakage inductance than the classical Litz one. Therefore, a prototype has been built for performing experimental verifications. Figure 10 presents this winding with 1 mm thick aluminum foils insulated by a 0.2 mm mylar sheet.

Accurate measurements are obtained when the recorded signal levels are significantly above the surrounding noise level. In the case of the short-circuit test of the transformer, the impedances are low; consequently, it is important to make measurements at high currents for obtaining voltages of the order of one volt. Considering the available current probe, the peak current was fixed to 100 A peak. At medium or high frequencies, the skin effect in the conductors is an important element that depends on frequency. Therefore, the measurements must be made with sine waves containing only one frequency. In practice, producing sinusoidal currents of such levels is very difficult. This problem has been solved using the actual voltage produced by an inverter that is not sine and decomposition in Fourier series for computing the resistance and the inductance with the first harmonic. Figure 11 presents the testing circuit; the feeding transformer provides high currents (100 A peak) with a standard inverter that operates up to 6000 Hz. It also provides galvanic insulation that connects the test point directly to the oscilloscope, accurately measuring the instantaneous voltage v_p. The Tektronix A621 precision AC probe is used for the current

i_p. This probe has a bandwidth of 50 kHz. For each command frequency, the oscilloscope records the instantaneous values with 100,000 points on several periods and provides the corresponding data files to a computer.

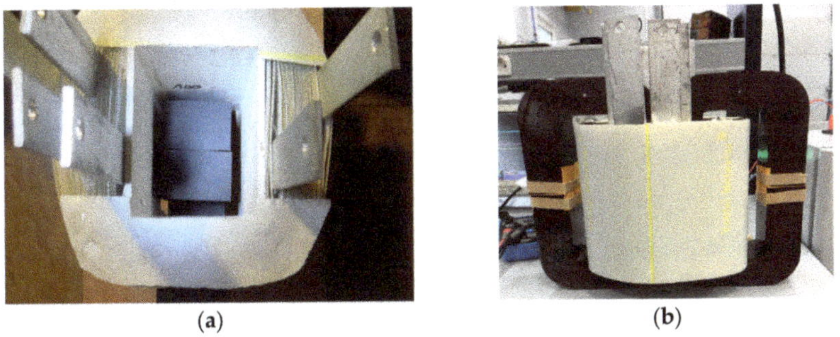

Figure 10. (**a**) The picture taken from above the interleave foil windings and (**b**) view of the prototype with its GOES wound core.

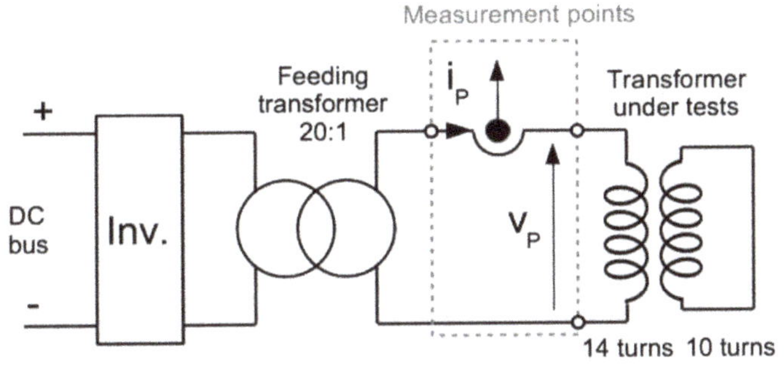

Figure 11. Short-circuit tests, measurement circuit [32].

The phase lag introduced by the limited bandwidth of the current probe is compensated by filtering the recorded data for the voltage: a Butterworth first-order low-pass numeric filter at 50 kHz is used for the voltage data, while the current data are used directly. This filter introduces the same phase lag on voltage as the current probe. It also eliminates the fast transients due to the classical high-frequency perturbations due to the inverter's fast switching.

Figure 12 presents an example of the recorded and filtered signal measured at 5 kHz. The current y-scale is on the left, and the voltage is on the right. High-frequency perturbations are observed on the recorded signal v_P (red curve), while the natural filtering of the current probe cleans the recorded current i_P (black curve). The blue curve shows the filtered voltage, and the high-frequency perturbations are eliminated.

A Fourier series of the recorded current and the filtered voltage is computed using an integer number of periods of the signals. The equivalent resistance and inductance are calculated with the first harmonic of the filtered voltage and current. Figures 13 and 14 show the recorded signal and the Fourier composition using ten harmonics. They also show the first harmonic of the voltage and od the current.

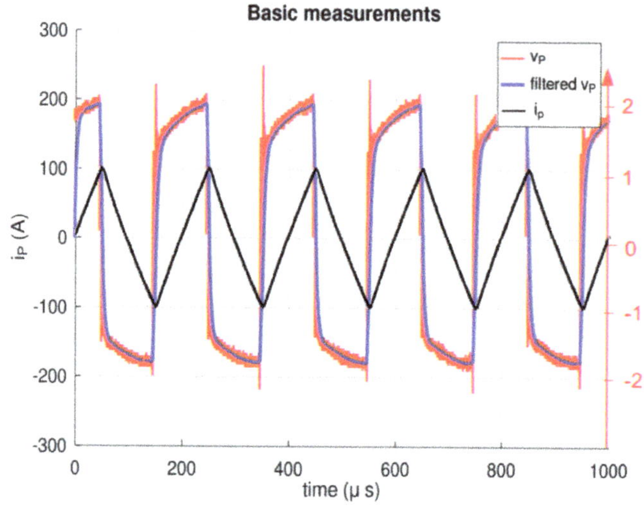

Figure 12. Example of instantaneous voltage and current.

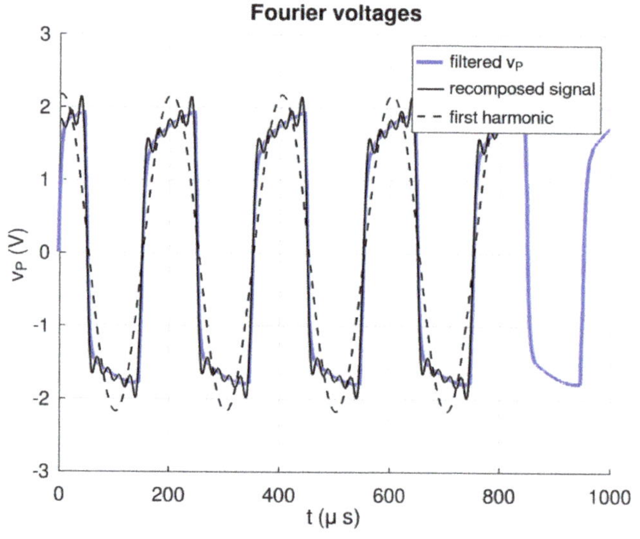

Figure 13. Filtered voltage, Fourier recomposed voltage with ten harmonics and the first one.

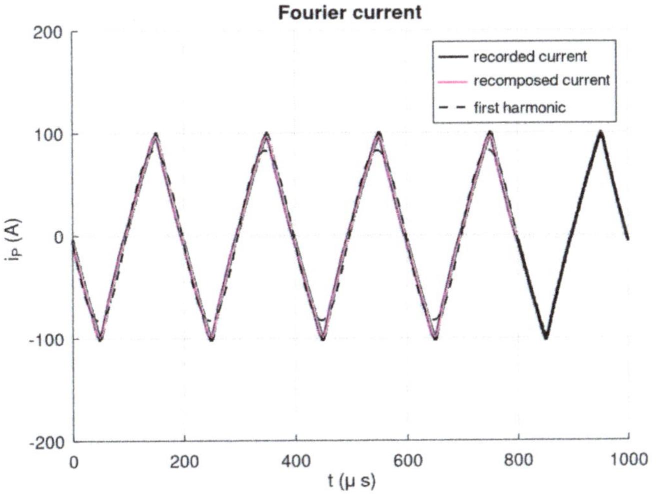

Figure 14. Recorded voltage, Fourier recomposed signal and first harmonic.

6. Comparison of Simulation and Experimental Results

Measurement errors are estimated considering that the oscilloscope (with the probes) acquires each signal with an error equal to ±1% of the channel's full range. Therefore, the 100 A probe provides an absolute error of ±1 A on each recorded point, corresponding to a relative error of ±1.21% for the peak value of the first harmonic. The voltage is measured using a 5 V full scale; the absolute error is ±0.05 V on each recorded point, corresponding to a relative error of ±2.31% for the first harmonic peak value. The relative error on the complex impedance is estimated to be ±3.52%, supposing that the numeric process that uses the recorded files is exact.

Results of the relative change of short-circuit resistance and leakage inductance versus frequency are presented in Figure 15a,b.

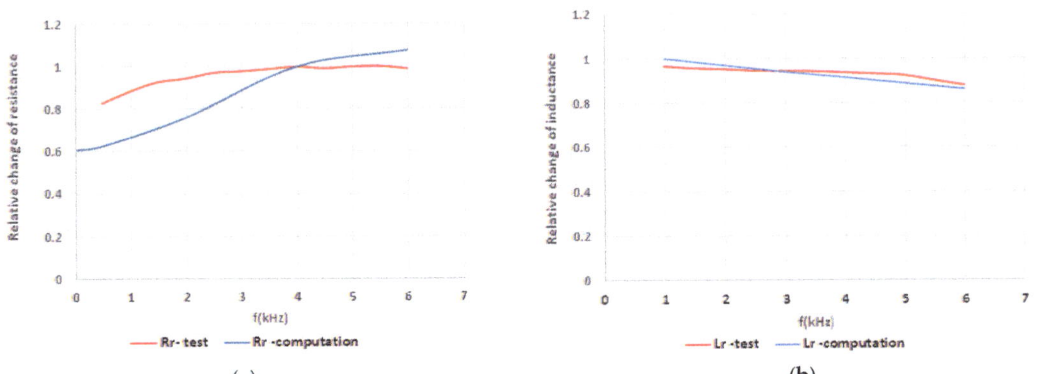

(a) (b)

Figure 15. Relative change of short-circuit (**a**) resistance and (**b**) leakage inductance versus frequency.

The results of measurements and calculations of the short-circuit resistance and leakage induction of the transformer differ by about 17%, but the values at 3000–4000 Hz are very similar.

The differences are much more significant for leakage inductance. The greater spacing between the aluminum foils can explain the higher leakage inductance measured on the prototype. During the winding process, the mechanical stresses of the aluminum foils wound on a square mandrel cause close contact with the 0.2 mm thick insulation layer only in the corner. Therefore, the actual shape differs from the theoretical; the distances between the foils are not constant but are much more significant in the middle of each side of the square. The leakage flux is greater due to the greater distance between the foils; therefore, the leakage inductance is higher.

There are considerable differences between simulation and measurement. In the 3D model, according to the design, the insulation between the foils was 0.2 mm, i.e., along the width of the windings along the core window, the insulation cross-section of the windings is 1.2 mm × 24 = 28.8 mm, and the insulation between the turns takes 4.8 mm and the same in the perpendicular direction (Figures 3 and 9).

However, in the prototype (Figure 10) in the same cross-section, it is approx. 38 mm, i.e., 38 mm ÷ 24 = 1.58 mm, and the insulation and air together give 13.98 mm. It should be added that six connection bars 32 × 5 mm are attached to the windings. Bars are connected perpendicularly on different sides and in other places of windings (Figure 10). Additional air gap dimensions due to connections can be estimated at approx. 5 mm, i.e., 18.98 mm in total. This is a significant difference. It causes a different value of the leakage inductance than assumed. The proportion of gaps between the prototype and the 3D model is about four times. Similarly, the measured value of the leakage inductance at 6000 Hz is 0.795 μH, and the calculated value is 0.19873 μH, nearly four times higher.

The above phenomenon is described in detail in the article [14] for a different type of transformer with foil windings. Of course, this is not a linear relationship.

In the case of resistance, the calculated resistance of the foil windings increases with increasing frequency due to the skin effect and proximity of foils, which is visible in Figure 9. However, the measurements on the prototype also show an almost constant value of resistance, which is related to significant air gaps and the lack of proximity effect.

The effect of frequency on changes in inductance is the same in both cases (Figure 15).

However, the measured leakage inductance remains very low (less than 1 μH), four times less than the value calculated for interlaced Litz coils. The inductance is, therefore, minimal, but it strictly depends on the spacing between successive layers of windings. In the target device, it will be necessary to use some technology to ensure the adhesion of the layers of the windings (probably gluing during winding).

7. Conclusions

For designing high-power SSTs, the MFT leakage inductance is a critical parameter.

As stated previously, the correct operation of the system requires obtaining the smallest value of the leakage inductance because the power that can be transferred through the transformer depends on it. So far, leakage inductance calculations have been based on approximate analytical formulas or 2D simulations. Unfortunately, in the 2D simulation, only a part of the windings in the transformer window is modeled, and the remaining significant portion of the winding, about 70% working in the air, is not modeled. Therefore, there are substantial differences between simulation and measurement. The second problem is the edge effect and proximity phenomena, which, apart from the skin effect, cause a considerable unevenness of the current distribution in the foil winding. As the analysis shows, the end effect (uneven current distribution at the height of the winding foil) is more important than the skin effect itself.

For this reason, only the 3D model allows us to consider all these phenomena (correct field distribution inside and outside the transformer window). For the over 2 million elements, it is practically impossible to 3D model the entire transformer under load. The control method would require a time-step solution with a sufficiently short time step, which exceeds the current calculation possibilities even when using parallel calculations. We tried to carry out such simulations in 2D, and the calculation time was very long. Therefore, the 3D harmonic variant using the complex representation of field quantities seems to be the best solution for now.

The use of 3D modeling was critical in determining the choice of winding type. This first step convinced the researchers that the chosen structure in the form of an interlaced aluminum foil winding could give good experimental results; this paved the way for the construction of a prototype. Measurements of the actual resistance and leakage inductance confirm this trend despite slightly higher leakage inductances (Figure 15), explained by the winding construction process. In addition, the experimental phase using the prototype identifies critical elements of the winding design that can be improved if lower leakage inductances are required.

Further work on the medium frequency transformer (SST) design operating at a frequency of 500 Hz to 6000 Hz will include an analysis of the interaction of the transformer with the converters and the load and, therefore, further optimization of the structure. The next step towards a more powerful SST will be a detailed study of its ability to operate at higher voltages compatible with a new generation of high-power electronic switches.

Author Contributions: Conceptualization: E.L. and D.R.; methodology: D.R. and E.L.; software: E.L.; validation: E.L. and D.R.; formal analysis: E.L.; investigation: D.R.; resources: E.L.; data curation: E.L. and D.R.; writing—original draft preparation: E.L. and D.R.; writing—review and editing: E.L.; supervision: E.L. All authors have read and agreed to the published version of the manuscript.

Funding: This research received no external funding.

Data Availability Statement: The data presented in this study are available on request from the corresponding author. Computer data is not publicly available because it is not suitable for use by other researchers.

Conflicts of Interest: The authors declare no conflict of interest.

References

1. Ortiz, G.; Leibl, M.; Kolar, J.W.; Apeldoorn, O. Medium Frequency Transformers for Solid-State-Transformer Applications—Design and Experimental Verification. In Proceedings of the IEEE 10th International Conference on Power Electronics and Drive Systems (PEDS), Kitakyushu, Japan, 22–25 April 2013; pp. 1285–1290. [CrossRef]
2. Chen, B.; Liang, X.; Wan, N. Design Methodology for Inductor-Integrated Litz-Wired High-Power Medium-Frequency Transformer with the Nanocrystalline Core Material for Isolated DC-Link Stage of Solid-State Transformer. *IEEE Trans. Power Electron.* **2020**, *35*, 11557–11573. [CrossRef]
3. Hannan, M.A.; Ker, P.J.; Lipu, M.S.H.; Choi, Z.H.; Rahman, M.S.A.; Muttaqi, K.M.; Blaabjerg, F. Blaabjerg State of the Art of Solid-State Transformers: Advanced Topologies, Implementation Issues, Recent Progress and Improvements. *IEEE Access* **2020**, *8*, 19113–19132. [CrossRef]
4. Bahmani, M.A.; Thiringer, T.; Kharezy, M. Design Methodology and Optimization of a Medium-Frequency Transformer for High-Power DC–DC Applications. *IEEE Trans. Ind. Appl.* **2016**, *52*, 4225–4233. [CrossRef]
5. Rylko, M.; Hartnett, K.; Hayes, J.; Egan, M. Magnetic material selection for high power high frequency inductors in dc-dc converters. In Proceedings of the 2009 Twenty-Fourth Annual IEEE Applied Power Electronics Conference and Exposition, Washington, DC, USA, 15–19 February 2009; pp. 2043–2049. [CrossRef]
6. Dujic, D.; Kieferndorf, F.; Canales, F.; Drofenik, U. Power electronic traction transformer technology. In Proceedings of the 7th International Power Electronics and Motion Control Conference, Harbin, China, 2–5 June 2012; Volume 1, pp. 636–642. [CrossRef]
7. Bahmani, M.A.; Thiringer, T.; Ortega, H. An accurate pseudo empirical model of winding loss calculation in HF foil and round conductors in switchmode magnetics. *IEEE Trans. Power Electron.* **2014**, *29*, 4231–4246. [CrossRef]
8. Meier, S.; Kjellqvist, T.; Norrga, S.; Nee, H.-P. Design considerations for medium-frequency power transformers in offshore wind farms. In Proceedings of the 2009 13th European Conference on Power Electronics and Applications, Barcelona, Spain, 8–10 September 2009; pp. 1–12.
9. Villar, I. Multiphysical Characterization of Medium-Frequency Power Electronic Transformers. Ph.D. Thesis, École Polytechnique Fédérale De Lausanne, Lausanne, Switzerland, 2010. [CrossRef]
10. Kumar, B.M.; Kumar, A.; Bhat, A.H.; Agarwal, P. Comparative study of dual active bridge isolated DC to DC converter with single phase shift and dual phase shift control techniques. In Proceedings of the 2017 Recent Developments in Control, Automation & Power Engineering (RDCAPE), Noida, India, 26–27 October 2017; pp. 453–458. [CrossRef]
11. Ruiz-Robles, D.; Ortíz-Marín, J.; Venegas-Rebollar, V.L.; Moreno-Goytia, E.; Granados-Lieberman, D.R.; Rodríguez-Rodriguez, J. Nanocrystalline and Silicon Steel Medium-Frequency Transformers Applied to DC-DC Converters: Analysis and Experimental Comparison. *Energies* **2019**, *12*, 2062. [CrossRef]
12. Roger, D.; Rossi, M.; Ichou, H.; Blaszkowski, J. Magnetic behavior of GOES wound cores of transformers fed by square or sine voltages. *J. Magn. Magn. Mater.* **2022**, *564*, 170032. [CrossRef]

13. Roger, D.; Napieralska, E.; Komeza, K.; Napieralski, P. Design of High−Power Solid−State Transformers with Grain−Oriented Electrical Steel Cores. *Electronics* **2022**, *11*, 2398. [CrossRef]
14. Diaz, G.A.; Mombello, E.E.; Venerdini, G.D.G. Calculation of Leakage Reactance in Transformers with Constructive Deformations in Low Voltage Foil Windings. *IEEE Trans. Power Deliv.* **2018**, *33*, 3205–3210. [CrossRef]
15. Ichou, H.; Roger, D.; Rossi, M. Thermal behavior of grain-oriented electrical steel wound core solid-state transformer. *COMPEL* **2022**, *41*, 732–751. [CrossRef]
16. Jezierski, E. *Power Transformer*; WNT: Warsow, Poland, 1983; pp. 1–650, ISBN 5905279053229. (In Polish)
17. Wiszniewski, A. *Instrument Transformers in the Power Industry*; WNT: Warsow, Poland, 1983; pp. 1–302, ISBN 8320403707. (In Polish)
18. Koszmider, A.; Olak, J.; Piotrowski, Z. *Current Transformers*; WNT: Warsow, Poland, 1985; pp. 1–283, ISBN 8320407109. (In Polish)
19. Asghari, B.; Dinavahi, V.; Rioual, M.; Martinez, J.A.; Iravani, R. Interfacing Techniques for Electromagnetic Field and Circuit Simulation Programs IEEE Task Force on Interfacing Techniques for Simulation Tools. *IEEE Trans. Power Deliv.* **2009**, *24*, 939–950. [CrossRef]
20. Amoiralis, E.I.; Georgilakis, P.S.; Tsili, M.A.; Kladas, A.G. Global transformer optimization method using evolutionary design and numerical field computation. *IEEE Trans. Magn.* **2009**, *45*, 1720–1723. [CrossRef]
21. Tong, Z.; Braun, W.D.; Rivas-Davila, J. Design and Fabrication of Three-Dimensional Printed Air-Core Transformers for High-Frequency Power Applications. *IEEE Trans. Power Electron.* **2020**, *35*, 8472–8489. [CrossRef]
22. Konrad, A.; Brudny, J.F. Virtual air gap length computation with the finite-element method. *IEEE Trans. Magn.* **2007**, *43*, 1829–1832. [CrossRef]
23. Lesniewska, E. Applications of the Field Analysis during Design Process of Instrument Transformers. In *Transformers. Analysis, Design, and Measurement*; Lopez-Fernandez, X.M., Ertan, B.H., Turowski, J., Eds.; CRC Press Taylor & Francis Group: Boca Raton, FL, USA; London, UK; New York, NY, USA, 2012; pp. 349–380.
24. Dems, M.; Komeza, K.; Kubiak, W.; Szulakowski, J. Impact of Core Sheet Cutting Method on Parameters of Induction Motors. *Energies* **2020**, *13*, 1960. [CrossRef]
25. Lesniewska, E.; Olak, J. Analysis of the Operation of Cascade Current Transformers for Measurements of Short-Circuit Currents with a Non-Periodic Component with a Large Time Constant of Its Decay. *Energies* **2022**, *15*, 2925. [CrossRef]
26. Lesniewska, E.; Kaczmarek, M.; Stano, E. 3D Electromagnetic Field Analysis Applied to Evaluate the Accuracy of a Voltage Transformer under Distorted Voltage. *Energies* **2021**, *14*, 136. [CrossRef]
27. Lesniewska, E.; Olak, J. Improvement the Insulation System of Unconventional Combined Instrument Transformer Using 3D Electric Field Analysis. *IEEE Trans. Power Deliv.* **2018**, *33*, 2582–2589. [CrossRef]
28. *KING MAGNETICS Catalogue*; Zhuhai King Magnetics Technology Co., Ltd.: Zhuhai, China. 2018. Available online: http://www.kingmagnetics.com/kingmagnetics_catalog.pdf (accessed on 31 July 2023).
29. Isolectra Martin Company Catalogue. Available online: https://www.isolectra.fr/catalogue_isolectra_en (accessed on 31 July 2023).
30. Markovic, M.; Perriard, Y. Eddy current power losses in a toroidal laminated core with rectangular cross section. In Proceedings of the 12th International Conference on Electrical Machines and Systems, Tokyo, Japan, 15–18 November 2009; pp. 1–4.
31. Zirka, S.E.; Moroz, Y.I.; Marketos, P.H.; Moses, A.J. Evolution of the loss components in ferromagnetic laminations with induction level and frequency. *J. Magn. Magn. Mater.* **2008**, *320*, e1039–e1043. [CrossRef]
32. Roger, D.; Napieralska, E.; Komeza, K.; Napieralski, P. Solid-state transformers of smart high-power battery charger for electric vehicles. *IEEE Trans. Ind. Appl.* **2023**, 1–10. [CrossRef]

energies

MDPI

Article

Field Test Method and Equivalence Analysis of Delay Characteristics of DC Electronic Current Transformer

Mengmeng Zhu [1], Hongda Tang [2], Zhaolei He [3], Yaohua Liao [1], Biao Tang [1], Qunqun Zhang [2], Hongchun Shu [2], Yaqi Deng [2], Fang Zeng [2] and Pulin Cao [2,]*

[1] Electric Power Research Institute of Yunnan Power Grid Co., Ltd., Kunming 650200, China; zhumengmeng@dlyjy.yn.csg.cn (M.Z.); liaoyaohui@dlyjy.yn.csg.cn (Y.L.); tangbiao@dlyjy.yn.csg.cn (B.T.)
[2] Faculty of Electric Power Engineering, Kunming University of Science and Technology, Kunming 650500, China; moguhuangei@163.com (Q.Z.); kmshc@sina.com (H.S.); dengyaqi1992@126.com (Y.D.); fanger1119@kmust.edu.cn (F.Z.)
[3] Measurement Verification Department of Measurement Center of Yunnan Power Grid Co., Ltd., Kunming 650200, China; archer_hzl@126.com
* Correspondence: pulincao_kust@sina.com

Abstract: DC electronic current transformer is the data source of the system control and protection device, so its measurement delay has always been concerned. In this paper, the DC current transformer is modelled equivalently and the key factors affecting the delay of shunt and the remote module are analyzed. A field test method for the delay characteristics of the DC electronic current transformer is proposed. Moreover, a complex multi-state large current generator and integrated test system are developed. On the basis of the IEC 60044-8 standard transmission protocol, the accuracy of standard signal acquisition with the shunt is better than 0.2%. The maximum output current of the established testing system is 600 A. Based on the field test in an actual HVDC project, transient step and various frequency signal components are applied for analyzing.

Keywords: DC electronic current transformer; group delay; field test

Citation: Zhu, M.; Tang, H.; He, Z.; Liao, Y.; Tang, B.; Zhang, Q.; Shu, H.; Deng, Y.; Zeng, F.; Cao, P. Field Test Method and Equivalence Analysis of Delay Characteristics of DC Electronic Current Transformer. *Energies* **2023**, *16*, 5727. https://doi.org/10.3390/en16155727

Academic Editors: Elzbieta Lesniewska, Xose Lopez-Fernandez and Pawel Witczak

Received: 2 May 2023
Revised: 13 June 2023
Accepted: 28 June 2023
Published: 31 July 2023

1. Introduction

Ultra (extra) high-voltage DC transmission (HVDC) is widely used in large-capacity and long-distance transmission projects, which has incomparable advantages in terms of controllability and flexibility. However, it also increases the risk of system resonance, especially in the asynchronous operation mode of the power grid. For example, the impact of harmonics on the safe and stable operation of the power grid is more prominent because harmonic resonance has a direct impact on transformer tap changers, stabilization devices, and other secondary equipment. In serious cases, it can even affect the isolation of the main equipment [1]. As a key device for the measurement data of the control and protective system, the measurement delay of DC current transformers is prone to the risk of broadband oscillations. Therefore, it is of great importance to understand the measurement delay of the direct current electronic current transformer to develop on-site testing systems for delay characteristics. It is crucial for adjusting control system circuit parameters, conducting rapid wideband oscillation suppression, and achieving precise measurements.

Nowadays, the widely used DC electronic current transformer is based on the shunt-sensing principle, which covers a whole data transmission chain, including primary sensing conversion, analog voltage acquisition, digital signal conversion, fiber optic transmission, merging unit (MU) processing, telegram transmission, control, and protection device reception processing [2]. The rated secondary signal of the DC electronic current transformer is a voltage signal of several tens of millivolts with digital information, which is different from the traditional instrumentation transformer with analog output.

In order to model and calibrate DC current transformers, the reference [3] proposes a laboratory calibration method for DC current transformers and conducts research on the uncertainty assessment of calibration results. Reference [4] presents a field calibration method for DCCT. This method synchronizes the DC current transformer field calibration by using GPS to measure the readings of the digital multimeter on the standard source side and the calibrated transformer's two ends. The measurement principle of this method and system is analyzed. If the calibrated DC current transformer has an analog output, GPS synchronization is used, and the data from the two-end digital multimeters are obtained. If the calibrated DC current transformer has a digital output, communication radios are used for synchronization. However, employing method, it is challenging to achieve on-site synchronization verification of the delay characteristics and transient characteristics of direct current electronic current transformers. This is because it relies on the stability of the current source and may not be suitable for achieving precise synchronization in the field. Reference [5] describes the design and setup of the new AC current transformer (CT) calibration system, which uses enhanced, actively compensated current comparators (CCs) for improving measurement accuracy. Reference [6] proposes a 3D magnetic-field model to analyze the impact of the magnetic sensor circular array position on current measurements, which successfully maintains accuracy class 1.0 under various position deviations in balanced or unbalanced three-phase power delivery systems. References [7,8] study from the perspective of shunt to improve measurement accuracy. Reference [9] proposes that the disparity in practical environmental conditions, which influences the proximity electric field, is a significant reason for active electronic voltage transformers exhibiting excessive errors at substation sites. The principles of selecting materials for capacitive voltage dividers and rationalization proposals for error calibration are given. In paper [10], a new method for performing power current measurement is proposed; this method involves the use of Hall sensors without iron cores, which greatly reduces the interference of environment on measurement. Currently, there is a lack of in-depth research on the delay characteristics of direct current electronic current transformers (DCCT). Due to limitations in testing methods and technical conditions, on-site calibration of delay testing under simulated operating conditions has not been effectively conducted. In particular, there is a lack of in-depth research and on-site test data analysis on the calibration of delay under different signal excitations. In DC systems, various signal waveforms occasionally appear, especially in wideband signal detection, relay protection, and control parameter adjustments, which requires more accurate measurement of the delay of direct current electronic current transformers.

The installation position of DC electronic current transformers in converter stations is dozens of meters high above the ground, so the disassembly and reconnection of high-voltage joints pose difficulties [11–14]. In order to complete the field test, this study first establishes an equivalent model for the delay characteristics of DC current transformers. Then, the impact of shunts and low-pass filters on the delay is analyzed. In the actual converter station, a field test method to measure the delay characteristics of direct current electronic current transformers is proposed. Furthermore, a complex polymorphic high-current generator and a related integrated testing system for delay characteristics are developed. Based on these devices, a field equivalent test circuit topology is constructed. Finally, the testing results in practical engineering are analyzed and compared.

2. Delay Analysis and Test Method Research

2.1. Causes of Delay

The DC electronic current transformer contains a remote module and an analog-to-digital conversion unit; hence, its structure can be simplified into three parts as shown in Figure 1. In Figure 1, R_k is the resistance of the shunt; L_k is the distribution inductance of the shunt; I_s the primary current flowing through the shunt; U_p is the secondary disconnect output voltage of the shunt.

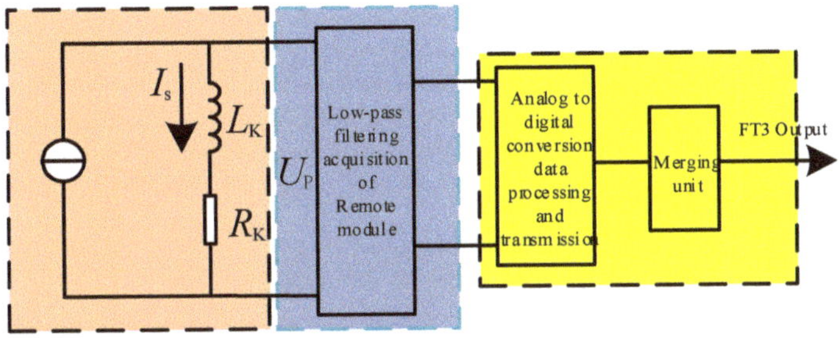

Figure 1. Schematic diagram of DC electronic transformer transmission link.

The voltage and current in Figure 1 adhere to the following equation:

$$U_P = (j\omega L_k + R_k)I_s. \tag{1}$$

L_k in Equation (1) has an impact on the transient characteristics of the shunt, not only on the phase shift of the transmitted high frequency signal but also on the rising edge of the step signal waveform. The shunt can have good transient response characteristics if L_k is reduced as much as possible. In addition, the low-pass filter in Figure 1 is generally composed of passive components. The filters commonly used in engineering are Bessel, Chebyshev, and Butterworth types, which present different transfer functions, resulting in different frequency delay characteristics.

The shunt and the low-pass filter together constitute the physical delay time part of the DC electronic current transformer. However, the delay time of the remote module after the AD acquisition is not related to input signal but to its sampling rate, processing method, and other factors, which can be called the rated delay time. The purpose of this paper is to accurately measure and calibrate the delay time of the DC electronic current transformer in actual operating conditions.

2.2. Field Equivalent Test Method for Delay Characteristics

Generally, the operating signals acting on the DC electronic transformer can be divided into two categories: steady state and transients. The steady-state signal is composed of DC component and high-frequency AC harmonics. The transient signal is the sudden change in electrical quantity after a fault or disturbance in the DC system. As a result, the delay characteristics field test circuit topology is designed as shown in Figure 2. It is mainly composed of current source, test object (DC electronic type current transformer), standard load (high precision resistor R_b), circuit control, and test system, etc.

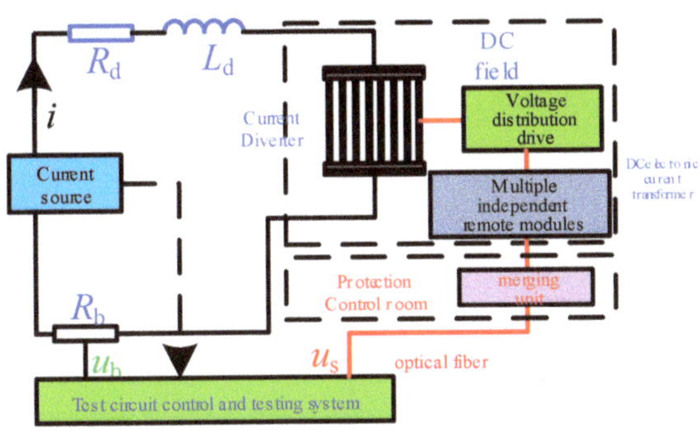

Figure 2. Equivalent test circuit topology.

The current source in Figure 2 can output many types of complex typical current waveforms. In Figure 2, R_d and L_d are the equivalent resistance and inductance of the test circuit, respectively; u_b is achieved by the high-precision resistor R_b; and u_s is the digital telegram signal output from the DC electronic current transformer under test. Moreover, the test system needs to synchronize the acquisition of two signals from the standard load and the test object to accurately quantify the delay.

3. Delay Characteristic Test System

3.1. Test System Design and Its Operating Mechanism

The core problem of field test is how to simulate the complex polymorphic signal waveforms in the actual operating conditions with a small engineering cost in the converter station. Thus, the main technical details are determined as follows:

(1) The step current signal is more or equal to 300 A (considered at 10% of the rated current of 3000 A) [15]; the rise time is less than 50 µs (one fifth of the step rise time of the DC transformer); and the duration is more or equal to 10 ms;
(2) The high-band steady frequency current amplitude is adjustable; the output maximum value is 300 A; and the frequency range covers at least 50–1200 Hz;
(3) It can output high-band frequency current with the DC component;
(4) The steady-state accuracy, frequency response, and transient step response characteristics must be tested;
(5) The digital interface of the delay testing system needs to comply with the transmission protocol in IEC 60044-8 standard;
(6) The testing system is small in size and easy to install at the site;
(7) The testing system complies with the international standard IEC 61869-14-2018.

Based on the above and the actual situation in the field, the complex multi-state current signal delay test system is established as shown in Figure 3. First, the current waveform required for the test is configured in the complex multi-state high-current source system to form a series of discrete sampling points as digital signals. Second, the signal generator converts the digital signal into an analog voltage and outputs the current through the power amplifier. Third, the P1 and P2 terminal of the high-precision resistor and the subject, in turn, receive the current. Driven by the synchronous clock, the calibrator receives the standard signal collected by the front unit and the FT3 digital signal output of the test product through two optical fibers. Finally, the calibrator completes the calculation of the delay.

A large capacity signal generator is home-made for testing the DC electronic current transformer. The output high-band frequency current of the complex high-current power supply is 600 A maximum; the frequency band covers 50 Hz~1.2 kHz; the output frequency accuracy is better than 0.01 Hz; the step signal rising time is less than 30 µs; and the steady step can last 100 ms. The sampling frequency is designed to be 500 kHz to meet the complex multi-state current signal sampling requirements. The measurement error of the calibrator does not exceed 5 µs. To verify the adaptability of the complex multi-state high current source and calibration system, the multi-state current signal source system generates step currents which are collected by the DC transformer calibrator. Figure 4 shows the current waveform at a step amplitude of 600 A with step width of 100 ms.

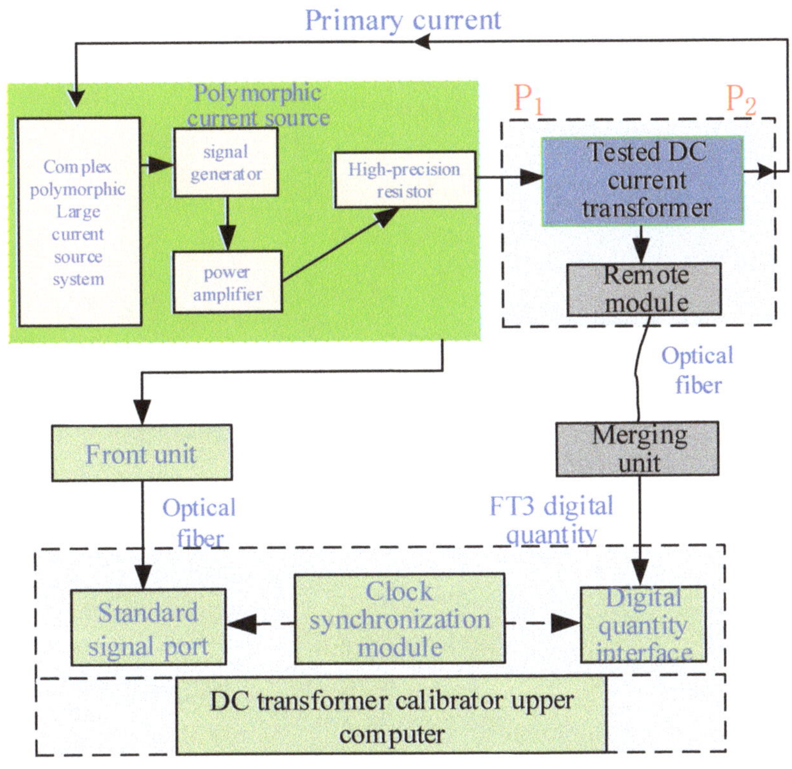

Figure 3. Time delay equivalent test system based on complex multi-state current signal.

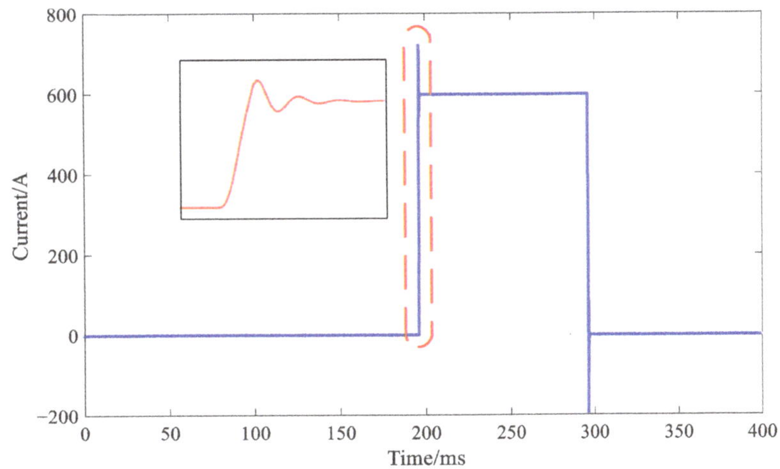

Figure 4. Transient step current waveform.

3.2. Key Parameters of Delay Characteristics

(1) Calculation method of delay of steady-state signals

The test current source outputs single frequency or DC component I_{dc} with a high-band frequency component I_{ac}, equivalent to the composition of the composite current source acting on the actual project. The expression is as follows:

$$i(t) = I_{dc} + \sum_{n=1}^{24} I_{ac}\sin(n\omega t + \varphi_n). \tag{2}$$

In this paper, the 50 Hz component, which is the power frequency in China, is applied as the fundamental frequency.

The essence of the delay calculation of the steady-state signal is to accurately calculate the phase error. In this test, the transformer calibrator is triggered by a synchronous clock with an error of less than 0.5 μs. The front unit of the standard voltage signal for high-precision acquisition extracts the phase φ_1 of the standard signal as the moment of appearance of the steady-state signal on the primary side of the DC electronic current transformer. The digital output (FT3 message) of the DC electronic current transformer under test is received in real time with the same synchronous clock. Then, the phase φ_2 of the measured transformer signal is extracted when the digital sampling value corresponds to the analog input. Assuming the current frequency is f, the absolute delay t_d is calculated by converting the phase error as follows:

$$t_d = (\varphi_2 - \varphi_1)/2\pi f. \tag{3}$$

(2) Calculation of delay of transient step signal

Figure 5 gives the extraction process of the delay of transient step signal. u(t) is the input step signal (standard acquisition signal); U_0 is the initial value of the input variable; V(t) is the output signal of DC electronic current transformer; V_0 is the initial value of the output variable before the step is applied; V_∞ is the steady-state value of the output variable after the step is applied; ΔV_s is usually 5% of V_∞; and T_{d1} and T_{d2} are the delay times at different parts of step signal, respectively.

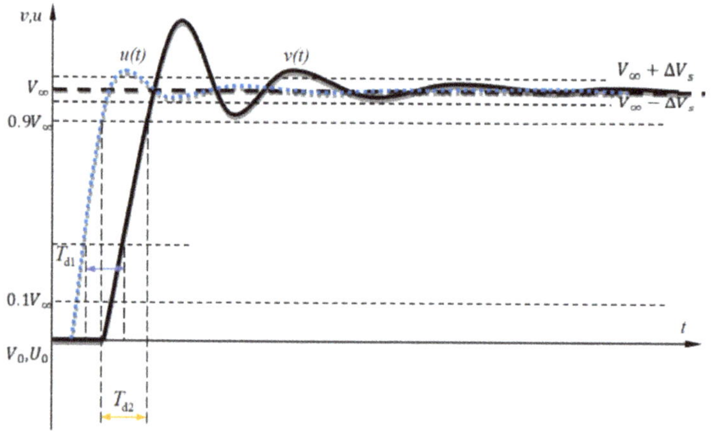

Figure 5. Schematic diagram of time delay extraction of typical transient step process.

The transient delay T_d of DC electronic current transformer is defined as

$$T_d = t_{bc} - t_{bz}, \tag{4}$$

where t_{bc} and t_{bz} are the moments when the tested DC electronic current transformer and the standard reach a certain value, respectively. First, the steady-state value, V_0, before the step and the steady-state value, V_∞, after the step are applied. Second, the moment corresponding to 10% or 90% of the variable reaching V_∞ is measured. Finally, the absolute delay of the DC current transformer is obtained.

4. Hardware Experiment

4.1. Field Test System Establishment

According to the requirements of GB/T 26216.1, IEC 61869, and other related standards, the field test system of the DC electronic transformer delay characteristics [6–10] is built as shown in Figure 6.

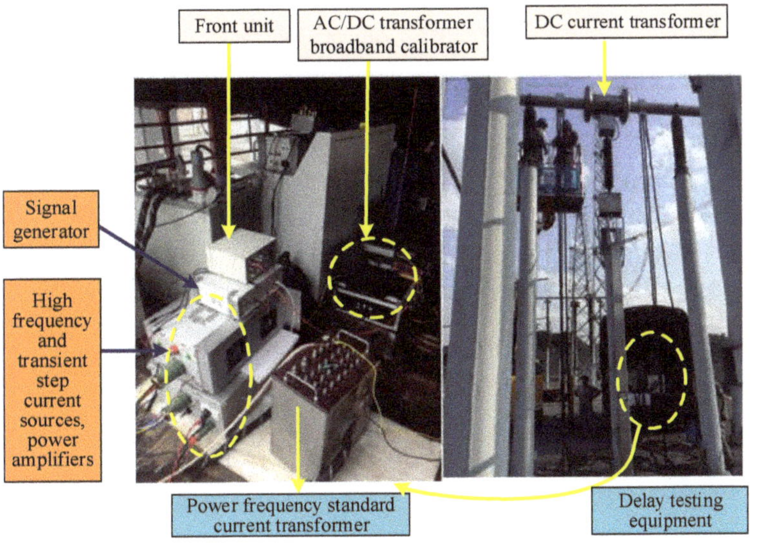

Figure 6. Schematic diagram of field test equipment.

This equipment is designed to obtain a greater primary current output with small impedance. In addition, this equipment has to be transported over a long distance of about 400 km, mostly on mountainous road, from the storage location to the convertor station. To ensure the accuracy and credibility of the field test, this equipment is tested in the field again, which verifies its effectiveness and accuracy.

The rated primary current of the DC current transformer under test is 3000 A with 10 kHz sampling rate output. First, the delay test under 50 Hz with the accuracy level of 0.01 S of 600 A current is carried out to the DC electronic current transformer and the load. The delay of 483 μs and 486 μs are calculated in two independent tests.

4.2. Field Test of Delay Time under the Action of Steady-State AC Signal

During the experiment, sinusoidal current signals from 50 Hz to 1200 Hz are sequentially applied to the DC current transformer. The calibration of the testing system is triggered and controlled by a high-precision synchronous clock. Figure 7 shows the variation of delays of different frequency components.

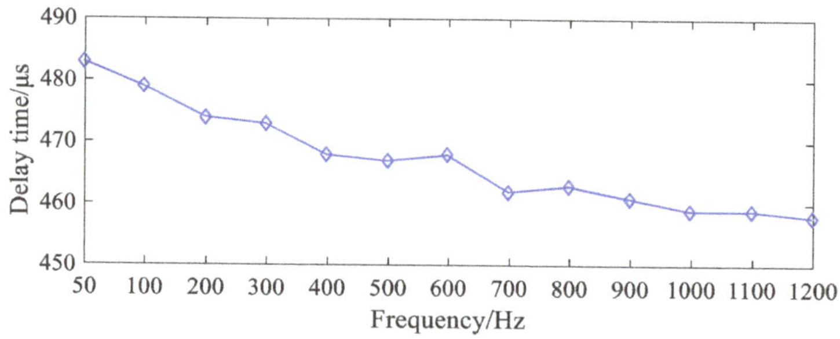

Figure 7. Curve of delay time versus frequency.

In Figure 8, it can be seen that the delay is gradually decreasing as the input current frequency increases, the value of which decreases from 483 μs to nearly 455 μs. In order to survey the influence of current amplitude, the current amplitude of 600 A, 410 A, and 365 A is applied on the primary side of the transformer, but the equivalent current does not have any significant effect on the delay.

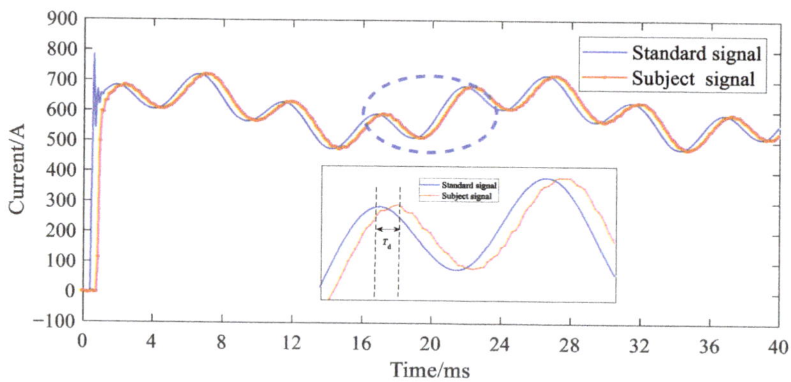

Figure 8. Curve of step signal.

In the case of HVDC system fault, AC components are able to transmit to HVDC transmission lines, which can induce large AC components superimposing DC component. Therefore, the waveform applied to the DC electronic current transformer in fault is not a single AC or DC but a composite current waveform. To be closer to the actual operating conditions in the field, waveforms with 600 A of DC, 50 A of 50 Hz, and 40 A of 200 Hz AC components are applied to the DC electronic current transformer. In Figure 8, the waveforms from the primary side and the secondary side are illustrated, and the calculated delay is 481 μs.

4.3. Field Test of Transient Step Signal

The field test result of the transient step signal on the DC electronic current transformer is shown in Figure 9.

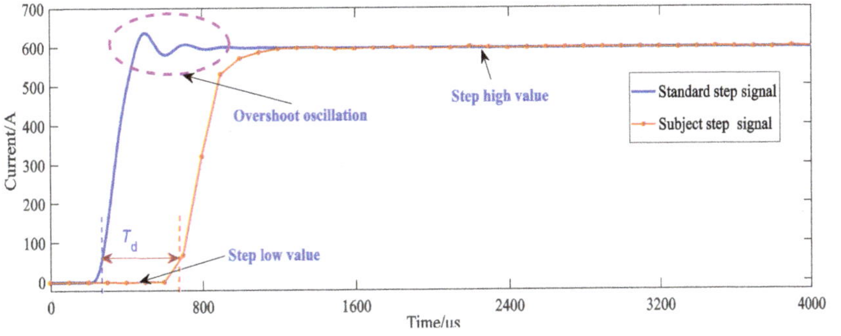

Figure 9. Field test delay waveform of transient step current.

The transient delay of the DC electronic current transformer is 415 μs at the start of rising time (the moment corresponding to 10% of step current amplitude) and 490 μs at the end of rising time (the moment corresponding to 90% of step current amplitude). There is 75 μs difference in step test, which may affect the low-pass filter by cutting off some high-frequency components.

4.4. Equivalent Analysis

This effectively analyzes the characteristic waveforms under different input to demonstrate the delay characteristics of DC electronic current transformers. These waveforms are applied to the tested DC electronic current transformers for conducting delay testing and calibration. It is necessary to ensure that the characteristic waveforms generated by the current source include single-frequency signals, DC overlaid with high-band frequency components, transient step signals, and other complex waveforms. Therefore, some requirements, e.g., 0~1500 Hz and 0~600 A current output, are imposed on the performance of the current source.

5. Conclusions

In this paper, the effects of shunt and low-pass filter on delay characteristics of DC electronic current transformers are studied. Furthermore, a field equivalent test method for the delay characteristics of DC electronic current transformers is established. Specific conclusions are as follows.

A complex multi-state high-current generation device and an integrated test system are developed, which support the IEC 60044-8 standard transmission protocol. It has a maximum output current of 600 A, with a measurement accuracy better than 0.01 Hz. Then, by simulating various signals to DC electronic current transformers, the accurate measurement and calibration of absolute delay are realized. Based on measurement results, the different frequency components are transmitted with various time delays via the transmitting of the DC electronic current transformer, which may result in the maloperation of the control system. Furthermore, the step response also reflects a nearly constant time delay of the DC electronic current transformer, even though the sampling rate of digital output of DC electronic current transformer is lower than the primary side.

Author Contributions: Conceptualization, M.Z.; formal analysis, H.T.; resources, Z.H.; methodology, Y.L.; software, B.T.; validation, Q.Z. and H.S.; investigation, Y.D. and F.Z.; project administration, P.C. All authors have read and agreed to the published version of the manuscript.

Funding: This work is supported by the Science and technology project of Yunnan Power Grid Co., Ltd. (No.YNKJXM20220161), and the Yunnan technological innovation talent training object project (No.202205AD160005).

Data Availability Statement: Data are unavailable due to privacy.

Conflicts of Interest: The authors declare no conflict of interest.

References

1. Yao, Z.; Wu, J.; Guo, X.; Zhang, C. Analysis and Suppression Control of High Frequency Resonance for MMC-HVDC System. *IEEE Trans. Power Electron.* **2017**, *32*, 3437–3452.
2. *GB/T 26216.1-2010*; DC Current Measuring Device for HVDC Transmission System—Part 1: Electronic DC Current Measuring Device. General Administration of Quality Supervision, Inspection and Quarantine of the People's Republic of China, Standardization Administration of China. Standards Press of China: Beijing, China, 2011.
3. Callegaro, L.; Cassiago, C.; Gasparotto, E. On the calibration of direct-current current transformers (DCCT). *IEEE Trans. Instrum. Meas.* **2015**, *64*, 723–727. [CrossRef]
4. Li, Q.; Li, H.; Zhou, Y.; Li, D.; Hu, H.; Zhan, S. On-site calibration technology for direct current transformers (DCCT) in ±800 kV DC transmission system converter stations. *High Volt. Eng.* **2011**, *37*, 3053–3058.
5. Mohns, E.; Roeissle, G.; Fricke, S.; Pauling, F. An AC Current Transformer Standard Measuring System for Power Frequencies. *IEEE Trans. Instrum. Meas.* **2017**, *66*, 1433–1440. [CrossRef]
6. Ma, X.; Guo, Y.; Chen, X.; Xiang, Y.; Chen, K.-L. Impact of Coreless Current Transformer Position on Current Measurement. *IEEE Trans. Instrum. Meas.* **2019**, *68*, 3801–3809. [CrossRef]
7. Zhang, J.; Pan, X.; Liu, W.; Gu, Y.; Wang, B.; Zhang, D. Determination of Equivalent Inductance of Current Shunts at Frequency Up to 200 kHz. *IEEE Trans. Instrum. Meas.* **2013**, *62*, 1664–1668. [CrossRef]
8. Pan, X.L.; Zhang, J.T.; Shao, H.M.; Liu, W.F.; Gu, Y.; Ma, X.F.; Wang, B.; Lu, Z.L.; Zhang, D.S. Measurement of the Phase Angle Errors of High Current Shunts at Frequencies up to 100 kHz. *IEEE Trans. Instrum. Meas.* **2013**, *62*, 1652–1657. [CrossRef]
9. Wu, W.; Xu, Y.; Xiao, X.; Hu, H. Research on Proximity Effect in Measuring Error of Active Electronic Voltage Transformers. *IEEE Trans. Instrum. Meas.* **2016**, *65*, 78–87. [CrossRef]
10. Chen, K.-L.; Chen, N. A New Method for Power Current Measurement Using a Coreless Hall Effect Current Transformer. *IEEE Trans. Instrum. Meas.* **2011**, *60*, 158–169. [CrossRef]
11. Jiang, J.; Zhao, M.; Zhang, C.; Chen, M.; Liu, H.; Albarracín, R. Partial Discharge Analysis in High-Frequency Transformer Based on High-Frequency Current Transducer. *Energies* **2018**, *11*, 1997. [CrossRef]
12. Chen, K.-L.; Wan, R.-S.; Guo, Y.; Chen, N.; Lee, W.-J. A Redundancy Mechanism Design for Hall-Based Electronic Current Transformers. *Energies* **2017**, *10*, 312. [CrossRef]
13. Kaczmarek, M.; Szczęsny, A.; Stano, E. Operation of the Electronic Current Transformer for Transformation of Distorted Current Higher Harmonics. *Energies* **2022**, *15*, 4368. [CrossRef]
14. Wang, D.; Liu, C. Combination Optimization Configuration Method of Capacitance and Resistance Devices for Suppressing DC Bias in Transformers. *Energies* **2019**, *12*, 1813. [CrossRef]
15. *GB/T 20840.14-2022*; Instrument Transformers—Part 14: Additional Requirements for Current Transformers for DC Applications. Standards Press of China: Beijing, China, 2022.

energies

MDPI

Review

SPICE-Aided Models of Magnetic Elements—A Critical Review

Krzysztof Górecki * and Kalina Detka

Department of Marine Electronics, Faculty of Electrical Engineering, Gdynia Maritime University, Morska 81-87, 81-225 Gdynia, Poland; k.detka@we.umg.edu.pl
* Correspondence: k.gorecki@we.umg.edu.pl

Abstract: This article analyzes the problem of modeling the properties of such magnetic elements as inductors, coupled inductors, and transformers using the SPICE software (version 17.2). Both the classical models of magnetic elements, built in this software, and the models implemented in the form of subcircuits are described. In particular, attention was paid to the possibility of taking into account the non-linearity of the characteristics of the considered elements and mutual couplings between electrical, magnetic, and thermal quantities. Using the results of thermographic measurements, the need to take into account the differences in temperature values between the individual windings and the core of inductors and transformers was justified. Selected models of the considered elements given in the literature are briefly characterized. The network structures of the electrothermal models of the considered elements elaborated at Gdynia Maritime University are presented. The results of calculations and measurements illustrating the correctness of the described models and their prac-tical usefulness for the elements of different structures are presented and discussed.

Keywords: inductors; transformers; coupled inductors; power losses; compact models; windings; ferromagnetic cores

check for updates

Citation: Górecki, K.; Detka, K. SPICE-Aided Models of Magnetic Elements—A Critical Review. *Energies* **2023**, *16*, 6568. https://doi.org/10.3390/en16186568

Academic Editor: Mario Marchesoni

Received: 2 August 2023
Revised: 3 September 2023
Accepted: 8 September 2023
Published: 12 September 2023

1. Introduction

Magnetic elements such as inductors, coupled inductors, and transformers are commonly used in different electronic and power electronic systems [1,2]. These elements perform different functions in the mentioned systems, which include, among others, electricity storage, voltage equalization and filtering, ensuring galvanic separation between circuits, etc. [3]. They are typically made of a ferromagnetic core and one or more windings wound on this core in the form of a winding wire, tape, or track on a printed circuit board [3,4].

Designers of electronic systems, before deciding on their structure, carry out proper computer analyses to design a device characterized by the most favorable values of operating parameters [5]. Thanks to computer simulations, it is possible to minimize or eliminate undesirable phenomena that may affect the correct operation of the designed systems already at the design stage [5].

When analyzing and designing such systems, computer programs supporting the work of engineers are commonly used [6]. One of the most popular programs of this type is SPICE [7,8], which has been used for several decades. It requires appropriate models of elements present in the analyzed systems. The manufacturer of this program implemented equations describing the most important electronic components, and in the attached libraries, the parameter values of such models for selected types of components, including magnetic elements such as inductors and transformers, are given. Of course, the credibility of the obtained calculation results depends on the accuracy of the models used, resulting from the consideration of the phenomena occurring in the considered elements.

According to the literature, the properties of magnetic elements significantly affect the characteristics of electronic systems [9,10]. In many papers concerning computer analysis of electronic or power electronic systems [11,12], the influence of the non-linearity of the characteristics of magnetic elements, the phenomenon of self-heating in these elements, and mutual couplings between their components are usually omitted. For example, inductors in the SPICE program are typically modeled using a linear coil model whose inductance is constant or a non-linear core model coupled with a linear winding model [13].

Despite the simple construction of magnetic elements, during their operation, electrical, magnetic, and thermal phenomena occur in them, causing deviations of these characteristics from the characteristics of ideal linear elements. These phenomena are characterized, e.g., in [14–16], by the non-linearity of the magnetization curve, the occurrence of power losses in the core and in the windings, thermal phenomena, i.e., self-heating in the core and in the windings, and mutual thermal couplings between them.

In the literature, there are also models of electronic elements, including magnetic elements, that are not dedicated to SPICE but are presented in an analytical form or are dedicated to another program, e.g., MATLAB [17–19] or others [20–22].

The paper [20] proposes a polynomial third-order model of an inductor used in electrical energy conversion systems. The modeling method proposed by the authors enables the estimation of the current up to saturation, understood as the point where the differential inductance decreases to half of its maximum value. The model also takes into account the influence of temperature on the inductor current. On this basis, the influence of the core temperature on the conduction time of the switch was determined. It was noticed from the research results that saturation requires an appropriate value of the temperature-dependent conduction time. The calculation results obtained using the non-linear model of the inductor in the DC–DC converter were experimentally verified for the boost converter.

In turn, the paper [21] proposes a method of modeling an inductor dedicated to operating in power electronic converters. The inductor model makes it possible to determine the reluctance of the equivalent circuit of this element. This model takes into account losses in the inductor core made of different ferromagnetic materials, with particular emphasis on a hysteresis loop and losses occurring in the core and the winding of the considered element. In addition, the influence of the frequency on the losses in the inductor was taken into account. Yet, this model does not take into account the influence of the magnetic flux density amplitude on the losses in the inductor core or the thermal phenomena occurring in this element.

On the other hand, the paper [22] describes behavioral modeling of the inductor used in switch-mode power supplies. The proposed modeling method takes into account the direct current flowing through the inductor and the switching frequency as input data for the model. This model takes into account current ripples, power losses, and temperature. It is based on simple analytical formulas formulated on the basis of measurements using a multi-object evolutionary algorithm. The resulting behavioral models define a "limited working area" of the inductor, which describes the boundary area in the three-dimensional domain of the switching frequency.

However, due to the wide range of material, this study focuses only on models dedicated to SPICE.

As mentioned earlier in the literature, many papers describe models of magnetic elements, but only a few are devoted to the review of such models. For example, the paper [23] is devoted to the modeling and design of transformers and coupled inductors only. The paper contains a short theoretical introduction concerning the physical phenomena occurring in the considered magnetic elements and the principles of formulating models of magnetic circuits. Magnetic and electrical models of transformers and coupled inductors proposed in the cited paper take into account phenomena such as core magnetization and stray flux. However, the considerations do not include the influence of many factors on the properties of the magnetic element, such as temperature, magnetic flux density amplitude, its shape, or the type of material from which the core was made.

On the other hand, in the papers [24,25], the review and comparison of non-linear behavioral models of the inductor dedicated to operating in power conversion systems are described. The paper [24] compares the models of the inductor by summarizing their main features, i.e., indicating the type of mathematical function used to formulate the model, the number of coefficients, and the nature of the resulting optimization problem. Typically, the models proposed in the cited paper show the method of defining the dependence of inductance on the current, i.e., using different functions and calculating its coefficients with different optimization methods. Because the described models relate to the description of flux density, they cannot be used to observe and determine the other characteristics of the inductor, for example, those related to its core and the influence of different phenomena on the core properties. It is not indicated whether the functions used to describe the dependence of inductance on different factors have the same character for the inductor, whose core is made of other materials.

In turn, the paper [25] compares two popular analytical models described in the literature. The first of these models is based on a polynomial, and the other uses the arctangent function to formulate a non-linear dependence of inductance on different factors. The efficiency of the models was compared, taking into account the method of determining the model parameters and evaluating the current profile using the characteristic equation of the inductor. Both models were implemented in switching power supply systems. The results presented in the paper take into account the time of calculations and the accuracy of modeling. The comparison of these models showed that both models ensured a good agreement between the results of calculations and measurements of the inductor operating in the power conversion system. It was noted that the duration of calculations using the polynomial model was 15% shorter than the time obtained using the model with the arctangent function.

The mentioned short reviews relate only to the behavioral models, which use different functions to describe the inductance of the inductor. In the literature, there is a lack of a review of magnetic element models in one paper (inductor, coupled inductor, transformer) that take into account their electric, magnetic, and thermal properties and can be applied to analyze the properties of magnetic elements containing different core materials.

The aim of the paper is a critical analysis of the models of magnetic elements available in the literature. Such models were reviewed, and models of three magnetic elements were considered, i.e., an inductor, a coupled inductor, and a transformer. This study deals with compact models dedicated to the SPICE program. When analyzing them, the models were divided into three groups. The first are the models built in SPICE. The second group includes non-linear isothermal models, i.e., those that ignore the phenomenon of self-heating and thermal couplings occurring between the components of the considered elements. In turn, the third group consists of non-linear electrothermal models of magnetic elements, taking into account thermal phenomena occurring in the considered elements. The descriptions of the presented models include both selected models available in the literature as well as the results of the authors' research on modeling the properties of magnetic elements, taking into account thermal phenomena occurring in them. The properties of the presented models were compared, and the ranges of their applicability were indicated. The results of the analyses of networks with magnetic elements were also shown and discussed.

Section 2 describes the models built in SPICE. Section 3 describes the factors causing the non-linearity of the characteristics of magnetic elements. Sections 4 and 5 describe non-linear isothermal and electrothermal models of the considered elements, respectively. In turn, exemplary results of the measurements and calculations obtained using the above-mentioned models are presented in Section 6.

2. SPICE Built-in Models

The user of SPICE can use the linear and non-linear models of the inductor and the transformer built into this program. These models are described, e.g., in [26].

The inductor can be represented by an ideal inductor, whose inductance is constant (linear model), or by an inductor, whose inductance L is described by the sum of square functions of current and temperature. The inductor models built in SPICE do not take energy losses in this element or the thermal phenomenon into account.

SPICE also includes a non-linear model of the ferromagnetic core, describing the dependence of magnetic flux density B on the magnetic force H. This description includes the hysteresis of the magnetization curve B(H) using the Jiles–Atherton model [27]. Using the mentioned core model, it is possible to model an inductor or a transformer, taking into account the non-linearity of the core magnetization curve. Unfortunately, the core model does not take into account the effect of temperature, which significantly affects the magnetization characteristics and magnetic permeability of the core [28].

In turn, the transformer can be modeled using coupled inductors. This coupling can be linear (then the described element behaves like an ideal transformer) or non-linear—using the ferromagnetic core model (then the non-linearity of the magnetization curve is taken into account). In the case of using linear coupling, the transformer voltage ratio is described by the classic formula of the form [4]:

$$\vartheta_U = k \cdot \sqrt{\frac{L_1}{L_2}} \tag{1}$$

where L_1 is the inductance of the primary winding of the transformer, L_2 is the inductance of the secondary winding of the transformer, and k is the magnetic coupling factor.

Winding losses are often taken into account when analyzing systems with a transformer. Figure 1 shows a network representation of a linear lossy transformer model. This model consists of the coupled inductors L_1 and L_2 and resistors describing losses in windings R_1 and R_2 and losses in the core R_C. The resistance values of these resistors and the inductance of the inductors depend on the material of the core used and the parameters of the windings.

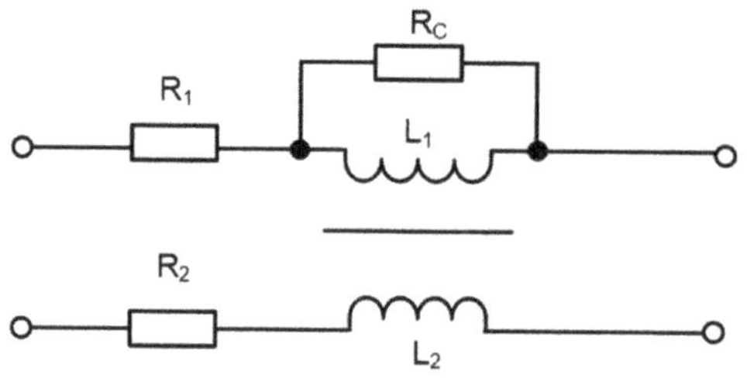

Figure 1. Network representation of a lossy model of the linear transformer.

In turn, in the case of non-linear coupling, the magnetic permeability of the core μ is determined from the inclination of the magnetization curve B(H) for inductors and the value of the induced electromotive force in the transformer windings based on the waveform of flux density B in the core and the number of turns z of the winding. In turn, the inductance L of the inductor depends on the number of turns z, the cross-sectional area of the core S_{Fe}, the relative magnetic permeability of the core μ_{Fe}, the length of the magnetic path in the core l_{Fe}, and the length of the air gap l_p in the core according to the relationship of the form [4].

$$L = \frac{\mu_0 \cdot z^2 \cdot S_{Fe} \cdot \mu_{Fe}}{l_{Fe} + l_p \cdot \mu_{Fe}} \tag{2}$$

where μ_0 denotes the magnetic permeability of free air.

This modeling method does not allow for winding or core losses, transformer inertia, or the influence of temperature on its properties. As mentioned above, the models of magnetic elements built in the SPICE program do not take into account many important phenomena occurring in these elements. Therefore, in the literature of the last dozen or so years, you can find information on macromodels of inductors and transformers for SPICE.

3. Factors Causing Non-Ideal Characteristics of Magnetic Elements

The properties of magnetic elements depend on a number of factors, such as temperature, frequency, the shape of the excitation signal, and the type of material used to build the core. The simple construction of these elements, containing only a ferromagnetic core and windings, often results in a very simplified description of their properties.

However, in reality, due to the properties of the core and the winding, the non-linear characteristics of these elements and the influence of thermal phenomena on their course are observed [14,16]. For example, there is a non-linear dependence of inductance L on current i and a non-linear dependence of the temperature of the core and windings on the power dissipated [29].

As mentioned earlier, the properties of the core of a magnetic element strongly depend on the type of material from which it was made [30–32]. It describes them, among others, in the magnetization curve B(H) that takes the form of a hysteresis loop. The shape of the hysteresis loop depends on many factors (temperature, frequency, magnetic field strength), and it also affects power losses in the core of the magnetic element.

An important parameter defining the material's ability to concentrate the magnetic force in it is the magnetic permeability μ, which is determined for various states of the core magnetization [4,28]. Magnetic permeability corresponds to the slope of the B(H) curve and affects, among others, the inductor inductance according to Formula (2). As can be seen, e.g., from [33], this permeability is a non-linear function of the magnetic force H. The relationship μ(H) has a maximum in the range of H values much lower than in the saturation range.

According to the catalog data of ferromagnetic core manufacturers, depending on the type of material, a strong influence of frequency and temperature on the magnetization curve B(H) and magnetic permeability is observed. Therefore, the catalog data also often presents the dependence of magnetic permeability μ on temperature T, frequency f, the magnetic force H, and the length of the air gap in the core l_p [28,34]. The use of an air gap in the core enables the linearization of the core characteristics, but at the same time it reduces magnetic permeability and thus the inductor inductance [28,35]. As it results from [16,29,31], the effect of temperature on the magnetization curve of the core is significant and noticeable for cores made of ferrite and nanocrystalline materials, while this effect is practically not observed for cores made of iron powder [29].

Figure 2 presents magnetization curves of a ferrite core made of EPCOS N27 material obtained at three different temperatures.

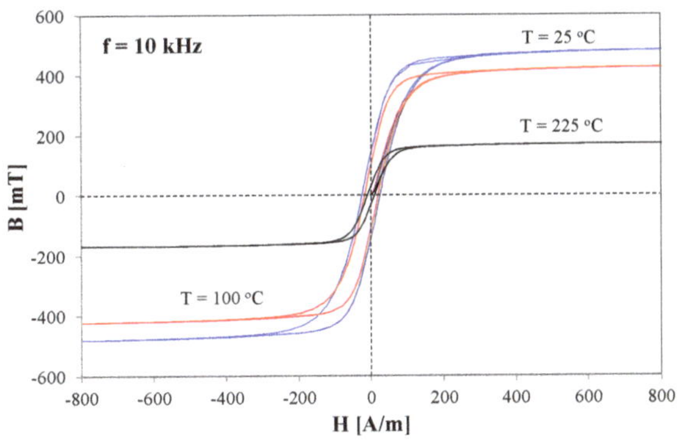

Figure 2. Magnetization curves of the ferrite core made of N27 material at selected values of temperature.

It can be seen that an increase in temperature causes changes in the course of the considered B(H) characteristics, and a decrease in the value of saturation flux density with an increase in temperature is particularly noticeable. An increase in this temperature to

225 °C causes a decrease in the value of saturation flux density to only 40% of the value of this parameter at the temperature of 25 °C.

In [15], it was shown that the magnetic permeability of ferromagnetic cores is a decreasing function of frequency, while the range of frequency, for which a decrease in the value of μ is small depends on the core material and ranges from a single kilohertz to even 1.5 MHz. In [29], it was shown that permeability depends on temperature, and after exceeding the Curie temperature, its value decreases rapidly. The cited paper also shows that at room temperature, the value of relative magnetic permeability, depending on the core material, ranges from 10 to 30,000.

During the operation of the magnetic element, power losses occur. These losses occur both in the core and in the windings. As a result of these losses, the efficiency of energy transfer between the windings of the transformer is below 100%. Core losses are related to, e.g., remagnetization of the core, and they depend on the amplitude of flux density, frequency, and temperature [15].

As mentioned before, the properties of magnetic elements are also significantly affected by temperature. As it follows from [15,31,36], depending on the ferromagnetic core used for the construction of the magnetic element, an increase in temperature reduces saturation flux density and also affects the values of power losses in the cores made of, e.g., ferrite materials.

Energy losses occurring in the components of inductors and transformers are converted into heat. The heat generated in the core and in the windings causes an increase in their temperature above the ambient temperature due to the phenomenon of self-heating and mutual thermal couplings between the core and the winding [28].

Thermal parameters are used to characterize the ability of an electronic element to dissipate the heat generated within it. In [37,38], it was proposed that transient thermal impedance $Z_{th}(t)$ be used for this purpose; the course of thermal impedance is approximated by the formula of the form [12]:

$$Z_{th}(t) = R_{th} \cdot \left[1 - \sum_{i=1}^{N} a_i \cdot \exp\left(\frac{-t}{\tau_{thi}}\right)\right] \tag{3}$$

where R_{th} is the thermal resistance of the element, equal to $Z_{th}(t)$ at the steady state, τ_{thi} is the i-th thermal time constant (i = 1, 2, …, N), a_i is the weighting factors, whose sum is 1, and N is the number of thermal time constants. In general, the temperature of each winding and core of the magnetic element may be different. For example, at the steady state, the inductor core temperature T_C can be described by a simplified formula of the form [9]:

$$T_C = T_a + R_{th} \cdot p_C + R_{thm} \cdot p_W \tag{4}$$

where T_a is the ambient temperature, R_{th} is the thermal resistance of the core, R_{thm} is the mutual thermal resistance between the core and the winding, p_C is the power dissipated in the core, and p_W is the power dissipated in the winding.

The model of an electronic element that takes into account the influence of the temperature of this element on its properties and the dependence of this temperature on the power dissipated in it is called the electrothermal model [39–41]. The graphic interpretation of this model is based on non-isothermal characteristics obtained as a result of electrothermal analysis [39,41–43]. The electrothermal model of an electronic component consists of three parts [29,39]:

(a) an electrical model describing the current–voltage–temperature characteristics of the considered element,
(b) thermal model describing the dependence of the temperature of the element on the power dissipated,
(c) power dependence of the quantities present in the electrical model.

4. Non-Linear Isothermal Models

This section describes models of inductors and transformers that take into account the non-linearity of physical phenomena occurring in these elements but ignore thermal phenomena. Such models are called isothermal models because, based on the characteristics obtained with their help, all the points correspond to one temperature of their components.

The paper [44] discusses a non-linear inductor model for SPICE, taking into account the hysteresis of the magnetization curve described by the Jiles–Atherton model [45], eddy currents in the core, and power losses in the core caused by the hysteresis phenomenon, as well as the presence of an air gap in the magnetic core. The calculation results presented in the cited paper showed the correctness of the shape of the obtained waveforms of voltages and clamping currents of the inductor, as well as the magnetization curves for inductors with a sheet metal core made of silicon steel at frequencies ranging from 50 Hz to 1 kHz.

In [45], two versions of the inductor macromodels for SPICE were proposed. The first of them describes a lossless inductor with a core with a non-linear magnetization curve devoid of hysteresis. This model makes it possible to obtain a non-linear relationship L(i) and to determine the value of the magnetic force, as well as the magnetic flux density. In turn, the other model is more complicated. It additionally takes into account the influence of the length of the air gap in the magnetic core on the characteristics L(i) and losses in the core caused by eddy currents. The results of calculations of the characteristics of the inductor with a sheet metal core made of silicon steel at frequencies of 50 and 250 Hz presented in the paper [46] have a qualitatively correct shape.

The paper [47] describes an inductor model containing only four passive elements: A capacitor representing inter-turn capacitance connected in parallel with a resistor representing core losses, and a series connection of a coil with inductance described by the quadratic function of the current with a resistor modeling winding losses. The model described here allows for a good agreement between the calculated and measured L(i) characteristics in the range of currents up to about 100 mA. Moreover, the frequency characteristics Z(f) of the inductor, describing the dependence of its impedance on frequency, show a good agreement between the results of calculations and measurements with the DC component of the current equal to zero.

In the paper [48], a non-linear inductor model was proposed using a multi-nominal description of the time course of the magnetic flux. In order to simplify the calculation procedure, a system of second-degree equations was used instead of a high-degree polynomial. The model presented in the cited paper takes into account the phenomenon of saturation in the core and energy losses in it. The results of the calculations of voltage and current waveforms at the transformer terminals presented in the cited paper, obtained using a non-linear inductor model in the form of magnetizing inductance, showed qualitatively correct shapes.

In [49], an inductor model was presented, taking into account the non-linearity of the characteristics of the ferromagnetic core and eddy currents in the core. This model uses the analogy of magnetic and electric quantities. In particular, the core magnetization characteristics were modeled by describing the core reluctance using a spline function (piecewise linear). The model does not include the hysteresis of the magnetization curve, but the winding resistance is included. The model was tested with voltage excitation of a low-frequency harmonic signal, obtaining the qualitatively correct time waveforms of the inductor current at the zero value of its DC component.

In [50], a non-linear macromodel of an inductor in the form of a gyrator loaded with non-linear capacitance was proposed. In this model, the hysteresis of the ferromagnetic core was omitted, and the primary magnetization curve was described using the piecewise linear function using the ETABLE-controlled voltage source available in SPICE. The considered model is suitable for use in the frequency small-signal analysis, while in the transient analysis, it shows significant discrepancies with the measurement results [50].

The paper [51] describes SPICE models of spiral inductors used in integrated circuits. The paper [52] presents the method of modeling the hysteresis loop of ferromagnetic cores, and it shows some results of calculations and measurements. The classic description of the B(H) curve is presented in [53]. In turn, the analytical description of the dependence of the inductor inductance on the DC current is presented in [47]. This description uses a polynomial.

The problem of modeling the properties of magnetic materials is also described in many other papers. For example, the approach to modeling the hysteresis loop of soft ferromagnetic materials is presented in the papers [54–56], whereas the method of modeling power losses in the mentioned materials is described in the papers [57,58].

Modeling inductors with non-linear saturable cores is the aim of the paper [46]. In turn, in the paper [59], the manner of modeling eddy currents and power losses in the core of the inductor, the transformer, or the electric machine operating with strong distortions is considered. In the paper [60], the manner of modeling magnetic cores in SPICE with the use of a subcircuit is described.

The papers [10,61] present the models of inductors that are dedicated to operating in electrical conversion systems. The model proposed in [61] is based on matching the dependence of inductance on the DC current of the inductor. To achieve good matching between the results of measurements and computations, a polynomial function was used. The common defect of the mentioned models is that they omit thermal phenomena occurring in the considered devices.

The paper [60] concerns modeling the properties of ferromagnetic cores in SPICE. It proposes a core model in the form of a subcircuit, which is a cascade combination of a static non-linear model of the magnetization curve and a linear low-pass filter of the second order. The core model formulated in this way was used as a component of the non-linear model of the inductor and the transformer. The results of the calculations of time waveforms of voltages and currents of the transformer and of the core hysteresis loop showed a good agreement with the measurement results in the frequency range up to several tens of kHz.

The paper [62] describes the method of modeling the hysteresis of the magnetization curve using SPICE, with the diode models built into this program and a set of controlled sources. This model has a complex structure, but it also allows modeling the inductor characteristics in terms of the core saturation.

In order to analyze the properties of the ferromagnetic material used to build the core of the inductor or transformer, models of the magnetization curve are used. Such models are the subject of many papers [63,64]. One type of literature model uses the laws of quantum mechanics. These are the most complex models, which require complete information on the material used. Additionally, they are described by complex mathematical formulas, which are difficult to implement in software [65]. Another type are physical models using non-linear formulas describing the relationships between such physical quantities as magnetic force and magnetic flux density [65].

There are many papers devoted to modeling the magnetization curve. The starting point for modeling the magnetization curve are models proposed by Rayleigh, Frolich, Hodgdon, Jiles–Atherton, or Preisach. These models take into account more physical phenomena or are dedicated to new ferromagnetic materials [39,65–68]. The paper [65] describes and compares six models. For example, model 1 presented in the cited paper only describes the primary magnetization curve. This is one of the simplest models discussed in the literature. Its advantage is that it uses the smallest number of parameters. The non-linear B(H) characteristics are described using an n-th root approximation.

Another model presented in the cited paper describes only the primary magnetization curve. The mentioned model uses the sum of a linear function and an arctangent cyclometric function to describe the magnetic flux density. The linear function describes the linear component of the magnetization curve, and the cyclometric function takes into account the non-linear nature of the ferromagnetic material and the saturation phenomenon.

The third model described in [65] represents one of the oldest magnetization curve models. The mentioned model is called the Rayleigh model. This model can be used to map the partial magnetization curve in the range of low magnetic force values. This model uses reversible and irreversible components related to the microstructure of the magnetic material.

Model 4, also described in the paper [65], is one of the first empirical models presented by Frolich [39]. The magnetic flux density in the Frolich model is determined using dependence [39]:

$$B = \frac{H - \delta \cdot H_C}{\alpha + \beta \cdot |H - \delta \cdot H_C|} \quad (5)$$

where α, β—model parameters, δ is equal to 1 when $dH/dt > 0$ and—to -1 when $dH/dt < 0$.

In turn, the modified Frolich model, called in the cited paper Model 5, is more widely described in the paper [69]. The modification of this model takes into account the influence of frequency and temperature on the calculated hysteresis loop. However, attention should be paid to the fact that the influence of these factors was not taken into account together; each selected factor required the formulation of other dependences to determine magnetic flux density. It is worth noting that model 5 allows determining magnetic flux density using only two factors, such as frequency and temperature of the core.

On the other hand, Model 6, also presented in [68], was described more widely in the paper [36]. The mentioned model takes into account the influence of temperature on the shape of the B(H) curve. The magnetic flux density is determined from the formula [32]:

$$B = \mu_0 \cdot (H + M \cdot y) \quad (6)$$

where μ_0 is the magnetic permeability of free air, M is magnetization, and y describes the influence of the Curie temperature on magnetic permeability. The magnetization M is expressed by the formula [32]:

$$M = -\frac{(M_a - M \cdot C_1) \cdot \mathrm{sgn}(dH/dt)}{C_1 \cdot H_C \cdot [1 + \alpha_{HC} \cdot (T_C - T_0)]} \cdot \frac{dH}{dt} - \frac{C}{C \cdot (1 + C)} \cdot \frac{dM_a}{dt} + M \quad (7)$$

where sgn(z) denotes the sign of z, C—the deformation parameter of the domain walls, and α_{HC}—the temperature coefficient of the coercive magnetic force. C_1 is a capacitor that describes capacitances occurring in the magnetic model of the core.

Many papers indicate that an important parameter determining the usefulness of the considered magnetic elements are power losses occurring in the components of the magnetic elements, such as in the winding and in the core. The power losses in the core depend on the type of material used to build it [4,28,70,71].

Power losses in ferromagnetic materials can be divided into three groups: Eddy current, hysteresis, and residual losses. Besides the type of magnetic material used to construct the core of the magnetic element, changes in frequency and the peak value of the waveform B(t) [4] also influence power losses in the core.

Hysteresis losses are typical for powdered materials used in devices operating in the frequency range of 10 kHz to 1 MHz. Eddy current losses dominate in the cores made of metallic materials (amorphous materials, steel sheet) used in low-frequency ($f < 10$ kHz) devices [4]. Typically, information about losses in ferromagnetic materials is given as values normalized to the unit of weight or volume.

In the literature, many papers are devoted to the calculation of power losses in magnetic elements and especially in their cores, for example [4,42,72]. The significant influence of the flux density amplitude, frequency, and temperature on power losses in the core is shown in the catalog data. The problem of formulating the universal formula describing power losses in different ferromagnetic materials is considered in many papers, e.g., [42,72,73].

In the literature, different models to determine power losses in ferromagnetic materials are proposed. For example, [28] considered power losses in the toroidal core of a transformer. It was noticed that the power losses occurring in the considered element are the sum of hysteresis and eddy current losses. The eddy current losses can be given by the equation [28]:

$$P_w = \frac{\pi}{6} \cdot \sigma_F \cdot d^2 \cdot {B_m}^2 \cdot f^2 \quad (8)$$

where σ_F is the conductivity of the material, B_m—amplitude of the sinusoidal waveform of B(t) at frequency f, d—thickness of a sheet.

In the cited paper, it is also noticed that the expression (8) is valid only for a sine wave of magnetic flux density. This means that it is necessary to use the correction coefficient for the rectangular waveform. In turn, to describe hysteresis losses, the modified Steinmetz equation was proposed. The Steinmetz formula has the form [28]:

$$P_v = k \cdot f^{\alpha} \cdot (B_m)^{\beta} \tag{9}$$

where k, α, β are material parameters.

On the other hand, the procedure for formulating a general model of power losses in the laminated sheets is presented in [74]. The mentioned model, based on a small amount of data, was proposed for frequencies up to 2 kHz. This model is a modified version of Bertotti's model. It takes into account hysteresis losses, residual losses, and eddy current losses dependent on frequency and the amplitude of magnetic flux density. The modification required consideration of the aforementioned factors and their application by the coefficients modifying Bertotti's model of the form [74]:

$$w_{Fe} = B_m^2 \cdot f \cdot [k_h(f, B_m) + f \cdot k_e(f, B_m)] \tag{10}$$

where k_h and k_e are the material coefficients dependent on f and B_m. The dependences $k_h(B_m)$ and $k_e(f)$, $k_e(B_m)$ were shown in the paper [37], but the authors of the cited paper did not present any description of the dependences $k_h(f,B_m)$ and $k_e(f,B_m)$.

In the paper [61], the description of power losses in ferromagnetic materials was given by the formula:

$$P_v = P_{v0} \cdot f^{\alpha} B_m^{\beta} \cdot (2 \cdot \pi)^{\alpha} \cdot \left(1 + D \cdot (T_C - T_m)^2\right) \cdot (0.6336 - 0.1892 \cdot \ln(\alpha)) \tag{11}$$

where D is the temperature coefficient of losses in the ferromagnetic material, T_m—temperature, at which material losses have the lowest value, and P_{v0}—the material parameter.

The important problem of modeling power losses in ferromagnetic materials using Equation (11) is omitting the influence of temperature and frequency on the material parameters (P_{v0}, α, β). In order to include the influence of temperature, frequency, and amplitude of magnetic flux density on the properties of ferromagnetic materials, the empirical equation describing power losses in the ferromagnetic material was formulated in the paper [15]. It is expressed by the formula [15]:

$$P_{v0} = a \cdot \exp\left(-\frac{f + f_0}{d \cdot (1 + d_1(T_C - T_0))}\right) + a_1 \cdot (b \cdot (T_C - T_M)) + c \cdot \exp\left(\frac{f - f_2}{f_1}\right) \tag{12}$$

where a, b, c, d, f_0, f_1, and f_2 are the parameters of the ferromagnetic material.

For ferrite materials, the parameter β occurring in Equation (11) depends on temperature and is expressed by the formula [15]:

$$\beta = \max(2 \cdot (1 - \exp(-(T_C - 273K)/\alpha_T)) + 1.5; 1.5) \tag{13}$$

α_T is the model parameter. The function $\max(x_1,x_2)$ has a higher value for arguments x_1 and x_2. From Equation (13), it is visible that the parameter β has a value in the range from 1.5 to 3.5.

5. Non-Linear Electrothermal Models

In [75], an electrothermal transformer model for SPICE was proposed. This model uses the method of modeling the magnetization curve proposed by Jiles and Atherton, additionally taking into account the influence of temperature on the course of the hysteresis loop. In this model, eddy currents and energy losses in the core and in the winding were also taken into account. Self-heating was considered using a compact thermal model in the form of the Cauer network, with mutual thermal couplings between the core and the windings omitted. The results of the analyses of a transformer with a ferrite core operating at a frequency of 80 Hz presented in the cited paper showed the correctness of modeling the core magnetization curve in a wide temperature range and small discrepancies between the calculated and measured voltage waveforms on the secondary winding of the transformer.

The authors' previous papers [29,30,37] presented electrothermal models of a transformer [30,37] and an inductor [29]. These models take into account self-heating in the core and in the windings as well as mutual thermal couplings between the core and the windings. The model described in the paper [29] does not take into account the hysteresis of the ferromagnetic core, while the models presented in the papers [30,37] contain a description of the hysteresis loop B(H) in accordance with the modified Jiles–Atherton model [26,45] taking into account the influence of the core temperature of this loop.

The electrothermal model of the inductor proposed in the paper [76] takes into account the self-heating phenomenon in the core only. It ignores self-heating in the windings and mutual thermal couplings between the windings. The skin effect is also omitted in the mentioned model.

In turn, the papers [10,61] describe a model of an inductor that takes into account thermal phenomena occurring in its components and between each pair of them. In the considered model, the parasitic winding capacitance and the skin effect are taken into account.

An important part of electrothermal models are thermal models. In many papers [9,31,77–80], thermal models of the transformer are described, but they have serious disadvantages. On the one hand, the models dedicated to the finite element method (FEM) make it possible to obtain the time–space distribution of temperature in the transformer, but on the other hand, such models assume the uniform distribution of the dissipated power density in the modeled element. Compact thermal models of the transformer are also presented in the literature [9,31,77], but typically these models use only one temperature of the whole transformer, neglecting the differences between the temperatures of each winding and of the core. Additionally, in the known models [9,79,80], the influence of the dissipated power on the efficiency of the heat removal is also neglected.

In the paper [75], an electrothermal model of the ferromagnetic core and the windings of magnetic elements was proposed. Figure 3 shows the circuit representation of the transformer model for SPICE, which is a combination of electro-thermal models of the ferromagnetic core and the windings presented in [75].

In this model, there are controlled voltage and current sources and passive elements. The first circuit, consisting of a series-connected controlled voltage source E_1 and a resistor R_1, is used to determine the effective magnetic force H_e. The circuit consisting of the controlled voltage source E_2 and resistance R_2 is used to determine the value of the irreversible magnetization M_{an}.

Another circuit consisting of the controlled voltage source E_3, capacitor C_1, resistor R_D, and voltage source V_1 is used to determine the time derivative of the magnetic field strength dH/dt. The circuit consisting of the controlled voltage source E_4, resistor R_2, and capacitor C_2 connected in series is responsible for delaying the signal of the analyzed system, slowing down the signal edges, or limiting the value of the dH/dt derivative. Four successive circuits consisting of the controlled voltage source E_6 and resistor R_5, a circuit with elements E_7 and R_6, and the controlled voltage source E_5 connected in series with diode D_1 and resistor R_4, as well as the controlled current source G_1, are used to determine the total magnetization M. The controlled voltage sources E_{21} and E_{22} are used to determine the B(H) core magnetization curve. The controlled current source G_1 is connected between the output terminals of the core model marked with numbers r_1 and r_2.

Another block consisting of voltage sources E_8, E_9, and E_{10} and the current source G_{10} is used to determine the power generated in the core and to determine the temperature transformer, corresponding to the voltage in node r3. An external RC network representing the transient thermal impedance of the transformer is connected to this node. Circuits consisting of series-connected controlled voltage sources E and resistors R are used to determine the values of the core model parameters MS, K, C, A, ECRATE, and ALPHA, taking into account the influence of temperature.

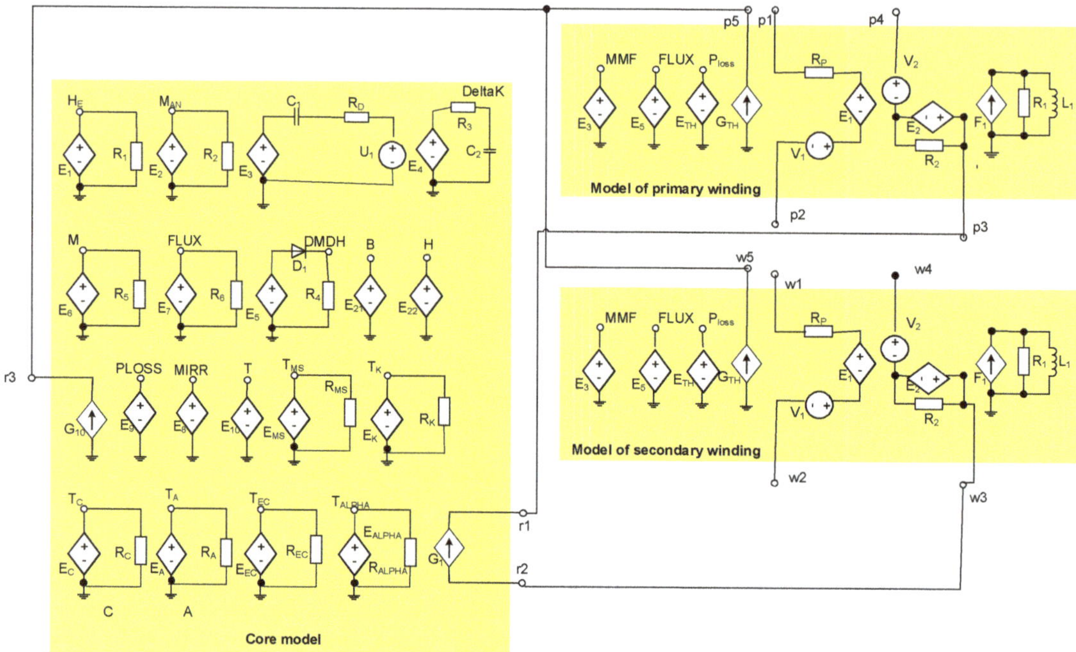

Figure 3. Network form of the electrothermal model of a transformer from the paper [75].

All the winding models have the same structure. They include circuits representing the electromotive force of induction and winding resistance, as well as circuits modeling the magnetomotive force generated by the current of each winding, as well as controlled sources enabling the determination of the winding loss power. In the considered model, the windings and the core of the transformer have the same temperature.

In the following subsections, the authors' electrothermal models of the inductor (Section 5.1), coupled inductors (Section 5.2), and transformers (Section 5.3) are presented. In each of these models, the non-linear thermal model is used. It is described in Section 5.4.

5.1. Electrothermal Model of an Inductor

The presented in this subsection electrothermal model of an inductor is a physical model widely described in the paper [29]. The network representation of the mentioned model corresponds to the parallel connection of capacitor C_W and the branch containing a non-linear inductor and a non-linear resistor connected in series.

In the presented model, the hysteresis of the magnetization curve is omitted due to its weak influence on the magnetic permeability of the core. On the other hand, the air gap in the magnetic core and the coupling coefficient with a value below 1 are taken into account. The discussed model has the form of a subcircuit dedicated to SPICE, as shown in Figure 4.

The considered model contains three blocks: Main circuit, auxiliary block, and thermal model. The skin effect, the triangular shape of the inductor current waveform, self-heating, and mutual thermal couplings between the core and the winding are included in this model. The influence of power losses in the core on the inductor series resistance and the dependence of the inductance of the inductor on frequency are also taken into account.

The main circuit contains five elements connected in series: Voltage source V_1, linear inductor L, controlled voltage source E_{LS}, resistor R_{SO}, and controlled voltage source E_{RS}. The voltage source V_1 has a zero value, and it is used to measure the inductor current. Inductor L is indispensable to calculate the time derivative of the inductor current used in the description of the voltage source E_{LS}. This source describes a voltage drop due to the non-linear inductance of the inductor. The output voltage of this source is given by the equation [29]:

$$E_{LS} = \omega_S \cdot \frac{V_L}{L \cdot (1 + f/f_g)} \cdot \frac{z^2 \cdot S_{Fe} \cdot B_{sat} \cdot A}{l_{Fe} \cdot (|H| + A)^2 + A \cdot B_{sat} \cdot l_p / \mu_0} \tag{14}$$

where z denotes the number of turns, V_L is the voltage drop on inductor L, S_{Fe} means the effective cross-sectional area of the core, B_{sat} is the saturation magnetic flux density, H—magnetic force, l_{Fe}—magnetic path, A—field parameter, l_p—air gap, f—frequency of the inductor current, f_g—frequency, at which the core magnetic permeability decreases twice.

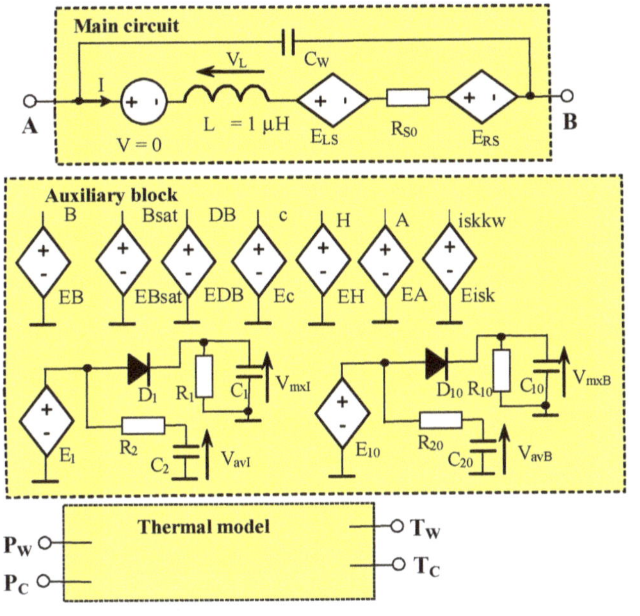

Figure 4. Network form of the electrothermal inductor model with the terminals A and B.

There are three factors in Equation (14). The first one is the coefficient of coupling ω_s, the second is equal to the time derivative of the inductor current, and the third describes the inductance L_s of the inductor. Capacitor C_W represents the parasitic winding capacitance, and resistor R_{SO}—DC series resistance—of the inductor at the ambient temperature T_a. The voltage source E_{RS} describes an additional voltage drop on the inductor series resistance caused by the skin effect and an increase in the winding temperature.

The auxiliary block contains voltage sources that represent the values of the magnetic quantities and selected parameters depending on temperature. To the first group of the mentioned quantities belong magnetic force H, magnetic flux density B, and dB, which is the time derivative of B. The second group contains the field parameter A, the maximum and average values of H and the inductor current, B_{sat} denoting the saturation value of B, and the coefficient α_{TC} describing the influence of the Curie temperature T_C on the value of B. To define the dependence of the flux density on the magnetic force, the following equation is selected [28,81]:

$$B = B_{sat} \cdot \frac{H}{|H| + A} \tag{15}$$

The value of H is given as follows [28,29]:

$$H = \frac{z \cdot I - \frac{B \cdot l_p}{\mu_0}}{l_{Fe} + l_p} \tag{16}$$

The field parameter A occurring in Equation (15) depends on the core temperature. This dependence is described using the exponential function. Moreover, the B_{sat} parameter strongly depends on the core temperature [32,37,82]. To describe this dependence, a linear function is used.

The average and peak values of the current and B are calculated in the auxiliary block using two networks containing two-terminal circuits R_1C_1, R_2C_2 and $R_{11}C_{11}$, $R_{21}C_{21}$ diodes D_1, D_{11}, and the controlled voltage sources E_1 and E_{11} representing the inductor current and the core magnetic flux density [29].

To calculate the temperatures of the core T_C and the winding T_W, the thermal model is used. The detailed form of this model is given in Section 5.4. Usually, the thermal model in all the electrothermal models of electronic elements has the same form. In the considered model, the controlled current sources represent the power dissipated in the core P_C and the power dissipated in the winding P_W. Losses in the winding are described by the following equation:

$$\begin{aligned} P_W = {} & \rho \cdot I^2 \cdot [1 + \alpha_\rho \cdot (T_U - T_0)] + l_d / d_d \cdot \sqrt{\mu_0 \cdot \rho \cdot f \cdot (1 + \alpha_\rho \cdot (T_U - T_0))} \cdot \\ & 2 \cdot \sum_{n=1}^{4} \sqrt{n \cdot f} \cdot (a_n \cdot \cos(2 \cdot \pi \cdot f \cdot t) + b_n \cdot \sin(2 \cdot \pi \cdot f \cdot t)) \cdot (I_{mx} - I_{av})^2 \end{aligned} \tag{17}$$

where I_{mx} is the maximum value of the inductor current.

In turn, the power dissipated in the core is given by the formula [61]:

$$P_C = V_e \cdot (B_m)^{\beta - \alpha} \cdot \left(1 + \alpha_P \cdot (T_R - T_m)^2\right) \cdot \frac{P_{V0}}{T} \int_0^T \left| \frac{dB}{dt} \right|^\alpha dt \tag{18}$$

where V_e denotes the equivalent core volume, P_{V0}—power losses per unit of volume at the nominal frequency and flux density, α_P—the temperature coefficient of power losses in the core, T_m—the temperature, at which P_C is the lowest, and T—the period of B, α and β are the model parameters.

The use of the model presented in this subsection requires the estimation of the values of the model parameters. The model parameters were calculated using the dedicated algorithm presented in [29].

5.2. Electrothermal Model of Coupled Inductors

In the paper [83], the authors proposed such a model for coupled inductors. The cited paper describes current applications and modeling methods for coupled inductors. It is stated that the T-type transformer model is commonly used when formulating a model of the coupled inductor.

In [84], it is shown that the T-type transformer model is asymmetric in contrast to the symmetrical structure of the actual transformer. This means that it is insufficient to be used in computer analyses of such power converters as a multi-phase DC–DC buck converter. This is due to the fact that in this model, the magnetizing inductance is only in one of the windings. The literature analysis shows that there is a lack of electrothermal models of coupled inductors that are dedicated to being used in power conversion systems.

This section describes an electrothermal model of the coupled inductors. It is dedicated to the analysis of power conversion systems and describes a coupled inductor consisting of any number of windings. The developed model takes into account such phenomena as the skin effect, self-heating in each component of the modeled element, and mutual thermal couplings between each pair of these components. In order to make it easier to understand the description of the model, considerations are presented for the coupled inductor with three windings. The described model has the form of a subcircuit for SPICE. Its network form is presented in Figure 5.

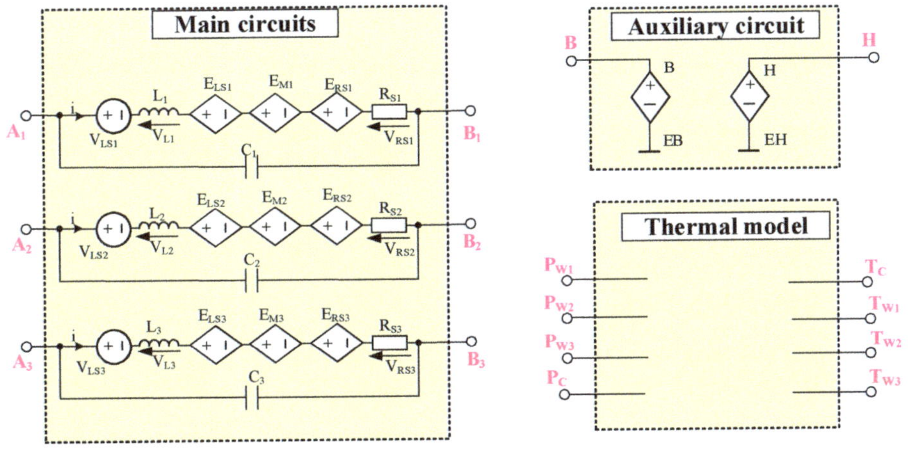

Figure 5. Network form of an electrothermal model of coupled inductors.

This model contains three blocks: main circuits, auxiliary circuit, and thermal model.

The main circuits model the properties of particular windings connected between terminals A_i and B_i (i = 1 ... n). Each winding is modeled using six series-connected elements: A voltage source V_{LSi} of the output voltage equal to zero, a linear inductor L_i, three controlled voltage sources E_{LSi}, E_{Mi}, E_{RSi}, and a resistor R_{Si}. Capacitor C_i represents the parasitic winding capacitance.

The inductance of the i-th inductor winding is described using the controlled voltage source $E_{LSi.}$ Its output voltage is given by the formula [83]:

$$E_{LSi} = a_i \cdot \frac{{z_i}^2 \cdot S_{Fe} \cdot B_{sat} \cdot A}{l_{Fe} \cdot (|H| + A)^2 + A \cdot B_{sat} \cdot l_p / \mu_0} \cdot \frac{V_{Li}}{L_i} \tag{19}$$

where z_i denotes the number of turns, a_i—the scaling factor, and V_{Li}—voltage on inductor L_i.

The mutual inductance between the i-th winding and the other windings is modeled using the controlled voltage source E_{Mi}. Its output voltage is given by the formula [83]:

$$E_{Mi} = \sum_{\substack{j=1 \\ j \neq 1}}^{n} k_j \cdot V_{ELSj} \tag{20}$$

where k_j represents the magnetic coupling coefficient, V_{ELSi}—voltage at the source E_{LSj}, whereas the number of the inductor windings is equal to n.

A voltage drop on the inductor series resistance is described by the controlled voltage source E_{Sri}. This description takes into account the skin effect and changes in the winding temperature. The output voltage of this source is given by the formula [83]:

$$E_{SRi} = V_{RSi} \cdot \alpha_\rho \cdot (T_{Wi} + T_a - T_0) + \frac{l_d}{d_d} \cdot \sqrt{\mu_0 \cdot \rho \cdot (1 + \alpha_\rho \cdot (T_{Wi} + T_a - T_0))} \cdot i_{Li} \tag{21}$$

where V_{RSi} represents the voltage on resistor R_{Si}, d_d denotes the diameter of the used winding wire, l_d—the length of this wire, ρ—copper resistivity, α_p—the temperature coefficient of ρ, T_{Wi}, T_a, and T_0 are winding, ambient, and reference temperatures, respectively, μ_0—magnetic permeability of free air, and i_{Li} current of the i-th inductor winding.

Resistor R_{Si} represents the DC series resistance of a single winding at temperature T_0.

The auxiliary circuit block contains the controlled voltage sources EB and EH. These sources make it possible to determine the values of B and H in the core. The output voltage of the source EB is given by Equation (15), whereas the source EH is described by the dependence [14,85]:

$$H = \frac{\sum_{i=1}^{n} z_i \cdot i_i}{l_{Fe} + l_p} \tag{22}$$

where i_i denotes the i-th winding current.

The hysteresis loops of the considered inductor core are very narrow. Therefore, in the presented model, the core losses are omitted. The thermal model of the coupled inductors makes it possible to compute the temperature of each winding (T_{W1}, T_{W2}, T_{W3}) and the temperature of the core (T_C). This model takes into account self-heating in each winding and mutual thermal couplings between each winding and the core as well as between each pair of windings.

5.3. Electrothermal Model of a Transformer

Figure 6 shows the circuit representation of the electrothermal model of a transformer for SPICE.

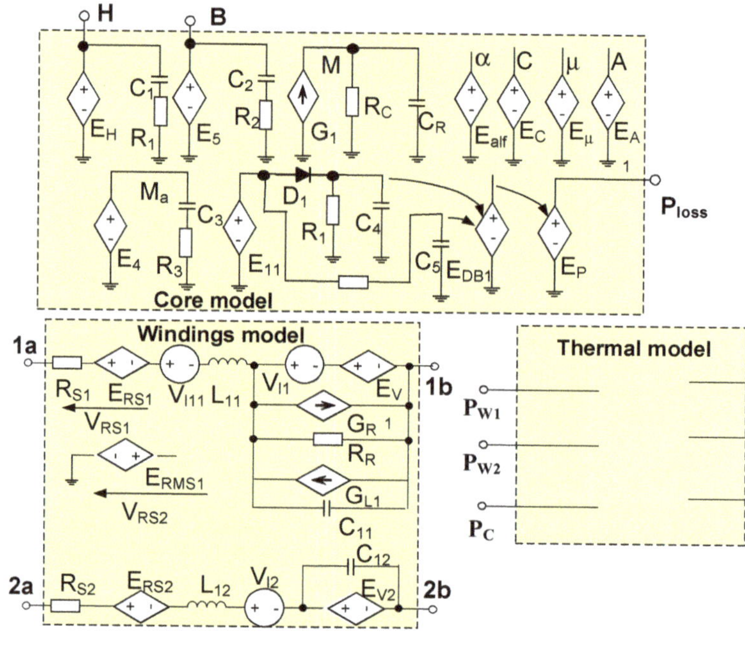

Figure 6. Network form of the electrothermal model of a transformer.

In this model, three blocks can be distinguished. Two blocks describe the properties of the core and the windings, whereas the third block is the thermal model. The model of the core is based on the equations proposed by Jiles and Atherton. In this model, the magnetic force H is computed in the network containing the controlled voltage source E_H, capacitor C_1, and resistor R_1. The voltage at the output terminal B represents the magnetic flux density, whereas the voltage at the P_{loss} terminal represents power losses in the core.

The winding model contains two networks representing the primary winding—with terminals 1a and 1b—and the secondary winding—with terminals 2a and 2b. Resistor R_{S1} represents the series resistance of the primary winding at temperature T_0. The controlled voltage source, E_{RS1}, models changes in this resistance when the temperature of the winding changes. The voltage induced in the primary winding is computed using the controlled voltage source E_V. The controlled current source G_{L1} models the magnetizing current, and the controlled current source G_R represents the power dissipated in the core. Voltage V_{RMS1} is calculated by the controlled voltage source E_{RMS1}. The voltage sources V_{L1} and V_{l11} of the zero value are used to monitor the value of their currents. Capacitor C_{11} represents the inter-turn capacitance of the primary winding.

On the other hand, the secondary winding model contains only the elements modeling the winding voltage (E_{V2}) and the series resistance of this winding (R_{S2} and E_{RS2}). Capacitor C_{12} represents the parasitic capacitance of the secondary winding. A detailed description of all the controlled sources occurring in this model is given in [30].

The thermal model makes it possible to calculate the core temperature T_C and the temperature of windings T_{W1} and T_{W2}, taking into account all the thermal phenomena occurring in the modeled transformer.

5.4. Non-Linear Thermal Model of Magnetic Elements

As can be seen in the paper [38], a compact thermal model can be used to describe the thermal properties of magnetic elements due to the practically uniform temperature distributions on each winding and on the core. Simultaneously, the temperatures of the core and each winding can differ significantly from one another [86]. Therefore, a compact thermal model of magnetic elements can be used. In each component of the magnetic element, the self-heating phenomenon occurs. Additionally, mutual thermal couplings are observed between each pair of such components.

The efficiency of the removal of the heat dissipated in electronic devices depends on such factors as [86–91], the construction of the cooling system, the values of the dissipated power, and the ambient temperature. A non-linear compact thermal model of the pulse transformer is proposed in the paper [38]. It is based on the model described in the paper [41] for the IGBT. The described model takes into account all the thermal phenomena occurring in all the transformer components and between each pair of them, as well as the dependence of the heat removal efficiency on the dissipated power.

The presented model enables calculating the waveforms of the core and each winding temperature, taking into account both of the mentioned thermal phenomena. The network form of the considered model has a different form for each of the considered elements. As an example, in Figure 7, the thermal model of a pulse transformer is shown.

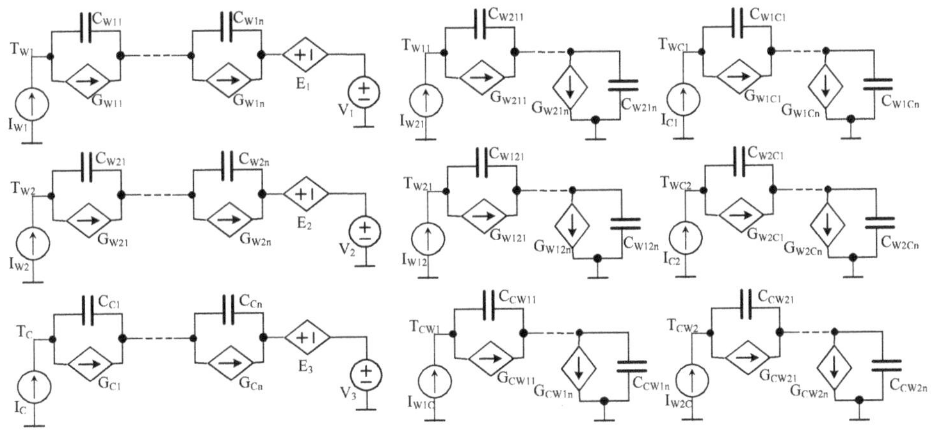

Figure 7. Network form of a non-linear compact thermal model of the pulse transformer.

This model consists of nine subcircuits. Three of them (on the left-hand side) are used to calculate the waveforms of temperature for each transformer component—for the primary winding T_{W1}, for the secondary winding T_{W2} and for the core T_C. The values of these temperatures correspond to the voltage in the nodes T_{W1}, T_{W2} and T_C. The powers dissipated in the core and the windings are represented by independent current sources I_C, I_{W1} and I_{W2}. The networks containing capacitors and the controlled current sources represent the self-transient thermal impedances of each of the transformer components.

The controlled current sources are used to describe the dependence of thermal resistances R_{th} on the dissipated power.

In order to take into account the influence of mutual thermal couplings between the components of each of the considered temperatures, the controlled voltage sources E_1, E_2 and E_3 are used. For example, the value of a voltage drop on the source E_1 is the sum of the voltages in nodes T_{W11} and T_{WC1}. The ambient temperature is represented by the voltage sources V_1, V_2 and V_3.

Six other subcircuits are used to model mutual thermal couplings in the transformer. The current sources describe the power dissipated in the transformer components. The networks containing capacitors and controlled current sources connected to these current sources represent the proper mutual transient thermal impedances.

The dependence of R_{th} on the dissipated power p is given by the empirical formula [38]:

$$R_{th} = R_{th0} \cdot \left[1 + a \cdot \exp\left(-\frac{p}{b}\right)\right] \tag{23}$$

where R_{th0} is the minimum value of R_{th}, whereas a and b are the model parameters.

The thermal model shown in Figure 7 can, after a simple modification, be used to model the thermal properties of the inductor and the coupled inductors. In the case of the inductor, instead of nine subcircuits, only four are used, visible in the upper left corner of Figure 7. In turn, for the coupled inductors containing three windings, in the thermal model, there will be 16 subcircuits describing self-heating in each winding and in the core and mutual thermal couplings between each pair of the windings and between the core and each winding.

The values of the model parameters are estimated using the measured waveforms of self and mutual transient thermal impedances occurring in the transformer thermal model. The proper manner of measuring is presented in the paper [37]. All the transient thermal impedances should be measured at different values of the power dissipated in the components of the transformer. In the next step, the method described in the papers [32] should be used for each measured waveform of transient thermal impedance. As a result of this step, the values of the parameters occurring in Equation (3) are obtained. Finally, the values of parameters a and b existing in Equation (23) are estimated using the idea of local estimation [32] for each thermal resistance existing in the transformer thermal model.

6. Results of Measurements and Computations

In this section, as an example, exemplary characteristics of the magnetic elements obtained using the previously discussed models are presented. In all the figures presented in this section, points denote the results of measurements, and solid lines are the results of calculations made using the non-linear compact electrothermal models.

Due to the non-linearity of the magnetization curve and the related dependence of magnetic permeability on the current, there is a non-linear characteristic L(i) of the inductor. Examples of the measured and calculated L(i) characteristics of inductors containing cores made of different materials are shown in Figure 8. The inductor containing the RTP powder core made of T106-26 material, the RTF ferrite core made of F867 material, and the RTN nanocrystalline core made of M070 material were considered. The investigations were carried out at a frequency of 100 kHz for an ambient temperature equal to T_a = 25 °C.

The obtained dependencies are monotonically decreasing functions, and the results of measurements and calculations are well matched. This proves the correctness and universality of the presented model.

For an inductor with a powder core of the same geometric dimensions, a constant value of inductance in the range of the DC current from 0 to 1.2 A is observed, which is the result of the presence of a longer air gap in this core than in the remaining considered cores. In addition, a decrease in the inductance of the inductor with the ferrite core (even two hundred times) is much greater than in the case of the inductor with the powder core (about 30%), but much smaller than in the case of the inductor with the nanocrystalline core.

The value of inductance of the inductor with the nanocrystalline core rapidly decreases at very low values of the DC current, from L = 1700 μH (I_{DC} = 0 A) to approx. 22 μH (I_{DC} = 0.3 A). The different course of the dependence L(i) of the investigated inductor results from the non-linear magnetization curve of the used ferromagnetic cores.

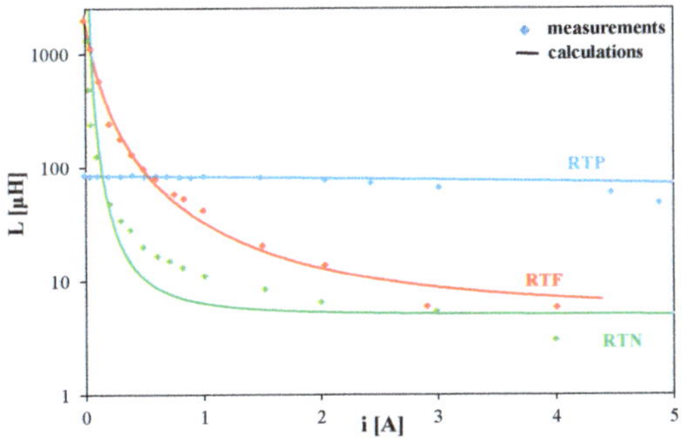

Figure 8. Measured and computed values of inductance of the inductors containing the core made of different ferromagnetic material in the function of the DC current.

Figure 9 shows the dependences of the impedance modulus of an inductor with the ferrite core (RTF F867) and the powder core (RTP T106-26) on frequency at the fixed values of the DC current of the inductor. The results of calculations obtained using the model mentioned in the paper [15] are marked with a solid line, and the results of measurements are marked with points. As can be seen, a good agreement is obtained between the measurements and the analysis results. By taking into account the winding capacitance in the electrothermal model of the inductor, resonance was obtained on the modeled characteristic, which corresponds to the obtained measurement results. The value of the resonant frequency for the inductor with the ferrite core increases with an increase in the DC current, while for the inductor with the powder core, it oscillates in the range from 1.3 MHz to about 2.3 MHz. It is also worth noting that the cited model is a universal model, which is proved by the good agreement between the results of measurements and calculations when considering the inductor containing various ferromagnetic materials.

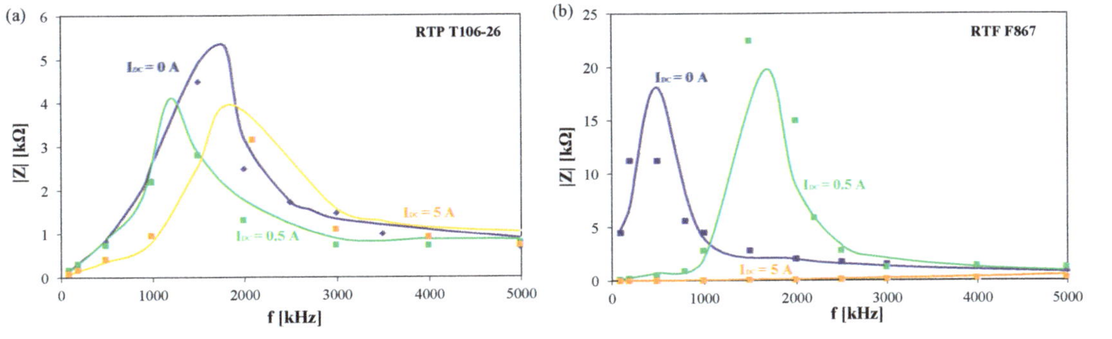

Figure 9. Measured dependences of the modulus of impedance of the inductor with the RTP T106-26 (**a**) and RTF F867 (**b**) cores on frequency.

Figure 10 presents the measured and computed waveforms of temperature for each transformer component: The primary winding T_{W1}, the secondary winding T_{W2} and the core T_C. The tested transformer contains the ferrite core RTF. In the investigations, both windings are simultaneously excited with a single rectangular pulse of different values of the pulse duration time. The values of the power amplitude are as follows: P_{W1} = 0.88 W, P_{W2} = 1.05 W (Figure 10a), and P_{W1} = 2.4 W, P_{W2} = 2.7 W (Figure 10b).

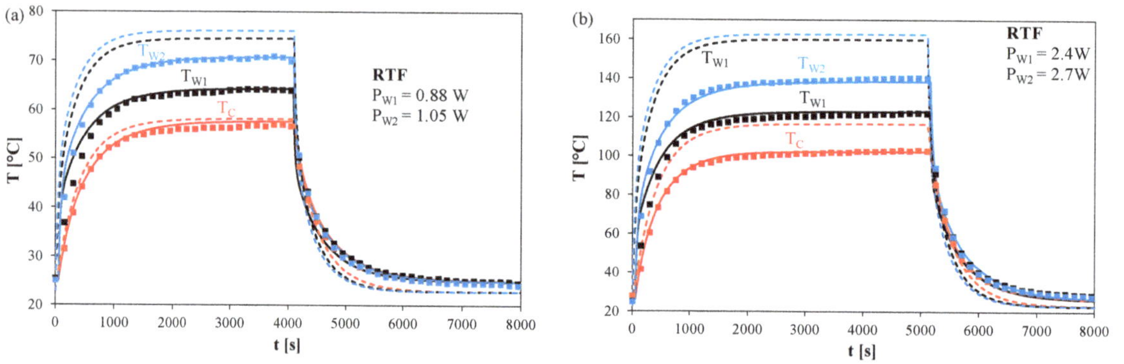

Figure 10. Measured and computed waveforms of the temperature of the windings and the core of the transformer containing the RTF core while dissipating the power of different values of amplitudes.

As it is visible, for the non-linear thermal model, the results of computations fit well with the results of measurements in both sets of power amplitude values. On the other hand, the computations performed using the linear thermal model give results that overestimate the measurement results by even 20%.

In turn, Figure 11 presents the peak-to-peak value of the input voltage of the transformer containing the RTP core as a function of frequency. The considered dependence was obtained at load resistance R_0 = 1 kΩ, and the peak-to-peak value of the excitation signal is 25 V. The considered function possesses a maximum voltage equal to 25 V at a frequency of 9 MHz.

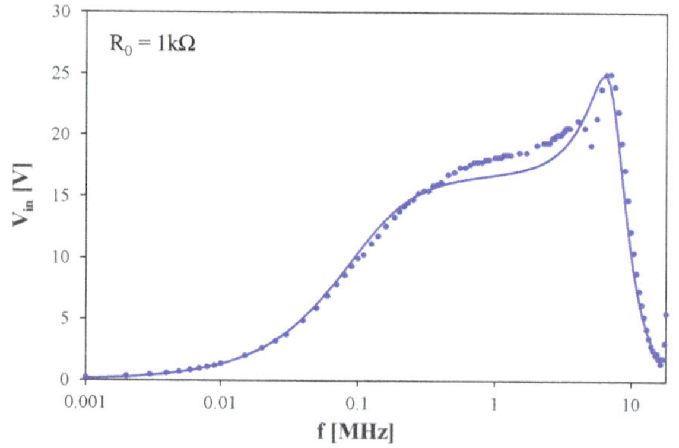

Figure 11. Measured and calculated dependences of the transformer input voltage on frequency.

In turn, Figure 12 presents the dependence of energy efficiency on load resistance R_0 for the transformer operating at a frequency f = 100 kHz. From the obtained results, it can be seen that the proposed model shows the best agreement between the results of calculations and measurements for load resistance R_0 higher than 40 Ω.

Figure 13 presents the results of calculations and measurements of the influence of the DC current on the temperatures of the components of the coupled inductor. The investigations were carried out for the coupled inductor containing the nanocrystalline core, on which three windings, eight turns each, were wound.

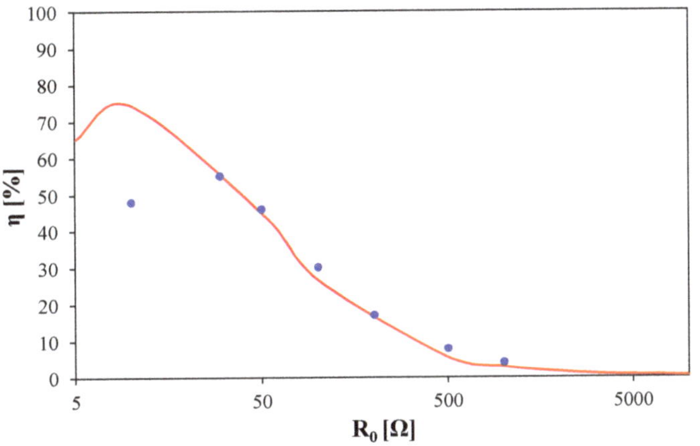

Figure 12. Measured and computed dependences of the transformer energy efficiency on load resistance.

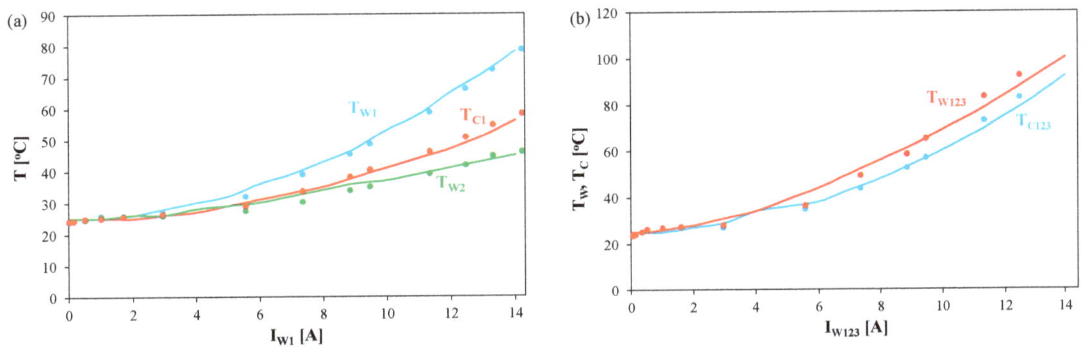

Figure 13. Courses of the inductor components temperature on the DC current at power dissipation in one winding only (**a**), in three windings connected in series (**b**).

It is visible, that all the curves shown in Figure 13 present monotonically increasing functions. The temperature of the power supplying the winding increases due to a self-heating phenomenon, but the temperature of the other components of the tested device increases due to mutual thermal couplings in this winding (Figure 13a). The values of the windings temperatures T_{W2} and T_{W3} at I_{DC} = 14 A exceed the ambient temperature even by 25 °C, whereas the core temperature T_C at the same value of I_{DC} is even by 33 °C higher than the ambient temperature. Figure 13b shows that the differences between the temperatures of the windings and of the core are clear if I_{DC} > 6 A. They can exceed even 10 °C, which means that the self-thermal resistance of each winding is higher than the transfer thermal resistance between the winging and the core.

The non-linear characteristics of inductors and transformers presented above cause changes in the course of the characteristics of switch-mode power conversion systems. Such characteristics were included in our previous papers. As examples, the relationships presented in Figures 14 and 15 are selected. They illustrate the effect of taking into account the non-linearity of the L(i) dependence for the inductor (Figure 14) and the coupled inductors (Figure 15) on the characteristics of the boost and flyback converters, respectively. In both figures, the solid lines denote the calculation results using the non-linear inductor model described in Section 5, the dashed lines—using the linear model (constant inductance value), and the points are the measurement results.

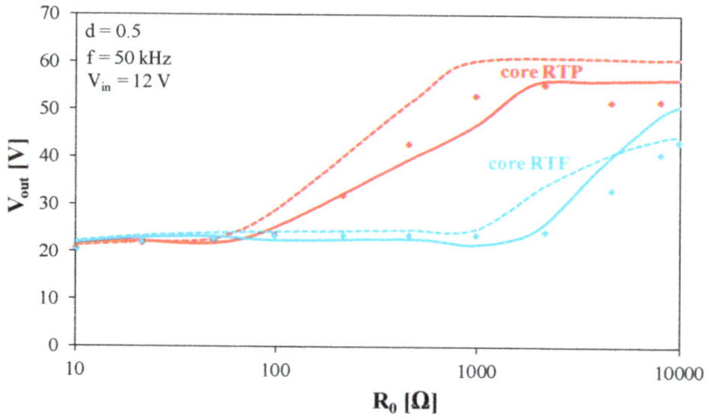

Figure 14. Calculated (lines) and measured (rhombus) dependences of the output voltage of the boost converter with different cores on load resistance.

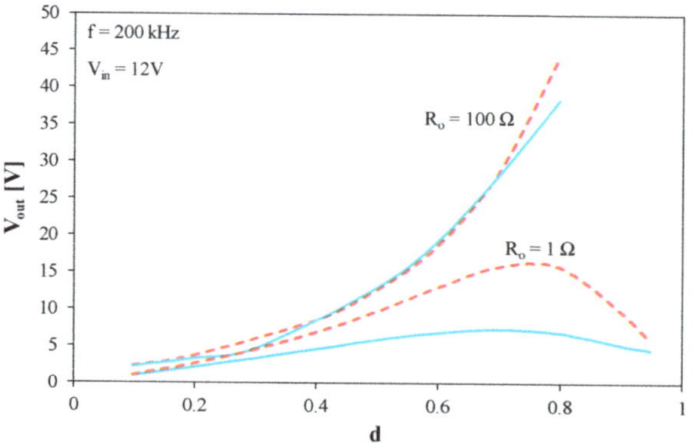

Figure 15. Dependences of the output voltage of a flyback converter on the duty cycle of the control signal for selected values of load resistance.

Figure 14 shows the characteristics of the boost converter operating at a frequency f = 50 kHz, the duty cycle d = 0.5 and an input voltage V_{in} = 12 V. The tests were carried out for the inductors with the RTF ferrite core (blue) and the RTP powder core (red).

As can be seen in Figure 14, the selection of the core material affects the shifts in the operating range of the tested converters between the CCM and DCM modes. For a converter containing an inductor with a ferrite core, CCM can be achieved with a load resistance as high as 1 kΩ, while for a converter containing an inductor with a powder core, the load resistance must be less than 80 Ω. When using the linear model, the output voltage of the converter in DCM mode is overestimated, and the linear model ensures a better agreement between the measurement and calculation results.

Figure 15 presents a flyback converter operating at an input voltage of V_{in} = 12 V, a control signal frequency f = 200 kHz, and coupled inductors with a nanocrystalline core. The dashed lines indicate the results obtained with the linear model, and the solid lines indicate the results obtained with the non-linear model described in the previous section.

In this case, for the higher of the considered load resistance values, there are practically no differences in the course of the V_{out}(d) dependence for both the considered models of the coupled inductors. In turn, clear differences between the V_{out}(d) dependence for both models are visible at low load resistance values. They are more than doubled at d = 0.8 and are caused by taking into account losses in the core and windings of the considered element in the non-linear model.

7. Final Remarks

The paper presents a review of literature models of such magnetic elements as inductors, coupled inductors, and transformers dedicated to SPICE. The described models are characterized by a different level of complexity and a set of physical phenomena taken into account. Particular attention was focused on electrothermal models of the considered elements. The advantages and disadvantages of the described models are given.

Table 1 collects the most important properties of the considered models of magnetic elements dedicated to SPICE.

Table 1. The most important properties of the considered models of magnetic elements for SPICE.

Source	Non-Linearity of L(i) Characteristic	Hysteresis Loop	Power Losses	Self-Heating	Mutual Thermal Couplings	Number of Modeled Temperatures
[10,47,50,51]	Yes	No	Yes	No	No	0
[23,25]	Yes	Yes	No	No	No	0
[17,26,28,44]	Yes	Yes	Yes	No	No	0
[75]	Yes	Yes	Yes	Yes	No	1
[9,15,29,61]	Yes	No	Yes	Yes	Yes	2
[32,34]	Yes	Yes	Yes	Yes	Yes	2
[30]	Yes	Yes	Yes	Yes	Yes	3
[83]	Yes	No	Yes	Yes	Yes	4

The forms of electrothermal models of the considered elements formulated by the authors are presented. A special feature of these models is that they take into account the non-linearity of electrical, magnetic, and thermal phenomena occurring in the modeled elements. In addition to determining the time courses of voltages and currents, these models also make it possible to determine the time courses of the temperature of each of the windings and the ferromagnetic core for each of the considered elements. The big advantage of the proposed model is its open structure, which makes it possible to add another winding to the described model.

The results of the experimental verification of the presented models prove their correctness and universality. They ensured a good agreement between the results of calculations and measurements for magnetic elements containing cores made of different ferromagnetic materials and different sizes of the cores. These results also show that the non-linearity of the characteristics of the considered devices can be effectively modeled using the proposed non-linear compact electrothermal models. The papers by the authors, in which it was proved that the described models can be successfully used for elements with the classical windings made of the winding wire, as well as for planar transformers with the windings in the form of PCBs, were also cited.

In our opinion, the new structures of magnetic elements will be elaborated and, in consequence, new models of these elements will have to be elaborated. In order to obtain better and better accuracy from these models, the number of phenomena taken into account in the model will increase. Two directions can be observed in the development of these models: (a) Compact models dedicated to the analyses of electronic networks and (b) detailed models dedicated to the analyses of the properties of magnetic elements only. We also hope that the description of power losses in the core that will be formulated will also characterize them more accurately over a wide range of frequencies. Due to the fast development of electrical vehicles, models of an air-core transformer with high accuracy will also be elaborated.

The presented research results may be useful for designers of electronic systems containing magnetic elements. The proposed models will enable more accurate analyses of the designed systems. These models can also be useful while educating students, as they enable the illustration of the phenomena occurring in these elements during their operation in selected electronic systems.

Author Contributions: Conceptualization: K.G. and K.D.; methodology: K.G. and K.D.; investigation: K.G. and K.D.; writing—original draft preparation: K.G. and K.D.; writing—review and editing: K.G. and K.D.; visualization: K.G.; supervision: K.G. All authors have read and agreed to the published version of the manuscript.

Funding: This research received no external funding.

Data Availability Statement: Data available for request.

Conflicts of Interest: The authors declare no conflict of interest.

References

1. Rashid, M.H. *Power Electronic Handbook*; Academic Press: Cambridge, MA, USA; Elsevier: Amsterdam, The Netherlands, 2007.
2. Billings, K.; Morey, T. *Switch-Mode Power Supply Handbook*; McGraw-Hill: New York, NY, USA, 2011.
3. Ericson, R.; Maksimovic, D. *Fundamentals of Power Electronics*; Kluwer Academic Publisher: Norwell, MA, USA, 2001.
4. Barlik, R.; Nowak, M. *Energoelektronika Elementy Podzespoły Układy*; Politechnika Warszawska: Warsaw, Poland, 2014.
5. Rashid, M.H. *SPICE for Power Electronics and Electric Power*; CRC Press: Boca Raton, FL, USA, 2017.
6. Maksimovic, D.; Stankovic, A.M.; Thottuvelil, V.J.; Verghese, G.C. Modeling and simulation of power electronic converters. *Proc. IEEE* **2001**, *89*, 898–912. [CrossRef]
7. Tuinenga, P.W. *SPICE a Guide to Circuit Simulation and Analysis Using PSpice*; Prentice-Hall, Inc.: Hoboken, NJ, USA, 1995.
8. Zhao, W.; Wei, P. PSpice system simulation application in electronic circuit design. In Proceedings of the 32nd Chinese Control Conference, Xi'an, China, 26–28 July 2013; pp. 8634–8636.
9. Detka, K.; Górecki, K.; Zarębski, J. Modeling single inductor DC–DC converters with thermal phenomena in the inductor taken into account. *IEEE Trans. Power Electron.* **2016**, *32*, 7025–7033. [CrossRef]
10. Detka, K.; Górecki, K.; Grzejszczak, P.; Barlik, R. Modeling and Measurements of Properties of Coupled Inductors. *Energies* **2021**, *14*, 4088. [CrossRef]
11. Vorperian, V. *Fast Analytical Techniques for Electrical and Electronic Circuits*; Cambridge University Press: Cambridge, MA, USA, 2002.
12. Górecki, P.; Górecki, K. Methods of Fast Analysis of DC–DC Converters—A Review. *Electronics* **2021**, *10*, 2920. [CrossRef]
13. Fitzpatrick, D. *Analog Design and Simulation Using OrCAD Capture and PSpice*; Newnes: Oxford, UK, 2017.
14. Tumański, S. *Handbook of Magnetic Measurements*; Taylor and Francis Group: Boca Raton, FL, USA, 2011.
15. Detka, K.; Górecki, K. Modelling the power losses in the ferromagnetic materials. *Mater. Sci.-Pol.* **2017**, *35*, 398–404. [CrossRef]
16. Kazimierczuk, M. *High-Frequency Magnetic Components*; Wiley: Hoboken, NJ, USA, 2014.
17. Le-Huy, H.; Sybille, G. MATLAB/Simulink and PSPice as modelling tools for power systems and power electronics. In Proceedings of the 2000 Power Engineering Society Summer Meeting, Seattle, WA, USA, 16–20 July 2000; Volume 2, pp. 766–767.
18. Yang, W.Y.; Kim, J.; Park, K.W.; Baek, D.; Lim, S.; Joung, J.; Park, S.; Lee, H.L.; Choi, W.J.; Im, T. *Electronic Circuits with MATLAB, PSpice, and Smith Chart*; John Wiley & Sons: Hoboken, NJ, USA, 2020.
19. Yi, R.; Chen, S.; Zhang, J.; Zhao, H.; Yu, Y.; Zhu, H. Research on Reverse optimized Control of Dual Active Bridge Based on PSpice-MATLAB/Simulink. In Proceedings of the 2022 IEEE 5th International Conference on Electronics Technology (ICET), Chengdu, China, 13–16 May 2022; pp. 243–248.
20. Scirè, D.; Lullo, G.; Vitale, G. Assessment of the Current for a Non-Linear Power Inductor Including Temperature in DC-DC Converters. *Electronics* **2023**, *12*, 579. [CrossRef]
21. Saeed, S.; Georgious, R.; Garcia, J. Modeling of Magnetic Elements Including Losses—Application to Variable Inductor. *Energies* **2020**, *13*, 1865. [CrossRef]
22. Stoyka, K.; Femia, N.; Di Capua, G. Power inductors behavioral modeling revisited. *IEEE Trans. Circuits Syst. I Regul. Pap.* **2020**, *67*, 5636–5649. [CrossRef]
23. Witulski, A.F. Introduction to modeling of transformers and coupled inductors. *IEEE Trans. Power Electron.* **1995**, *10*, 349–357. [CrossRef]
24. Oliveri, A.; Lodi, M.; Storace, M. Nonlinear models of power inductors: A survey. *Int. J. Circuit Theory Appl.* **2022**, *50*, 2–34. [CrossRef]
25. Scirè, D.; Lullo, G.; Vitale, G. Non-Linear Inductor Models Comparison for Switched-Mode Power Supplies Applications. *Electronics* **2022**, *11*, 2472. [CrossRef]
26. Wilamowski, B.M.; Jaeger, R.C. *Computerized Circuit Analysis Using SPICE Programs*; McGraw-Hill: New York, NY, USA, 1997.
27. Jiles, D.C.; Atherton, D.L. Theory of ferromagnetic hysteresis. *J. Appl. Phys.* **1984**, *55*, 2115–2120. [CrossRef]
28. van den Bossche, A.; Valchev, V.C. *Inductors and Transformers for Power Electronics*; CRC Press, Taylor & Francis Group: Boca Raton, FL, USA, 2005.
29. Górecki, K.; Detka, K. The parameters estimations of the electrothermal model of inductors. *J. Microelectron. Electron. Compon. Mater.* **2015**, *45*, 29–38.
30. Górecki, K.; Godlewska, M. Modelling characteristics of the impulse transformer in a wide frequency range. *Int. J. Circuit Theory Appl.* **2020**, *48*, 750–761. [CrossRef]

31. Detka, K.; Górecki, K. Influence of the size and the material of the magnetic core on thermal properties of the inductor. *Microelectron. Reliab.* **2022**, *129*, 114458. [CrossRef]
32. Górecki, K.; Rogalska, M.; Zarębski, J. Parameter estimation of the electrothermal model of the ferromagnetic core. *Microelectron. Reliab.* **2014**, *54*, 978–984. [CrossRef]
33. Sumarac, D.; Medjo, B.; Trisovic, N. Hysteretic behaviour modeling of elastplastic materials. *Theoret. Appl. Mech.* **2008**, *35*, 287–304. [CrossRef]
34. McLyman, C.W.T. *Transformer and Inductor Design Handbook*; CRC Press: Boca Raton, FL, USA, 2017.
35. Lv, G.; Zhi, R.; Cui, L.; Zhang, Z.; Liu, Y. Analysis on the Air-Gap Magnetic Flux Density and Propulsion of the TFLSM Considering Cogging Effect. *IEEE Trans. Magn.* **2022**, *59*, 1–7. [CrossRef]
36. Chwastek, K. Modelling of dynamic hysteresis loops using the Jiles-Atherton approach. *Math. Comput. Model. Dyn. Syst.* **2009**, *15*, 95–105. [CrossRef]
37. Górecki, K.; Zarębski, J. Electrothermal analysis of the self-excited push-pull dc-dc converter. *Microelectron. Reliab.* **2009**, *49*, 424–430. [CrossRef]
38. Górecki, K.; Detka, K.; Górski, K. Compact thermal model of pulse transformer taking into account nonlinearity of heat transfer. *Energies* **2020**, *13*, 2766. [CrossRef]
39. Fröhlich, H. Electrons in lattice fields. *Adv. Phys.* **1954**, *3*, 325–361. [CrossRef]
40. Szewczyk, R. Modelling of the magnetic and magnetostrictive properties of high permeability Mn-Zn ferrites. *Pramana-J. Phys. India Acad. Sci.* **2006**, *67*, 1165–1171. [CrossRef]
41. Górecki, K.; Górecki, P. Non-linear compact thermal model of the IGBT dedicated to SPICE. *IEEE Trans. Power Electron.* **2020**, *35*, 13420–13428. [CrossRef]
42. Parchimiuk, M. Badania przekształtnika DC/DC z izolacją transformatorową przeznaczonego do zasilania potrzeb własnych pojazdów trakcyjnych. *Inform. Autom. Pomiary Gospod. Ochr. Sr.* **2014**, *4*, 113–116. [CrossRef]
43. Goodenough, J.B. Summary of losses in *magnetic* materials. *IEEE Trans. Magn.* **2002**, *38*, 3398–3408. [CrossRef]
44. Mandache, L.; Topan, D.; Sirbu, I.G. Accurate time-domain simulation of nonlinear inductors including hysteresis and eddy-current effects. In Proceedings of the World Congress on Engineering WCE 2011, London, UK, 6–8 July 2011; Volume 2, pp. 1327–1332.
45. Jiles, D.C.; Thoelke, J.B.; Devine, M.K. Numerical determination of hysteresis parameters for the modeling of magnetic properties using the theory of ferromagnetic hysteresis. *IEEE Trans. Magn.* **1992**, *28*, 27–35. [CrossRef]
46. Mandache, L.; Topan, D.; Iordache, M.; Dumitriu, L.; Sirbu, I.G. On the time-domain analysis of analog circuits containing nonlinear inductors. In Proceedings of the 20th European Conference on Circuit Theory and Design ECCTD 2011, Linkoping, Sweden, 29–31 August 2011; pp. 98–101.
47. O'Hara, M. *Modeling Non-Ideal Inductors in SPICE*; Newport Components: Milton Keynes, UK, 1993.
48. Stefopoulos, G.K.; Cokkinides, G.J.; Meliopoulos, A. Quadratized model of nonlinear saturablecore inductor for time-domain simulation. In Proceedings of the 2009 IEEE Power & Energy Society General Meeting, Calgary, AB, Canada, 26–30 July 2009; pp. 1–8.
49. Mandache, L.; Al-Haddad, K. An accurate design tool for filter inductors. In Proceedings of the 33rd Annual Conference of the IEEE Industrial Electronics Society IECON, Taipei, Taiwan, 5–8 November 2007; pp. 1414–1419.
50. Rozanov, E.; Ben-Yaakov, S. A SPICE behavioral model for current-controlled magnetic inductors. In Proceedings of the 23rd IEEE Convention of Electrical and Electronics Engineers in Israel, Tel-Aviv, Israel, 6–7 September 2004; pp. 338–341.
51. Hristov, M.H.; Gadjewa, E.D.; Pukneva, D.I. Computer Modelling and Geometry Optimilization of Spiral Inductors for RF Application Using SPICE. In Proceedings of the 10th International Conference Mixed Design of Integrated Circuits and Systems MIXDES 2003, Łódz, Poland, 25–27 June 2003; pp. 413–418.
52. Fuzi, J.; Helerea, E.; Ivanyi, A. Experimental construction of Preisach models for ferromagnetics cores. In Proceedings of the International Conference ZM Communications GmbH (PCIM 1998), Power Conversion, Nurnberg, Germany, May 1998; pp. 661–666.
53. Jiles, D.C.; Atherton, D.L. Theory of ferromagnetic hysteresis. *J. Magn. Magn. Mater.* **1986**, *61*, 48–60. [CrossRef]
54. Faiz, J.; Saffari, S. A new technique for modeling hysteresis in soft magnetic materials. *Electromagnetics* **2010**, *30*, 376–401. [CrossRef]
55. Mészáros, I. Magnetization curve modelling of soft magnetic alloys. *J. Phys. Conf. Ser.* **2011**, *268*, 012020. [CrossRef]
56. Zidari, B.; Miljavec, D. A new ferromagnetic hysteresis model for soft magnetic composite materials. *J. Magn. Magn. Mater.* **2011**, *323*, 67–71. [CrossRef]
57. Willard, M.A.; Francavilla, T.; Harris, V.G. Core-loss analysis of an (Fe, Co, Ni)-based nanocrystalline soft magnetic alloy. *J. Appl. Phys.* **2005**, *97*, 10F502. [CrossRef]
58. Cardelli, E.; Giannetti, R.; Tellini, B. Numerical characterization of dynamic hysteresis loops and losses in soft magnetic materials. *IEEE Trans. Magn.* **2005**, *41*, 1540–1543.
59. Mandache, L.; Topan, D. Managing Eddy Current. Losses and Ferromagnetic Material Nonlinearities in Distorting Regimes. In Proceedings of the IEEE International Electric Machines and Drives Conference IEMDC 2009, Miami, FL, USA, 3–6 May 2009; pp. 1449–1454.

60. Ngo, K.D.T. Subcircuit modeling of magnetic cores with hysteresis in PSpice. *IEEE Trans. Aerosp. Electron. Syst.* **2002**, *38*, 1425–1434. [CrossRef]
61. Górecki, K.; Zarębski, J.; Detka, K. Application of the electrothermal average inductor model for analyses of boost converters, In Proceedings of the 22nd International Conference Mixed Design Integrated Circuits Systems Mixdes, Toruń, Poland, 25–27 June 2015. pp. 417–421.
62. Brachtendorf, H.G.; Eck, C.; Laur, R. Macromodeling of hysteresis phenomena with SPICE. *IEEE Trans. Circuits Syst. II Analog. Digit. Signal Process.* **1997**, *44*, 378–388. [CrossRef]
63. Mazgaj, W. Properties of magnetic hysteresis models. *Arch. Electr. Eng.* **2006**, *55*, 147–161.
64. Ossart, F.; Meunier, G. Comparison between various hysteresis models and experimental data. *IEEE Trans. Magn.* **1990**, *26*, 2837–2839. [CrossRef]
65. Detka, K.; Górecki, K. Validation of selected models of the magnetization curve of ferrite cores. In Proceedings of the 2022 IEEE 16th International Conference on Compatibility, Power Electronics, and Power Engineering (CPE-POWERENG), Birmingham, UK, 29 June–1 July 2022; pp. 1–6. [CrossRef]
66. Rayleigh, L. On the behavior of iron and steel under the operation of feeble magnetic forces, XXV. Notes on electricity and magnetism.—III. *Philos. Mag. J. Sci.* **1887**, *23*, 225–245. [CrossRef]
67. Hodgdon, M.L. Mathematical theory and calculations of magnetic hysteresis curves. *IEEE Trans. Magn.* **1998**, *24*, 3120–3122. [CrossRef]
68. Iványi, A. *Hysteresis Models in Electromagnetic Computation*; Akadémiai Kiadó: Budapest, Hungary, 1997.
69. Wilson, P. Modeling and Simulation of Magnetic Components in Electric Circuits. Ph.D. Thesis, Department of Electronics and Computer Science, University of Southampton, Southampton, UK, 2001.
70. Bozorth, R. *Ferromagnetism*; D. Van Nostrand Company: Princeton, NJ, USA, 1951.
71. Cullity, B.D.; Graham, C.D. *Introduction to Magnetism Material*; John Wiley and Sons: Hoboken, NJ, USA, 2011.
72. Bertotti, G.; Fiorillo, F.; Soardo, G. The Prediction of Power Losses in Soft Magnetic Materials. *J. Phys. Colloq.* **1988**, *49*, C8-1915–C8-1919. [CrossRef]
73. Wciślik, M.; Suchenia, K. Modeling of core losses of switched reluctance motor. *Przegląd Elektrotechniczny* **2016**, *92*, 152–155. [CrossRef]
74. Popescu, M.; Miller, T.; McGlip, M.; Ionel, D.; Dellinger, S.; Heidemann, R. On the Physical Basis of Power Losses in Laminated Steel and Minimum-Effort Modeling in an Industrial Design Environment. In Proceedings of the Conference Record of the 2007 IEEE Industry Applications Conference Forty-Second IAS Annual Meeting, New Orleans, LA, USA, 23–27 September 2007; p. 60.
75. Wilson, P.R.; Ross, J.N.; Brown, A.D. Simulation of magnetic component models in electric circuits including dynamic thermal effects. *IEEE Trans. Power Electron.* **2002**, *17*, 55–65. [CrossRef]
76. Lullo, G.; Sciere, D.; Vitale, G. Non-linear inductor modelling for a DC/DC Buck converter. In Proceedings of the International Conference on Renewable Energies and Power Quality (ICREPQ'17), Malaga, Spain, 4–6 April 2017. [CrossRef]
77. Magambo, J.S.N.T.; Bakri, R.; Margueron, X.; Le Moigne, P.; Mahe, A.; Guguen, S.; Bensalah, T. Planar Magnetic Components in More Electric Aircraft: Review of Technology and Key Parameters for DC–DC Power Electronic Converter. *IEEE Trans. Transp. Electrif.* **2017**, *3*, 831–842. [CrossRef]
78. Valchev, V.C.; Van den Bossche, A. *Inductors and Transformers for Power Electronics*; Informa UK Limited: London, UK, 2018.
79. Górecki, K.; Rogalska, M. The compact thermal model of the pulse transformer. *Microelectron. J.* **2014**, *45*, 1795–1799. [CrossRef]
80. Penabad-Duran, P.; López-Fernández, X.; Turowski, J. 3D non-linear magneto-thermal behavior on transformer covers. *Electr. Power Syst. Res.* **2015**, *121*, 333–340. [CrossRef]
81. Buschow, K.H.J. *Handbook of Magnetic Materials*; Elsevier: Amsterdam, The Netherlands; Berlin, Germany, 2015.
82. Wilson, P.R.; Ross, J.N. Definition and application of magnetic material metrics in modeling and optimization. *IEEE Trans. Magn.* **2001**, *37*, 3774–3780. [CrossRef]
83. Detka, K.; Górecki, K. Electrothermal Model of Coupled Inductors with Nanocrystalline Cores. *Energies* **2021**, *15*, 224. [CrossRef]
84. Zhu, G.; Wand, K. Modeling and Design Consideration of Coupled Inductor Converters. In Proceedings of the 2010 Twenty-Fifth Annual IEEE Applied Power Electronics Conference and Exposition (APEC), Palm Springs, CA, USA, 21–25 February 2010. [CrossRef]
85. Bahman, A.; Ma, K.; Blaabjerg, F. A lumped thermal model including thermal coupling and thermal boundary conditions for high power IGBT modules. *IEEE Trans. Power Electron.* **2018**, *33*, 2518–2530. [CrossRef]
86. Swift, G.; Molinski, T.S.; Bray, R. A fundamental approach to transformer thermal modeling. Part I: Theory and equivalent circuit. *IEEE Trans. Power Deliv.* **2001**, *16*, 171–175. [CrossRef]
87. Gamil, A.; Al-Abadi, A.; Schatzl, F.; Schlucker, E. Theoretical and Empirical-Based Thermal Modelling of Power Transformers. In Proceedings of the 2018 IEEE International Conference on High Voltage Engineering and Application (ICHVE), Athens, Greece, 10–13 September 2018; IEEE: Piscataway, NJ, USA, 2018; pp. 1–4.
88. Souza, L.; Lemos, A.; Caminhas, W.; Boaventura, W. Thermal modeling of power transformers using evolving fuzzy systems. *Eng. Appl. Artif. Intell.* **2012**, *25*, 980–988. [CrossRef]
89. Tang, W.H.; Wu, Q.H.; Richardson, Z.J. A simplified transformer thermal model based on thermal–electric analogy. *IEEE Trans. Power Deliv.* **2004**, *19*, 1112–1119. [CrossRef]

90. Tang, W.; Wu, Q.H.; Richardson, Z. Equivalent heat circuit based power transformer thermal model. *IEE Proc.-Electr. Power Appl.* **2002**, *149*, 87. [CrossRef]
91. Lopez, G.S.; Exposito, A.D.; Muñoz-Antón, J.; Ramirez, J.A.O.; Lopez, R.P.; Delgado, A.; Oliver, J.A.; Prieto, R. Fast and Accurate Thermal Modeling of Magnetic Components by FEA-Based Homogenization. *IEEE Trans. Power Electron.* **2020**, *35*, 1830–1844. [CrossRef]

MDPI
St. Alban-Anlage 66
4052 Basel
Switzerland
www.mdpi.com

Energies Editorial Office
E-mail: energies@mdpi.com
www.mdpi.com/journal/energies

www.ingramcontent.com/pod-product-compliance
Lightning Source LLC
Chambersburg PA
CBHW061518210926
51850CB00008B/1961
9783725802524